AF400831

Systems & Control: Foundations & Applications

More information about this series at http://www.springer.com/series/4895

Alexander Poznyak • Andrey Polyakov

Vadim Azhmyakov

Attractive Ellipsoids in Robust Control

Alexander Poznyak
Automatic Control Department
Centro de Investigacion y Estudios
    Avanzados
México, Distrito Federal
Mexico

Vadim Azhmyakov
Faculty of Electronic and Biomedical
    Engineering
University of Antonio Nariño
Neiva, HUILA
Colombia

Andrey Polyakov
Non-A
INRIA-LNE
Villeneuve d'Ascq
Nord, France

ISSN 2324-9749   ISSN 2324-9757 (electronic)
ISBN 978-3-319-09209-6   ISBN 978-3-319-09210-2 (eBook)
DOI 10.1007/978-3-319-09210-2
Springer Cham Heidelberg New York Dordrecht London

Library of Congress Control Number: 2014946581

Mathematics Subject Classification (2010): 93D09, 93D21

Printed on acid-free paper

Springer is part of Springer Science+Business Media (www.springer.com)

To Russia with love

Preface

The material discussed in this monograph is a result of our research program at the Mexican Center for Advanced Studies and Research (CINVESTAV, Mexico City, Mexico) and the Institute of Control Problems, Russian Academy of Sciences (IPU RAN, Moscow, Russia). The main purpose of this book is to provide an advanced account of a newly developed robust control design technique for a wide class of continuous-time dynamical systems. We call the approach under discussion the "Attractive Ellipsoid Method." As the general methodology of the stabilization methods develops, more and more recent results are filtering through to graduate courses. Therefore, our book also contains a coherent introduction to the proposed control design technique and related topics.

We study nonlinearly affine control systems in the presence of uncertainties and are interested in a constructive and easily implementable control strategy that guarantees in a practical sense some stability properties of the closed-loop realizations. In fact, we deal with a linear-type feedback control synthesis in the context of the above-mentioned nonlinear uncertain systems of an affine structure. Throughout, the emphasis is on understanding and the ability to apply the developed theory to examples rather that on rigorous mathematical development. Although there are theorems proved in a systematic way, the level of rigor is not that of a purely mathematical text. Nonetheless, appreciating the limitations of any method is vital, and so we have stated all results in a precise form. The choice of topics has also been influenced by a desire to cover different dynamical systems and consider possible applications. In particular, this monograph contains some illustrative examples and applications of the attractive ellipsoid method to mechanical and electromechanical systems.

We expect that this book will be useful to interested graduate students and advanced undergraduate students with sufficient knowledge of modern systems theory as well as to researchers in the fields of control engineering and applied mathematics. The book can be also considered a complementary text to graduate courses in advanced robust nonlinear control. We also assume that the reader has a rudimentary knowledge of analysis and linear algebra, while a little more is presupposed from the theory of ordinary differential equations and Lyapunov stability theory. We have attempted throughout to provide detailed and transparent proofs of the main results.

Of course, the book claims to present only an introduction to the theory and applications of stable operators in infinite-dimensional spaces. We have made an attempt to unify, simplify, and relate many scattered results in the literature. Some of the topics discussed here are new; others are not. Therefore, the book is not a collection of research papers, but is rather a monograph whose aim is to present recent developments of the theory that might constitute a foundation for further development.

The book contains this preface and twelve chapters and is organized as follows.

- **Chapter 1 (Introduction)** presents motivation and intuitive concepts.
- **Chapter 2 (Mathematical Background)** contains a short collection of necessary mathematical facts from classical analysis and related areas, namely a description of the class of nonlinear uncertain models (Quasi-Lipschitz dynamical systems and examples thereof), differential inclusions and their general solution concept, the Filippov regularization procedure, the Lyapunov approach to Quasi-Lipschitz dynamical systems, elements of linear matrix inequalities (LMIs) including the existence of solutions and some numerical approaches, the duality for LMI-constrained problems, the S-lemma, and more.
- **Chapter 3 (Robust State Feedback Control)** establishes the main concepts of linear (proportional to the current state) feedback design using an S-procedure-based approach, discusses the storage function method providing the boundedness of all possible trajectories of a controlled system from a given class, and presents a technique of minimization of the attractive ellipsoid containing all bounded trajectories. Aspects of practical stabilization are also discussed.
- **Chapter 4 (Robust Output Feedback Control)** is devoted to direct feedback control design and considers two feedback structures:

  - observer-based feedback,
  - full-order linear dynamic controllers,

  and to both of these, the attractive ellipsoid method is applied and analyzed.
- In **Chapter 5 (Control with Sample-Data Measurements),** the main problem is formulated, and some necessary mathematical concepts are discussed related to the feedback control design for nonlinear systems under sample-data output measurements. Then we present a theoretical analysis of an extended version of the invariant ellipsoid method. Then two types of feedback are analyzed:

  - a linear feedback proportional to the current state estimate obtained by a Luenberger-type estimator,
  - a full-order linear dynamic controller governed by a linear ODE with available sample data as input.

Then we construct a minimal attractive ellipsoid that guarantees stability of the system in a practical sense, varying all parameters of the suggested feedbacks. An associated numerical techniques is also presented. Implementable algorithms for the constructive treatment of the robust control design problem are proposed.

- **Chapter 6 (Sample Data and Quantifying Output Control)** considers the analysis and design of an output feedback controller for a perturbed nonlinear system in which the output is sampled and quantized. Using the invariant ellipsoid method, which is based on Lyapunov analysis techniques, together with the relaxation of a nonlinear optimization problem, sufficient conditions for the design of a robust control law are obtained. Since the original conditions result in nonlinear matrix inequalities, a numerical algorithm to obtain the solution is presented. The obtained control ensures that the trajectories of the closed-loop system will converge to a minimal (in a sense to be made specific) ellipsoidal region. Finally, numerical examples are presented in order to illustrate the applicability of the proposed design method.
- **Chapter 7 (Robust Control of Implicit Systems)** focuses on the analysis and synthesis of robust feedback for a class of implicit systems whose state derivatives cannot be expressed analytically as functions of its state coordinates. The transformation to differential–algebraic form is presented, and the attractive ellipsoid method is designed for such systems. The reduction of bilinear matrix inequalities to linear inequalities is presented in detail, and some specific numerical aspects are also discussed.
- **Chapter 8 (Attractive Ellipsoids in Sliding Mode Control)** deals with the minimization-of-unmatched-uncertainties effect in sliding mode control. In particular, LMI-based sliding mode control design is considered, and the optimal sliding surface is constructed. In addition, gain matrix tuning in dynamic actuators is analyzed, and the sliding mode control of time-delay systems with a predictive control is discussed in detail.
- **Chapter 9 (Robust Stabilization of Time-Delay Systems)** considers the class of uncertain time-delay affine-controlled systems whereby a delay is admitted to be in state variables as well as in their derivatives (neutral systems), and shows that the attractive ellipsoid method makes it possible to create feedback that provides the convergence of any state trajectory of the controlled system from a given class to an ellipsoid whose "size" depends on the parameters of the applied feedback. Finally, we present a method for numerical calculation of these parameters providing the "smallest" zone convergence for controlled trajectories.
- **Chapter 10 (Robust Control of Switched Systems)** deals with robust control problems in which a structure of the controlled dynamics may vary in time according to some fixed program strategy or in which the controlled trajectories cross some given surfaces. All nonlinearities of each structure are admitted to be uncertain but to belong a wide class of quasi-Lipschitz functions. The corresponding switching of the applied feedback is shown to be much more effective than a "solid structure" of the feedback providing a smaller convergence zone.

- **Chapter 11 (Bounded Robust Control)** describes the application of the attractive ellipsoid method to controlled systems in which the control actions are a priori bounded, so that a current control at each time is the projection of a linear function of the state or its estimate. This constraint certainly makes the convergence zone a bit larger in comparison to the unconstrained case, but our approach allows us to make this zone as small as possible.
- **Chapter 12 (Attractive Ellipsoid Method with Adaptation)** deals with designing a state estimator and adaptive controller for a class of uncertain nonlinear systems having "quasi-Lipschitz" nonlinearities as well as external perturbations. The set of stabilizing feedback matrices is given by a specific matrix inequality including the characteristic matrix of the attractive ellipsoid, which contains all possible bounded trajectories around the origin. Here we present two modifications of the attractive ellipsoid method that allow us

  - to use online information obtained during the process,
  - to adjust matrix parameters participating in some constraint that characterizes the class of adaptive stabilizing feedbacks.

The proposed approach guarantees that under a specific persistent excitation condition, the controlled system trajectories converge to an ellipsoid of a "minimal size" having a minimal trace of the corresponding inverse ellipsoidal matrix, which turns out to be significantly smaller than one without adaptation.

Standard notation is used throughout the book. The opening chapter contains a brief collection of necessary mathematical facts that will be useful for a deeper understanding of what follows.

Many individuals have influenced the content and presentation of this book, and we are grateful to all of them. We would like to thank especially Professor A.B. Kurzhanski (Lomonosov State University, Moscow, and University of California, Berkeley) and Professor F.L. Chernousko (Institute for Problems in Mechanics, Russian Academy of Sciences) as the pioneers of this approach. The authors are grateful to all participants of the seminars at the Mexican Center for Advanced Studies and Research (CINVESTAV, Mexico City) and the Institute of Control Problems the Russian Academy of Sciences (IPU RAN, Moscow, Russia) headed by Professor B.T. Polyak. We also would like to express our gratitude to Professor A.P. Kurdyukov (IPU RAN, Moscow) and to Professor S. Mondie (CINVESTAV, Mexico City) for their critical comments and suggestions.

We wish to thank Dr. I Chairez (UPIBI-IPN, Mexico) and Dr. F. Castanos (CINVESTAV, Mexico City) for reading the manuscript and noting both substantive and typographical errors.

Finally, we wish to express our appreciation to the staff of Birkhäuser Science for their accomplished handling of the manuscript as well as their understanding and patience.

México, Mexico
Nord, France
Neiva, HUILA, Colombia
2014

Alexander Poznyak
Andrey Polyakov
Vadim Azhmyakov

Contents

List of Figures

Chapter 1
Introduction

Abstract This introductory chapter briefly reviews the evolution of optimal control design. First, it considers the classical control principles of optimal design for ideal completely known systems. Then, the case of incomplete information is studied. The ideas of robust control design and related optimization issues are discussed. The general principles of ellipsoid-based control design are introduced.

Keywords Optimal control • Robust control • Ellipsoid-based feedback control design

1.1 Complete Information Case: Classical Control Approaches

Optimal control is a rapidly expanding field developed during the last half-century to analyze optimal behavior of a constrained process that evolves in time according to prescribed laws. Its applications now embrace a variety of new disciplines such as economics and production planning. The main supposition of *classical optimal control theory* (OCT) is that the mathematical techniques especially designed for analysis and synthesis of an optimal control of dynamic models are based on the assumption that a designer (or an analyst) possesses *complete information* on the model under consideration as well as on the environment in which this controlled model will evolve.

There exist two principal approaches in solving *optimal control problems* (OCPs) in the presence of complete information on considered dynamic models:

- The first is the *maximum principle* (MP) of Pontryagin (Boltyanski, Gamkrelidze, & Pontryagin 1956).
- The second is the *dynamic programming method* (DPM) of Bellman (Bellman 1956).

Formally, a description of the notion of OCP in its classical form is as follows.

A. Poznyak et al., *Attractive Ellipsoids in Robust Control*, Systems & Control: Foundations & Applications, DOI 10.1007/978-3-319-09210-2_1

1.1.1   System Description

- *Controlled plant dynamics* are given by the following system of ordinary differential equations (ODEs):

$$\left.\begin{array}{l} \dot{x}\,(t) = f\,(x\,(t)\,,u\,(t)\,,t)\,, \quad \text{a.e. } t \in [0,T]\,, \\[2mm] x\,(0) = x_0, \end{array}\right\} \tag{1.1}$$

where $x = \left(x^1,\dots,x^n\right)^T \in \mathbb{R}^n$ is its state vector, $u = \left(u^1,\dots,u^r\right)^T \in \mathbb{R}^r$ is a control, which may run over a given control region $U \subset \mathbb{R}^r$.

- *Cost functional* is defined as

$$J\,(u\,(\cdot)) := h_0(x\,(T)) + \int\limits_{t=0}^{T} h_1\,(x\,(t)\,,u\,(t)\,,t)\,dt. \tag{1.2}$$

It contains an integral term as well as a terminal term, and the time process or *horizon* T is supposed to be fixed or not and may be finite or infinite.

- *The terminal set* $\mathcal{M} \subseteq \mathbb{R}^n$ is given by the inequalities

$$\mathcal{M} = \{x \in \mathbb{R}^n : g_l(x) \leq 0 \quad (l = 1,\dots,L)\}. \tag{1.3}$$

- The function (1.2) is said to be given in *Bolza form.* If in (1.2), we have

$$h_0(x) = 0,$$

then we obtain the cost functional in the *Lagrange form*, that is,

$$J\,(u\,(\cdot)) = \int\limits_{t=0}^{T} h_1\,(x\,(t)\,,u\,(t)\,,t)\,dt \tag{1.4}$$

If in (1.2), we have

$$h_1(x,u,t) = 0,$$

then we obtain the cost functional in *Mayer form*, that is,

$$J\,(u\,(\cdot)) = h_0(x\,(T)). \tag{1.5}$$

Usually, the following assumptions are assumed to be in force:

(A1) (U,d) is a separable metric space (with metric d) and $T > 0$.

(A2) The maps

$$\left.\begin{aligned} f &: \mathbb{R}^n \times U \times [0, T] \to \mathbb{R}^n \\[2mm] h_1 &: \mathbb{R}^n \times U \times [0, T] \to \mathbb{R} \\[2mm] h_0 &: \mathbb{R}^n \times U \times [0, T] \to \mathbb{R} \\[2mm] g_l &: \mathbb{R}^n \to \mathbb{R}^n \ (l = 1, \ldots, L) \end{aligned}\right\} \tag{1.6}$$

are measurable, and there exist a constant L and a continuity modulus

$$\bar{\omega} : [0, \infty) \to [0, \infty)$$

such that for

$$\varphi = f(x, u, t), h_1(x, u, t), h_0(x, u, t), g_l(x) \ (l = 1, \ldots, L),$$

the following inequalities hold:

$$\left.\begin{aligned} \|\varphi(x, u, t) - \varphi(\hat{x}, \hat{u}, t)\| &\leq L \|x - \hat{x}\| + \bar{\omega}(d(u, \hat{u})) \\ \forall t \in [0, T], \ x, \hat{x} \in \mathbb{R}^n, \ u, \hat{u} \in U & \\[3mm] \|\varphi(0, u, t)\| \leq L \quad \forall u, t \in U \times [0, T] & \end{aligned}\right\} \tag{1.7}$$

(A3) The maps

$$f, h_1, h_0 \text{ and } g_l \ (l = 1, \ldots, L)$$

are from C^1 in x, and there exists a continuity modulus

$$\bar{\omega} : [0, \infty) \to [0, \infty)$$

such that for

$$\varphi = f(x, u, t), h_1(x, u, t), h_0(x, u, t), g_l(x) \ (l = 1, \ldots, L),$$

the following inequalities hold:

$$\left.\begin{aligned} \left\|\frac{\partial}{\partial x}\varphi(x, u, t) - \frac{\partial}{\partial x}\varphi(\hat{x}, \hat{u}, t)\right\| &\leq \bar{\omega}(\|x - \hat{x}\| + d(u, \hat{u})) \\ \forall t \in [0, T], \ x, \hat{x} \in \mathbb{R}^n, \ u, \hat{u} \in U & \end{aligned}\right\} \tag{1.8}$$

1.1.2 Feasible and Admissible Control

A function $u(t)$, $t_0 \leq t \leq T$, is said to be a *feasible control* if it is measurable and $u(t) \in U$ for all $t \in [0, T]$. Denote the set of all feasible controls by

$$\mathcal{U}[0, T] := \{u(\cdot) : [0, T] \to U \mid u(t) \text{ is measurable}\}. \tag{1.9}$$

The control $u(t)$, $t_0 \leq t \leq T$ is also said to be *admissible* or to *realize the terminal condition (1.3)* if the corresponding trajectory $x(t)$ satisfies the terminal condition, that is, if it satisfies the inclusion $x(T) \in \mathcal{M}$. Denote the set of all admissible controls by

$$\mathcal{U}_{admis}[0, T] := \{u(\cdot) : u(\cdot) \in \mathcal{U}[0, T], \ x(T) \in \mathcal{M}\}. \tag{1.10}$$

In view of the theorem on the existence of the solutions to an ODE (see Coddington & Levinson 1955 or Poznyak 2008), it follows that under the assumptions **(A1)–(A2)**, for every $u(t) \in \mathcal{U}[0, T]$ the equation (1.1) admits a unique solution $x(\cdot) := x(\cdot, u(\cdot))$, and the functional (1.2) is well defined.

1.1.3 Problem Setting in the General Bolza Form

Based on the definitions given above, the classical OCP can be formulated as follows.

Problem 1.1 (OCP in Bolza Form).

$$\text{Minimize (1.2) over } \mathcal{U}_{admis}[0, T]. \tag{1.11}$$

Problem 1.2 (OCP with a Fixed Terminal Term). If in the problem (1.11),

$$\mathcal{M} = \{x_f \in \mathbb{R}^n\} =$$
$$\{x \in \mathbb{R}^n : g_1(x) = x - x_f \leq 0, \ g_2(x) = -(x - x_f) \leq 0\} \tag{1.12}$$
$$\text{or equivalently, } x = x_f,$$

then it is called an **OCP** *with fixed terminal term* x_f.

Every control $u^*(\cdot) \in \mathcal{U}_{admis}[0, T]$ satisfying

$$J\left(u^*(\cdot)\right) = \min_{u(\cdot) \in \mathcal{U}_{admis}[0,T]} J\left(u(\cdot)\right) \tag{1.13}$$

is called an *optimal control*; the corresponding state trajectories $x^*(\cdot) := x^*(\cdot, u^*(\cdot))$ and $(x^*(\cdot), u^*(\cdot))$ are called an *optimal state trajectory* and an *optimal pair*.

1.1.4 Specific Features of Classical Optimal Control

The main specific features of both the MP and DPM approaches are as follows:

1. The cost functional defining the quality of the applied control is given in the Bolza form (1.2) containing terminal term as well as the integral term characterizing the cost functional of a designer during all time of the control process.
2. The function f in (1.1) is known a priori and may be used in the control-designing process. In other words, the right-hand side of the dynamic equation (1.1) does not contain any uncertainty or disturbances that are unavailable during the control process.
3. The state vector $x(t)$ is assumed to be available for control designing.

The solution of the classical control problem can be found, for example, in Boltyanski and Poznyak (2012). If one of this three features does not hold, then the classical Optimal Control approach is not applicable.

1.2 Case of Incomplete Information

When we have complete information on a dynamic model to be controlled, the main problem consists in designing an acceptable control that remains "close to the optimal or desired one" (having small sensitivity with respect to every unknown (unpredictable) factor from a given set of possibilities). In other words, the desired control should be *robust* with respect to unknown factors. In the presence of any sort of uncertainty (e.g., parametric type, unmodeled dynamics, external perturbations), the main methodology applied in this book for obtaining a solution suitable for a class of given models is to formulate a corresponding *tracking control* problem, where we are interested in the "best approximation" to a desired trajectory. In other words, we are interested in a zone stabilization or in the practical stability of the deviation of the trajectories of the given system from the desired one. The *robust stabilization problem* considered for different classes of nonlinear systems has been a hot topic of research over the past two decades (Ioannou & Sun 1996; Narendra & Annaswamy 2005; Schweppe 1973; Utkin 1992).

1.2.1 Robust Tracking Problem Formulation

Formally, the robust tracking problem can be described as follows:

- *The controlled plant dynamics* are given by

$$\dot{\bar{x}}(t) = \bar{f}(\bar{x}(t)) + B\bar{u}(t) + \bar{\xi}_x(t), \quad \text{a.e. } t \in [0, T]$$

$$\bar{x}(0) = \bar{x}_0 \tag{1.14}$$

$$y(t) = \bar{h}(\bar{x}(t)) + \xi_y(t)$$

where

$\bar{x} = \left(\bar{x}^1, \ldots, \bar{x}^n\right)^T \in \mathbb{R}^n$ is its state vector.
$\bar{u} = \left(\bar{u}^1, \ldots, \bar{u}^r\right)^T \in \mathbb{R}^r$ is the control to be designed.
$y = \left(y^1, \ldots, y^m\right)^T \in \mathbb{R}^m$ is the measurable output of the system available for a designer at any time $t \geq 0$.

The functions $\bar{\xi}_x(t)$ and $\xi_y(t)$ represent external perturbations that are not measurable (unavailable) for a designer.

- The *desired dynamics* $x^*(t)$ are governed by the following *reference model*:

$$\dot{x}^*(t) = \varphi\left(x^*(t), t\right), \tag{1.15}$$

where $x^* = \left(x^{*1}, \ldots, x^{*n}\right)^T \in \mathbb{R}^n$, and $x^*(t)$ is supposed to be measurable (available) at every time $t \geq 0$. The matrix $B \in \mathbb{R}^{n \times r}$ characterizing the actuator properties is also assumed to be known.

- The tracking error $x(t)$ is defined as

$$x(t) = \bar{x}(t) - x^*(t). \tag{1.16}$$

So the ODE describing the dynamics of the tracking error is

$$\left. \begin{aligned} \dot{x}(t) = f\left(x(t), t\right) + B\bar{u}(t) - \varphi\left(x^*(t), t\right) + \bar{\xi}_x(t) \\ \text{a.e. } t \in [0, T] \\ y(t) = h\left(x(t), t\right) + \xi_y(t) \end{aligned} \right\} \tag{1.17}$$

where

$$f\left(x(t), t\right) := f\left(x(t) + x^*(t)\right), \quad x(0) = \bar{x}_0 + x^*(0)$$

$$h\left(x(t), t\right) := \bar{h}\left(x(t) + x^*(t)\right) \tag{1.18}$$

- The control action $\bar{u}$ consists of two terms:

$$\bar{u}(t) := u(t) + u_{comp}(t), \tag{1.19}$$

where the *compensating control* $u_{comp}(t)$ is selected in such a way that the effect of the dynamics $\varphi\left(x^*(t), t\right)$ of the desired trajectory will be compensated or minimized, namely,

$$u_{comp}(t) = \underset{u_{comp}}{\arg\min} \left\| Bu_{comp} - \varphi\left(x^*(t), t\right) \right\|^2$$

$$\tag{1.20}$$

$$= B^+\varphi\left(x^*(t), t\right)$$

where

$$B^+ := \left(B^\mathsf{T} B\right)^{-1} B^\mathsf{T}$$

if we assume that

$$B^\mathsf{T} B > 0.$$

In view of this, the dynamics of the tracking error can be represented as

$$\left. \begin{array}{l} \dot{x}(t) = f\left(x(t), t\right) + Bu(t) + \xi_x(t), \text{ a.e. } t \in [0, T] \\[2ex] y(t) = h\left(x(t), t\right) + \xi_y(t) \end{array} \right\} \tag{1.21}$$

where

$$\xi_x(t) := \bar{\xi}_x(t) + \left(BB^+ - I\right)\varphi\left(x^*(t), t\right). \tag{1.22}$$

For the tracking error dynamics (1.21), the following assumptions usually are supposed:

(B1) The dynamic plant (1.21) is *controllable* and *observable* (see, for example, Isidori 1995).

(B2) The functions f and h may be unknown, but they belong to the *given classes* $\mathcal{C}_f$ and $\mathcal{C}_h$ of nonlinear functions, respectively. In this book, both classes are the classes of the *quasi-Lipschitz functions*, whose exact definition is given in the next chapter.

(B3) The unmeasured functions $\xi_x(t)$ and $\xi_y(t)$ are *bounded*.

(B4) The control $u(t)$ is designed as a feedback (static or dynamic) of a given structure containing the set of parameters $\mathcal{P}$, that is,

$$u(t) = u\left(y(\tau)\,|_{0 \leq \tau \leq t}, t, \mathcal{P}\right), \tag{1.23}$$

so that $u(t)$ depends on all measurable data $y(\tau)$ in the time interval $[0, t]$.

1.2.2   What Is the Effectiveness of a Designed Control in the Case of Incomplete Information?

If we have the nonzero terms $\xi_x(t)$ (1.22) and $\xi_y(t)$ (1.21), which are unmeasurable during the control process, then obviously, the application of the classical optimal control approach (as described above) is impossible. The situation looks much more difficult if the functions $f \in C_f$ and $h \in C_h$ describing the dynamic process are unknown a priori. In this case, the following questions seem to be important:

How can one formulate the problem of control in the uncertain model case, and based on what performances we can estimate the effectiveness of an applied control strategy?

Several approaches may be considered in this situation.

One suggests that we formulate the corresponding control problem as a *min–max optimal control*, where the *maximum* is taken over all existing uncertainties, and the *minimum* is realized within an admissible control set (see, for example, the H^∞ approach Zhou, Doyle, & Glover 1996 and robust maximum principle Boltyanski & Poznyak 2012). Such min–max consideration work successfully if the set of uncertain terms has a sufficiently simple structure (external perturbations are quadratically integrable, parametric uncertainties are from finite sets or belong to a measurable compact set of a simple nature).

Here we present another approach, referred to below as the *attractive ellipsoid method (AEM)*, which turns out to be workable for a significantly wide spectrum of uncertainties participating in a model description.

1.3 Ellipsoid-Based Feedback Control Design

The main features of AEM are as follows:

- Since in the uncertain case, the optimization of the cost functional [such as (1.2)] cannot be realized exactly because of uncertain factor participation, the control problem is formulated as a *tracking problem*, which equivalently is reduced to the minimization of the vector trajectory $x\,(t)$ (1.21) by an adequate selection of control strategies $u\,(t)$.
- The set of considered control strategies is suggested to belong to a *parameterized class of nonlinear (perhaps nonstationary) feedbacks* (1.23)

$$u\,(t) = u\,(y\,(\tau)\,|_{0\leq\tau\leq t}, t, \mathcal{P})\,,$$

  whose parameters $\mathcal{P}$ are selected in such a way that all possible trajectories $x\,(t)$ of the closed controlled systems remain bounded and closed to the origin.
- Taking into account that every set of bounded trajectories may be imposed within a convex bounded set, and particularly within an ellipsoid, the AEM suggests that we select the feedback parameters $\mathcal{P} = \mathcal{P}^*$ providing a minimal "size" of this ellipsoid containing all possible bounded trajectories of every dynamical system from the considered class of dynamics containing uncertain elements. In this case, we talk about *zone convergence* or "practical stability" (with a prescribed convex convergence zone) if the size of the convergence zone is of a predetermined value, so that *the effectiveness of such robust control strategies is associated with the "size" of the corresponding attractive ellipsoid set.*

- During the control process, these "optimal" parameters may be adjusted online (*learning* or *adaptive* version of AEM), making the attractive ellipsoid of a smaller size.

1.4 Overview of the Book

In general, in this book we describe a methodology that provides successful designing of feedbacks for tracking or stabilization of nonlinear systems in the presence of uncertainties or disturbances. The class of stabilizing feedbacks is given by the corresponding BMIs (bilinear matrix inequalities) or LMIs (linear matrix inequalities). If they are satisfied, then one may guarantee that all possible trajectories of the considered systems are bounded. Since bounded dynamics may be imposed inside an ellipsoid, we associate the *"best parameters"* of the feedback with the minimal size of this ellipsoid. Unfortunately, this finite-dimensional optimization problem with matrix constraints cannot be resolved analytically. Therefore, we suggest an associated numerical procedure for designing robust and adaptive-robust feedbacks for a wide class (quasi-Lipshitz) of nonlinear uncertain systems with

- state and output feedbacks,
- sample-data and quantized output feedbacks,
- robust feedback design for time delay and for implicit and switched structure systems.

All these subclasses of uncertain systems are treated by a unique methodology based on *AEM*. Such a unified approach has not been previously considered. However, this book has been preceded by many other fine books on modern control theory, such as the following:

- Maciejowski, J. (1989). *Multivariable feedback design*. New York: Addison Wesley. (Maciejowski 1989).
- Grimble, M. (1994). *Robust industrial control*. Hemel Hempstead, UK: Prentice Hall International. (Grimble 1994).
- Zhou, K., Doyle, J., & Glover, K. (1996). *Robust and optimal control*. Upper Saddle River, NJ: Prentice Hall. (Zhou et al. 1996).
- Kurzhanski, A., & Valyi, I. (1997). *Ellipsoidal calculus for estimation and control*. Boston, MA: Birkhauser. (Kurzhanski & Valyi 1997).
- Mahmoud, M. S. (2000). *Robust control and filtering for time-delay systems*. New York: Marcel Dekker. (Mahmoud 2000).
- Blanchini, F. & Miani, S. (2008). *Set theoretic methods in control. Systems and control: Foundations and applications*. Boston, MA: Birkhäuser. (Blanchini & Miami 2008).
- Haddad, W., & Chellaboina, V. (2008). *Nonlinear dynamical systems and control*. Princeton: Princeton University Press. (Haddad & Chellaboina 2008).

Of course, the authors hope that this book may serve as a complement to the excellent books mentioned above and provide a constructive instrument for feedback designing in practical situations.

Chapter 2
Mathematical Background

Abstract This chapter is of a tutorial character. The aim here is to introduce some main concepts and auxiliary facts related to the class of control systems and problems under consideration. Moreover, we also give a first abstract problem formulation in the framework of robust and practically stable control design. Some conventional and advanced results related to ordinary differential equations in the framework of the examined class of dynamical systems are also included. Moreover, we take a quick look at the basic computational tool for our problems, namely, at linear matrix inequality techniques. We focus our attention primarily on motivations of the proposed "attractive ellipsoid" method and illustrate it in some simple situations.

Keywords Quasi-Lipschitz systems • Lyapunov functions • Attractive sets

This chapter is of a tutorial character. The aim here is to introduce some main concepts and auxiliary facts related to the class of control systems and problems under consideration. Moreover, we also give a first abstract problem formulation in the framework of robust and practically stable control design. Some conventional and advanced results related to ordinary differential equations (ODEs) in the framework of the examined class of dynamical systems are also included. Moreover, we take a quick look at the basic computational tool for our problems, namely, at linear matrix inequality (LMI) techniques. We focus our attention primarily on motivations of the proposed "attractive ellipsoid" method and illustrate it in some simple situations.

2.1 The Class of Nonlinear Uncertain Models

2.1.1 Quasi-Lipschitz Dynamical Systems

The basic inspiration for classic control theory is the state equation parameterized by an "input parameter"

© Springer International Publishing Switzerland 2014
A. Poznyak et al., *Attractive Ellipsoids in Robust Control*, Systems & Control:
Foundations & Applications, DOI 10.1007/978-3-319-09210-2_2

$$\dot{x}(t) = g(t, x(t), u(t)), \quad t \geq 0$$
$$x(0) = x_0 \in \mathbb{R}^n \tag{2.1}$$

where $g : \mathbb{R}^n \times \mathbb{R}^m \to \mathbb{R}^n$ is a suitable right-hand side, and the "parameter" $u(t)$ is chosen from a control set $U \subseteq \mathbb{R}^m$.

Parametric initial value problems of the type (2.1) have been objects of the theory of ODEs for a long time. In particular, a very important question of stabilizability of various classes of systems (2.1) by an appropriate feedback control $u : \mathbb{R}^n \to U$ has been asked by many authors (see, e.g., Isidori 1995; Khalil 2002; Sastry 1999; Sontag 1998; Zabczyk 1995).

Let us denote by $\mathcal{U}_h$ a class of admissible control functions of type $u(x)$ such that the closed-loop system

$$\dot{x}(t) = g(t, x(t), u(x(t))), \quad t \geq 0$$
$$x(0) = x_0 \in \mathbb{R}^n \tag{2.2}$$

has a well-defined solution. It is well known that a general approach to the stability of (2.2) is based on the celebrated *Lyapunov method*. We refer to Zubov (1962), Michel, Hou, and Liu (2007), Blanchini and Miami (2008), and Haddad and Chellaboina (2008) for traditional and advanced Lyapunov techniques.

During the last few years, there has been a revival of the theoretical and practical methodologies associated with stabilizing and robust control design (Burton 2006; Fridman & Orlov 2007; Liberzon 2003; Michel et al. 2007; Naghshtabrizi, Hespanha, & Teel 2008; Polyak, Nazin, Durieu, & Walter 2004; Polyak and Topunov 2008; Polyakov & Poznyak 2011; Teel, Nesic, & Kokotovic 2001). This is because of their relatively simple verifiability in combination with the existence of well-established analytic results. Moreover, modern computational technologies make it possible to obtain an adequate and effective numerical implementation for a concrete Lyapunov-based method.

At this point, let us recall the well-established approaches that reduce stabilizability problems to certain auxiliary LMI-constrained problems.

The ensuing stabilizability techniques discussed in this book are restricted to a specific class of systems (2.1), namely, the so-called *quasi-Lipschitz dynamic models* with bounded uncertainties.

The formal definition of the class of the quasi-Lipschitz functions $g(x)$ is formulated in the definition below.

Definition 2.1. A vector function $g : \mathbb{R}^n \to \mathbb{R}^k$ is said to be of the class $\mathcal{C}(A, \delta_1, \delta_2)$ of quasi-Lipschitz functions if there exist a matrix $A \in \mathbb{R}^{k \times n}$ and nonnegative constants δ_1 and δ_1 such that for every $x \in \mathbb{R}^n$, the following inequality holds:

$$\|g(x) - Ax\|^2 \leq \delta_1 + \delta_2 \|x\|^2.$$

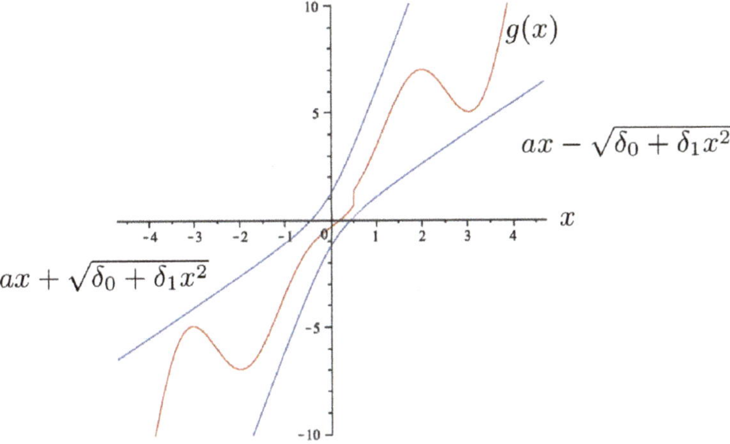

Fig. 2.1 A quasi-Lipschitz function: the single-dimensional case $n = k = 1$, $a > c_1 > 0$

This implies that the growth rates of $g(x)$ as $\|x\| \to \infty$ are not faster than linear (see Fig. 2.1, illustrating the single-dimensional case $n = k = 1$, $a > \sqrt{\delta_2} > 0$).

We now give a formal characterization of the specific class given by (2.1):

(A1)   The piecewise continuous function $g : \mathbb{R} \times \mathbb{R}^n \times \mathbb{R}^m \to \mathbb{R}^n$ is defined in $\mathbb{R} \times \mathbb{R}^n \times \mathbb{R}^m$, and for some $A \in \mathbb{R}^{n \times n}$ and $B \in \mathbb{R}^{n \times m}$ the inequality

$$\|g(t, x, u) - Ax - Bu\|_{Q_f}^2 \leq f_0 + \|x\|_{Q_x}^2 + \|u\|_{Q_u}^2 \tag{2.3}$$

holds almost everywhere in $\mathbb{R} \times \mathbb{R}^n \times \mathbb{R}^m$, where

$$f_0 \geq 0, \ Q_f \in \mathbb{R}^{n \times n} : Q_f \geq 0$$

$$Q_x \in \mathbb{R}^{n \times n} . \ Q_x \geq 0, \ Q_u \in \mathbb{R}^{m \times m} : Q_u \geq 0$$

The above assumption **(A1)** is compatible with several widely used techniques of linear approximation related to plant models. Note that similar linearization-like ideas are common in the theoretical and numerical practice of control engineering (see, e.g., Zabczyk 1995; Khalil 2002).

Let us also introduce the main hypothesis for the class of feedback controls we apply.

(A2)   The class of admissible feedbacks

$$u = u(x)$$

consists of functions piecewise continuous in $\mathbb{R}^n$ such that for all $x \in \mathbb{R}^n$, the inequality

$$\|u(x)\|_{Q_u}^2 \leq u_0 + \|x\|_{Q_{xu}}^2$$

holds almost everywhere in $\mathbb{R}^n$, where

$$u_0 \in \mathbb{R}_+ \cup \{0\}, \quad Q_{xu} \geq 0.$$

In fact, we can extend our consideration to the wide class of nonlinear control systems that also contain measurable bounded disturbances. By $\|\cdot\|_{Q_j}$, $j = f, x, u$ (where Q_j is a given suitable symmetric positive definite matrix) we denote a weighted Euclidean norm. Throughout this book, the integrable functions $g(\cdot, \cdot, \cdot)$ that satisfy the basic hypothesis indicated below are called *quasi-Lipschitz functions*. We easily obtain an alternative description of the quasi-Lipschitz model as follows:

$$\dot{x} = Ax + Bu + Df(t, x, u), \tag{2.4}$$

where

$$f : \mathbb{R} \times \mathbb{R}^n \times \mathbb{R}^m \to \mathbb{R}^k$$

$$x \in \mathbb{R}^n, \ u \in \mathbb{R}^m, \ A \in \mathbb{R}^{n \times n}, \ B \in \mathbb{R}^{n \times m}, D \in \mathbb{R}^{n \times k}$$

and

$$f^T(t, x, u) Q_f f(t, x, u) \leq \delta + x^T Q_x x + u^T Q_u u. \tag{2.5}$$

It is easy to see that the quasi-Lipschitz model introduced above covers a large number of existing control systems. Evidently, systems of the type (2.2) with quasi-Lipschitz right-hand sides also contain conventional ODEs with Lipschitz continuous and locally Lipschitz right-hand sides. Let us note that the class of functions $g(\cdot)$ introduced above is different from the quasi-Lipschitz concept used in the theory of partial differential equations. In fact, the basic dynamic model (2.4) can also be interpreted as a "perturbed" variant of (2.1). Moreover, the condition (2.3) can be considered from two points of view:

- as a kind of a "linearization" procedure applied to a known function $g(\cdot)$;
- as an a priori estimate for of the perturbation associated with a given system (2.4).

In this book, we will consider both of these as interpretations of the basic quasi-Lipschitz condition. As we can see, the class of the right-hand sides introduced above also contains some possible discontinuous functions. In general, continuous and discontinuous affine control systems and dynamic models of type (2.4) have become a modern application focus of practical control theory; cf. (Bartolini, Fridman, Pisano, & Usai 2008; Poznyak 2008; Utkin 1992). We next discuss the solution concepts of (2.1) with discontinuous right-hand sides.

2.1.2 Examples of Quasi-Lipschitz Systems

Let us continue with a brief discussion of some examples of dynamical systems that allow the quasi-Lipschitz modeling framework. We consider some well-known models from conventional control theory.

Example 2.1 (LPV, LTV and Uncertain Systems).   Consider the linear model with time-varying parameters given by

$$\dot{x}(t) = (A + \Delta A)x(t) + (B + \Delta B)u + f(t). \tag{2.6}$$

Here we assume that unknown variations ΔA and ΔB possibly depend on time and are bounded as

$$\|\Delta A\| \le a_+, \quad \|\Delta B\| \le b_+.$$

If $f(t)$ is also bounded, i.e., $\sup_{t \ge 0} \|f(t)\| \le f_+$, then the right-hand side of (2.6) can be represented in the form (2.4):

$$(A + \Delta A)x(t) + (B + \Delta B)u + f(t) \\ = Ax(t) + Bu + Df(t, x, u)$$

with $(D = I)$,

$$Df(t, x, u) = f(t, x, u) = \Delta Ax(t) + \Delta Bu + f(t),$$

satisfying (2.5):

$$\|Df(t, x, u)\|^2 = \|f(t, x, u)\|^2 \le$$

$$3\,(a_+)^2\,\|x\|^2 + 3\,(b_+)^2\,\|y\|^2 + (f_+)^2$$

Example 2.2 (Affine Control System).   If in the ODE

$$\dot{x}(t) = f(t, x) + Bu, \tag{2.7}$$

the function $f(t, x)$ is quasi-Lipschitz satisfying

$$\|f(t, x)\|^2 \le f_0 + f_1 \|x\|^2,$$

then the right-hand side of (2.7) also satisfies (2.5).

Example 2.3 (Nonlinear Mechanical Systems).   By the main theorem of mechanics, in each mechanical model given by

$$J(q(t))\ddot{q}(t) + Q(t, q(t), \dot{q}(t)) = Bu, \tag{2.8}$$

the matrix $J(q(t))$ is nonsingular, that is, $\det J(q(t)) \ne 0$, and hence (2.8) can be rewritten as

$$\ddot{q}(t) + J^{-1}(q(t))Q(t, q(t), \dot{q}(t)) = J^{-1}(q(t))Bu,$$

or equivalently, in standard Cauchy form, as

$$q^{(1)} = q, \; q^{(2)} = \dot{q}$$

$$\dot{q}^{(1)} = q^{(2)}$$
$$\dot{q}^{(2)} = -J^{-1}(q^{(1)})Q(t, q^{(1)}, q^{(2)}) - J^{-1}(q^{(1)})Bu \tag{2.9}$$

In general, $J(q(t))$ is bounded, i.e.,

$$\|J(q)\| \leq J^{+} < \infty,$$

and $Q(t, q(t), \dot{q}(t))$ is Lipschitz continuous and does not increase faster than linear in the second and third arguments. Therefore, the model (2.9) also satisfies (2.5) and is quasi-Lipschitz.

Example 2.4 (Relay and Sliding Mode Control Systems). If

$$\dot{x} = g(t, x, u)$$

$$u = K_1 x + K_2 \mathrm{sign}(Sx) \tag{2.10}$$

the function $g(t, x, u)$ satisfies (2.3), then the controlled system (2.10), closed by the control u, also satisfies (2.3), and hence is quasi-Lipschitz too.

Descriptively, the given class of uncertain systems of type (2.4) is quite general and contains locally Lipschitz (on an interval) and possibly some discontinuous right-hand sides in (2.4). For the ODEs of type (2.4) discontinuous right-hand side, one needs to extend a classical solution concept in order to develop a consistent modeling framework for the systems under consideration. Later, we will examine briefly a common technique for dealing with discontinuous dynamic behavior. Since the class of quasi-Lipschitz functions is sufficiently rich (in particular, it contains the locally Lipschitz functions), the set of solvable closed-loop realizations (2.3) is nonempty. Recall that the continuity of the feedback $u(\cdot)$ and function $g(\cdot, \cdot, \cdot)$ together with their derivatives

$$\frac{\partial u(x)}{\partial x}, \; \frac{dg(t, x, u)}{dx}, \; \frac{dg(t, x, u)}{dx}$$

guarantees the local Lipschitz property of the right-hand side of (2.3).

2.1.3 Differential Inclusions and General Solution Concept

In this section, we discuss briefly some existence and prolongability results for the dynamical systems of type (2.4) under the main technical assumptions from Sect. 2.1. Let us also present a standard prolongability result for a related class of dynamical systems (see, e.g., Hale 1969; Poznyak 2008).

Recall that a set $A \subset \mathbb{R}^n$ is closed if $A = \overline{A}$. We next give a formulation of the well-known Carathéodory's theorem (see, e.g., Aliprantis & Border 1999; Deimling 1992; Rockafellar 1970).

Theorem 2.1 (Carathéodory's Theorem). *Let $M \subset \mathbb{R}^n$ be an arbitrary set. Every point $x \in \overline{\mathrm{co}}M$ can be represented in the form*

$$x = \alpha_1 x_1 + \alpha_2 x_2 + \ldots + \alpha_k x_k$$

$$\alpha_i \geq 0, \alpha_1 + \alpha_2 + \ldots + \alpha_k = 1$$

where $x_i \in M$ and $k \leq n + 1$.

Next let us consider the celebrated Hausdorff semidistances (see, e.g., Aliprantis & Border 1999; Atkinson & Han 2005; Edwards 1995), formally given as

$$\rho(x, B) = \inf_{y \in B} \|x - y\|, \quad \beta(A, B) = \sup_{x \in A} \rho(x, B),$$

where

$$x, y \in \mathbb{R}^n, \ A, \ B \subset \mathbb{R}^n.$$

The following set-valued mapping (function) gives rise to a formal definition of the differential inclusions associated with the dynamic models from the previous section:

$$F : \mathbb{R}^n \to 2^{\mathbb{R}^m}.$$

Here $2^{\mathbb{R}^m}$ is a standard power set consisting of all subsets of $\mathbb{R}^m$. We continue with some auxiliary concepts and facts.

Definition 2.2. The set-valued function $F : \mathbb{R}^n \to 2^{\mathbb{R}^m}$ is said to be **upper semicontinuous** at the point x_0 if

$$\beta(F(x'), F(x)) \to 0$$

results from $x' \to x$. The set-valued function F is said to be upper semicontinuous in some domain if it is upper semicontinuous at each point of that domain.

The following simple example illustrates this definition.

Example 2.5. The set-valued extension

$$\mathrm{sign}\,(x) = \begin{cases} 1 & \text{for } x > 0 \\ \{-1, 1\} & \text{for } x = 0 \\ -1 & \text{for } x = 0 \end{cases}$$

of the sign function is upper semicontinuous.

Consider the differential inclusion associated with the set-valued function F:

$$\dot{x}(t) \in F(t, x), \ t > t_0, \tag{2.11}$$

where

$$F : \mathbb{R} \times \mathbb{R}^n \to 2^{\mathbb{R}^n}, \ x \in \mathbb{R}^n.$$

The inclusion (2.11) is usually considered with the initial condition

$$x(t_0) = x_0, \tag{2.12}$$

where $x_0 \in \mathbb{R}^n$ is some given vector. A suitable solution concept to the initial value problem (2.11)–(2.12) is determined in the wide class of absolutely continuous functions.

Definition 2.3. The absolutely continuous function $x : \mathbb{R} \to \mathbb{R}^n$ is called a **solution of the differential inclusion** (2.11) if it satisfies this inclusion almost everywhere on some segment or a time interval $\mathcal{I}$.

The Cauchy problem (2.11)–(2.12) constitutes a mathematically formal representation of the general control system given by (2.1) or (2.2). We are interested in giving a consistently general existence result that also includes the case of quasi-Lipschitz right-hand sides of the dynamic models under consideration. This requirement and as a consequence, some stability results constitute a significant reason for considering differential inclusions of the type (2.11).

Theorem 2.2 (Filippov 1988). *Let*

1. the set-valued function $F : \mathbb{R}^{n+1} \to \mathbb{R}^n$ is upper semicontinuous on t and x in the domain

$$G = \{(t, x) \in \mathbb{R}^{n+1} : a < t < b \ \text{and} \ \|x - x_0\| < c\}.$$

2. the set $F(t, x)$ is nonempty, compact, and convex for every $(t, x) \in \mathbb{R} \times \mathbb{R}^n$.
   Then the Cauchy problem (2.11)–(2.12) has an absolutely continuous solution $x(t)$ determined at least up to the exit to the boundary of G.

Analogously to the classic theory of ODEs, the prolongation of solutions to (2.11)–(2.12) requires some additional properties of the right-hand side. We give here only a semiconventional result in that direction.

Theorem 2.3. *Let the set-valued function $F : R^{n+1} \to \mathbb{R}^{n+1}$ be defined to be upper semicontinuous, compact, and convex-valued at each point of $\mathbb{R}^{n+1}$. If there exists a real-valued function*

$$L : \mathbb{R}^+ \cup \{0\} \to \mathbb{R}^+ \cup \{0\}$$

such that

$$\rho(0, F(t, x)) \leq L(\|x\|) \quad and \quad \int_0^{+\infty} \frac{1}{L(r)} dr = +\infty,$$

then for every $(t_0, x_0) \in \mathbb{R}^{n+1}$, each solution of the Cauchy problem (2.11)–(2.12) is defined for all $t \in \mathbb{R}$.

2.1.4   The Filippov Regularization Procedure

Modern control of networks with sophisticated dynamic topologies also gives rise to discontinuous systems of type (2.2) (Boiko 2009; Orlov 2008). In this case, the vector field defining the resulting dynamical system is a discontinuous function of the state, and hence system stability can be analyzed using nonsmooth Lyapunov theory involving concepts such as weak and strong stability notions, differential inclusions, and generalized gradients. In many applications of discontinuous dynamical systems, for example mechanical systems having rigid-body modes and consensus protocols for dynamical networks, the system dynamics give rise to a continuum of equilibria. Under such dynamics, the limiting system state achieved is not determined completely by the dynamics, but depends on the initial system state as well. In our book, we follow the general theoretic concept of differential inclusions, which provides a universal modeling idea and also includes the case of control systems (2.2) with quasi-Lipschitz characterization. According to the Filippov regularization approach mentioned earlier, let us introduce the set-valued functions (see Deimling 1992)

$$U(x) = \bigcap_{\delta \in \mathbb{R}_+} \bigcap_{\mu_{\mathbb{R}^n}(\mathcal{N})=0} \overline{co}\, u(\mathcal{B}(x, \delta) \backslash \mathcal{N})$$

$$G(t, x, u) := \bigcap_{\delta \in \mathbb{R}_+} \bigcap_{\mu_{\mathbb{R}^n}(\mathcal{N})=0} \bigcap_{\mu_{\mathbb{R}^m}(\mathcal{M})=0} \overline{co}\, g(t, \mathcal{B}(x, \delta) \backslash \mathcal{N}, \mathcal{B}(u, \delta) \backslash \mathcal{M})$$

where $\mathcal{N} \subset \mathbb{R}^n$ ($\mathcal{M} \subset \mathbb{R}^m$) is a set of measure zero, i.e., $\mu_{\mathbb{R}^n}(\mathcal{N}) = 0$, $\overline{co}$ denotes convex closure, and $\mathcal{B}(x, \delta)$ is the ball of radius δ with the center at x.

Let us now suppose that assumptions (**A1**) and (**A2**) introduced above hold. In that case, the set-valued functions $U(x)$, $G(t, x, u)$, and $\overline{co}G(t, x, U(x))$ are defined, upper semicontinuous, compact, and convex-valued for every $t \in \mathbb{R}$, $x \in \mathbb{R}^n$, and $u \in \mathbb{R}^m$. These properties of these set-valued functions are direct corollaries of the Filippov regularization procedure applied to the initial (nonconvex) differential inclusion (2.11) with quasi-Lipschitz right-hand side (Aubin & Celina 1984; Berkovitz 1974; Deimling 1992; Filippov 1988).

Definition 2.4. An absolutely continuous vector function $x : \mathbb{R} \to \mathbb{R}^n$ is called a **solution of the closed-loop quasi-Lipschitz system** (2.2) if it satisfies the differential inclusion

$$\dot{x}(t) \in \overline{\mathrm{co}}\, G(t, x, U(x))$$

almost everywhere on some time interval or segment $\mathcal{I}$.

The above definition gives rise to the corresponding existence result.

Theorem 2.4. *Under assumptions (A1) and (A2), for every $x_0 \in \mathbb{R}^n$ and $t_0 \in \mathbb{R}$, the Cauchy problem*

$$\begin{cases} \dot{x} = g(t, x, u(x)), \ t > t_0 \\[2ex] x(t_0) = x_0 \end{cases}$$

has at least one solution, and each solution is defined for all $t > t_0$.

The proof of this theorem is an immediate consequence of Theorems 2.1 and 2.2 and the following fundamental lemma.

Lemma 2.1. *Under assumptions (A1) and (A2), the following hold:*

- *For every $x \in \mathbb{R}^n$ and $u \in U(x)$, the inequality*

$$\|u\|_{Q_u}^2 \leq u_0 + \|x\|_{Q_{xu}}^2$$

  holds.
- *For every $x \in \mathbb{R}^n$ and $u \in \mathbb{R}^m$, the inequality*

$$\|z - Ax - Bu\|_{Q_f}^2 \leq f_0 + \|x\|_{Q_x}^2 + \|u\|_{Q_u}^2$$

  holds for all $z \in G(t, x, u)$ and $t \in \mathbb{R}$.
- *There exist $k_0 \geq 0$ and $k_1 \geq 0$ such that for every $x \in \mathbb{R}^n$, the inequality*

$$\|y\| \leq k_0 + k_1 \|x\| \tag{2.13}$$

  holds for all $y \in \overline{\mathrm{co}}\, G(t, x, U(x))$ and $t \in \mathbb{R}$.

Proof.   I. Let $x \in \mathbb{R}^n$. According Carathéodory's theorem (Theorem 2.1) and the definition of the set $U(x)$, every $u \in U(x)$ can be represented in the form

$$u = \sum_{i=1}^{k} \alpha_i u_i, \ \alpha_i \geq 0, \ \sum_{i=1}^{k} \alpha_i = 1, \ k \leq m + 1,$$

where

$$u_i = \lim_{x_j \to x} u(x_j), u_i \in U(x),$$

and $x_j \in \mathbb{R}^n$, $\{x_j\}_{j=1}^\infty$ is some sequence converging to x (see Aubin & Celina 1984; Deimling 1992; Filippov 1988; Himmelberg 1975 for details). Assumption **(A2)** implies

$$u_i^T Q_u u_i \leq u_0 + x^T Q_{xu} x$$

for all $i = 1, 2, \ldots, k$. Hence

$$\sum_{i=1}^k \alpha_i u_i^T Q_u u_i \leq u_0 + x^T Q_{xu} x.$$

Finally, using Jensen's inequality (Atkinson & Han 2005; Poznyak 2008) we obtain

$$\left(\sum_{i=1}^k \alpha_i u_i \right)^T Q_u \left(\sum_{i=1}^k \alpha_i u_i \right) \leq \sum_{i=1}^k \alpha_i u_i^T Q_u u_i.$$

II. Similar considerations prove the second statement of the lemma.

III. By Carathéodory's theorem, every $y \in \overline{co} G(t, x, U(x))$ can be represented as

$$y = \sum_{i=1}^k \alpha_i z_i, \quad \alpha_i \geq 0, \sum_{i=1}^k \alpha_i = 1, k \leq n + 1,$$

where

$$z_i \in G(t, x, U(x)).$$

Then there exists $u_i \in U(x)$ such that $z_i \in G(t, x, u_i)$. For

$$w_i = z_i - Ax - Bu_i,$$

we evidently have

$$\|w_i\|_{Q_f}^2 \leq f_0 + \|x\|_{Q_x}^2 + \|u_i\|_{Q_u}^2 \leq f_0 + u_0 + \|x\|_{Q_x}^2 + \|x\|_{Q_{xu}}^2.$$

Finally, taking into account that

$$y = Ax + B \sum_{i=1}^k \alpha_i u_i + \sum_{i=1}^k \alpha_i w_i,$$

we complete the proof.   ∎

Corollary 2.1. *If the feedback control $u(x)$ is a continuous function, then*

$$\overline{co}\,G(t, x, U(x)) = G(t, x, u(x)),$$

and for every $x \in \mathbb{R}^n$, the inequality

$$\|z - Ax - Bu(x)\|_{Q_f}^2 \leq f_0 + \|x\|_{Q_x}^2 + \|u(x)\|_{Q_u}^2$$

holds for every $z \in G(t, x, u(x))$ and for almost all $t \in \mathbb{R}$.

This last result establishes a useful relationship between the derivative of the differential inclusion under consideration and the linear part of the initial control system.

2.2 The Lyapunov Approach to Quasi-Lipschitz Dynamical Systems

The methodology proposed in this book has essential formal interconnections with the conventional Lyapunov-like technique from the set-stability approach. We refer to Zubov (1962) and Poznyak (2008) for some basic definitions and constructive results. However, the numerically oriented robust control design concept we follow here allows us to establish "practical stability" not only for a concrete uncertain system but also for a wide family of dynamical systems. This family is naturally given by the quasi-Lipschitz condition for the right-hand sides of the resulting dynamical systems of type (2.2). In that sense, the approach we develop here is more general, and moreover, it initially provides direct access to the powerful computational LMI-related methods. In order to give consistent consideration to the numerical methods for the proposed robust control schemes, we need to introduce some basic ideas from set-stability theory.

Definition 2.5 (Positive Invariant Set). The set Ω is said to be **positive invariant** for the system (2.11) if every solution of the Cauchy problem (2.11), (2.12) with $x_0 \in \Omega$ satisfies the inclusion

$$x(t) \in \Omega \quad \text{for all } t > t_0.$$

Note that there are many theoretical and practical examples of invariant sets for dynamical systems scattered in the modern applied mathematical literature. Roughly speaking, a set in the state space is said to be positively invariant (for a given dynamical system) if every trajectory initiated in this set remains inside the set at all future time. The theoretical questions related to the existence and possible constructive characterizations of an invariant set are very sophisticated mathematical questions. Under some structural assumptions, it is possible to apply the celebrated invariant ellipsoid method and to determine an invariant set constructively. This set will usually be geometrically chosen in the form of an ellipsoid in the given state space of the system. Our book essentially extends the classic invariant ellipsoid approach to a class of dynamical systems with a wide variety of right-hand sides. This class (as we have seen above) also includes the representative family of discontinuous dynamical systems.

Definition 2.6 (Attractive Set). The set Ω is said to be Lyapunov asymptotically attractive for the system (2.11) if every solution of the Cauchy problem (2.11), (2.12) with $x_0 \notin \Omega$ tends to Ω as t tends to infinity, i.e.,

$$\rho(x(t), \Omega) \to 0 \ \text{ if } t \to +\infty.$$

Note that the attractivity property mentioned above does not imply in general the Lyapunov asymptotic stability of the invariant set under consideration. Well-known counterexamples are given in Blanchini (1999). Recall that the simple Lyapunov stability of a set Ω usually can be defined as the following requirement: for all $\epsilon > 0$, there exists $\delta_1 > 0$ such that

$$\text{dist}\{x(0), \Omega\} < \delta_1$$

implies

$$\text{dist}\{z(t), \mathcal{D}\} < \epsilon, \ t \geq 0.$$

It is notable that the various versions of the basic Lyapunov function method provide the main tools for stability and robustness analysis and the corresponding control design for nonlinear control systems. In that connection, let us recall the fundamental Lyapunov function concept.

Definition 2.7. The function $V : \mathbb{R}^n \to \mathbb{R}$ is said to be **proper** if it satisfies the following conditions:

- Iy id continuously differentiable in $\mathbb{R}^n$.
- It is positive finite ($V(x) > 0$ for $x \neq 0$ and $V(0) = 0$).
- Iy is radially unbounded ($\|x\| \to +\infty$ implies $V(x) \to +\infty$).

Let us now continue by considering an ellipsoidal set

$$\mathcal{E} := \{z \in \mathbb{R}^n \mid z^T P^{-1} z \leq 1\}$$

associated with the closed-loop system (2.2). Here P is a symmetric positive definite $n \times n$ matrix, called the shape (or configuration) matrix of the ellipsoid. Our aim is to generate a simple feedback-type control strategy $u(x)$ such that $\mathcal{E}$ is a globally asymptotically stable positively invariant set for the realization (2.3). Based on the classical concepts mentioned above, we now introduce our local definition of the attractive ellipsoid.

Definition 2.8. We call $\mathcal{E}$ an **attractive ellipsoid** for the closed-loop system (2.2) if it is a globally asymptotically attractive invariant set of a system (2.3).

The analytic background of the attractive ellipsoid method we developed for the class of systems with quasi-Lipschitz right-hand sides is given by the following simple conceptual result.

Theorem 2.5. *Let assumptions* **(A1)**, **(A2)** *hold, and let* $u : \mathbb{R}^n \to \mathbb{R}^m$ *be a continuous function. If there exists a proper function*

$$V : \mathbb{R}^n \to \mathbb{R}_+ \cup \{0\}$$

such that

$$\frac{\partial V(x)}{\partial x}(Ax + Bu(x) + w) < 0$$

$$\text{for all } x, w \in \mathbb{R}^n : \left\{ \begin{array}{c} \|w\|_{Q_f} \leq \delta + \|x\|_{Q_x} + \|u(x)\|_{Q_u} \\ V(x) > 1 \end{array} \right.$$

(2.14)

then the set

$$\Omega = \{x \in \mathbb{R}^n : V(x) \leq 1\}$$

is **asymptotically attractive** *and the* **invariant set** *of the quasi-Lipschitz system (2.2) with feedback control* $u = u(x)$.

Proof. By Theorem (2.5), the quasi-Lipschitz system (2.3) has solutions for every set of initial conditions. Let $x(t)$ be an arbitrary solution. Corollary 2.1 implies that the relations

$$\dot{V}(x(t)) = \frac{\partial V(x)}{\partial x}\dot{x}(t) = \frac{\partial V(x)}{\partial x}y(t)$$

$$y(t) \in G(t, x(t), u(x(t)))$$

hold for almost all t. Observe that by the same corollary, we have

$$\|y(t) - Ax(t) - Bu(x(t))\|^2_{Q_f} \leq$$

$$f_0 + \|x(t)\|^2_{Q_x} + \|u(x(t))\|^2_{Q_u}$$

Hence, condition (2.14) implies that

$$\dot{V}(x(t)) < 0$$

for $V(x(t)) > 1$. So the set Ω is invariant and asymptotically attractive. ∎

Theorem 2.5 makes it possible to specify constructively an attractive invariant set not only for a concrete system (2.3), but also for the class (family) of corresponding dynamic processes that possess quasi-Lipschitz right-hand sides. The result presented provides a basic theoretical tool for a possible numerical (algorithmic) scheme for robust control design and analysis of the quasi-Lipschitz systems introduced here. Particularly, constructive algorithms for robust control design can be obtained if the function $V(x)$ in Theorem 2.5 is quadratic. The corresponding robust and/or optimal control design schemes become LMI constraints in this case.

Let us suppose here (in a specific case) that the resulting system (2.3) is closed by a linear-type feedback

$$u(x) := Kx, \tag{2.15}$$

where $K \in \mathbb{R}^{m \times n}$ is an appropriate gain matrix. In that case, an attractive ellipsoid $\mathcal{E}$ (introduced above) that is associated with the given system (2.2) is characterized by a matrix pair $\{P, K\}$. Clearly, the chosen controller matrix K from that pair determines a dynamic behavior of the state vector $x(t)$ such that the ellipsoid inequality

$$x^T(t)P^{-1}x(t) \leq 1$$

holds in an exact or approximate sense. We call this dynamic behavior a "practically stable" system realization (2.3).

From the point of view of implementable control applications, we are strongly interested in constructing a concrete attractive ellipsoid $\mathcal{E}$ of minimal size (in some suitable sense). This requirement can be formalized in the form of a specific minimization problem related to a characteristic parameter of $\mathcal{E}$. We follow the robust concept and define P such that the size of the attractive ellipsoid $\mathcal{E}$ under construction will be minimal. This minimizing problem evidently includes some natural additional restrictions for the "free" parameters, namely for P and for the gain matrix K. The above "minimality" property can be mathematically formalized by the following optimization problem (see also Azhmyakov 2011; Poznyak, Azhmyakov, & Mera 2011):

$$\text{minimize tr}\{P\} \text{ by } \{P, K\} \in \Gamma \tag{2.16}$$

$$\text{subject to } P > 0, P = P^T$$

Here Γ is a set of restrictions that determines a class of admissible matrices $P \in \mathbb{R}^{n \times n}$ and $K \in \mathbb{R}^{m \times n}$ such that $\mathcal{E}(P)$ has the property of being attractive and/or invariant for the corresponding closed-loop system (2.4), (2.15).

The trace $\text{tr}\{P\}$ defines the sum of the squares of the ellipsoid's semiaxes. The minimization problem (2.16) guarantees the minimization of the "size" of the desired invariant ellipsoid $\mathcal{E}$.

We also need to assume that (2.16) has an optimal solution $(\hat{P}, \hat{K})$. An ellipsoid determined by $\hat{P}$ is called a minimal attractive ellipsoid associated with (2.16). It is clear that the resulting control strategy closed by the specific (linear) feedback control $u(x) = \hat{K}x$ will possess some good robustness properties with respect to the above-mentioned attractive ellipsoidal region in the state space of (2.4). A constructive solution of (2.16) constitutes our generic approach to the robust control design for the class of quasi-Lipschitz uncertain control systems of type (2.4).

The main problem, then, is to give a concrete constructive characterization of the above set of restrictions Γ. We use here the LMI technique for this purpose.

2.3 Elements of LMIs

2.3.1 Main Concepts

This section is devoted to a brief overview of LMIs (see, e.g., Ghaoui & Niculescu 2000; Scherer & Weiland 2000).

Recall that an LMI is a specific expression of the form

$$F(x) := F_0 + x_1 F_1 + \ldots + x_n F_n < 0, \tag{2.17}$$

where

- $x = (x_1, \ldots, x_n)$ is a vector of n real numbers, called the decision variables;
- $F_0, \ldots, F_n$ are real symmetric matrices, i.e.,

$$F_j = F_j^T$$

  for $j = 0, \ldots, n$;
- the inequality $\prec 0$ in (2.17) means "negative definite," that is,

$$w^T F(x) w < 0$$

for all nonzero real vectors w. Because all eigenvalues of a real symmetric matrix are real, it follows that (2.17) is equivalent to saying that all eigenvalues $\lambda(F(x))$ are negative. Equivalently, the maximal eigenvalue also is negative:

$$\lambda_{\max}(F(x)) < 0.$$

Recently, the LMI concept has played a key role in the theoretical and computational treatment of some modern robust control design techniques. We refer to Boyd, Ghaoui, Feron, and Balakrishnan (1994) for some basic facts related to LMI methods in control. In connection with the further treatment of LMIs, it is convenient to introduce some classical notation associated with symmetric and Hermitian matrices. A matrix A is Hermitian if it is square and

$$A = A^* = \bar{A}^T,$$

where $\bar{A}$ denotes the matrix each entry of which is the complex conjugate of the corresponding entry in A. If A is real, then this amounts to saying that $A = A^T$, and we call A symmetric. The sets of all $m \times m$ Hermitian and symmetric matrices will be denoted by $\mathbb{H}^m$ and $\mathbb{S}$, respectively, and we will omit the superscript m if the dimension is not relevant for the context.

Definition 2.9 (Linear Matrix Inequality). A linear matrix inequality (LMI) is an inequality

$$F(x) < 0, \tag{2.18}$$

where F is an affine function mapping a finite-dimensional vector space $\mathbb{X}$ to either $\mathbb{H}$ or $\mathbb{S}$.

Remark 2.1. An affine mapping

$$F : \mathbb{X} \longrightarrow \mathbb{S}$$

necessarily takes the form

$$F(x) = F_0 + T(x),$$

where $F_0 \in \mathbb{S}$ (i.e., F_0 is real symmetric) and $T : \mathbb{X} \to \mathbb{S}$ is a linear transformation. Thus if $\mathbb{X}$ is finite-dimensional, say of dimension n, and $\{e_1, \dots, e_n\}$ constitutes a basis for $\mathbb{X}$, then every $x \in \mathbb{X}$ can be represented as

$$x = \sum_{j=1}^{n} x_j e_j,$$

and we can write

$$T(x) = \sum_{j=1}^{n} x_j e_j = \sum_{j=1}^{n} x_j F_j,$$

where $F_j = T(e_j) \in \mathbb{S}$. We next easily obtain (2.17) as a special case.

Note that in most control applications, LMIs arise as functions of matrix variables rather than scalar-valued decision variables. This means that we consider inequalities of the form (2.18), where $\mathbb{X} = \mathbb{R}^{m_1 \times m_2}$. A simple example with $m_1 = m_2 = m$ constitutes the celebrated Lyapunov inequality

$$F(X) = A^T X + XA + Q < 0.$$

Here, $A, Q \in \mathbb{R}^{m \times m}$ are assumed to be given, and X is the unknown matrix variable of dimension $m \times m$. Note that this defines an LMI only if $Q \in \mathbb{S}^m$. We can view this LMI as a special case of (2.17) by defining an arbitrary basis $e_1, \dots, e_n$ of $\mathbb{X}$ and writing

$$X = \sum_{j=1}^{n} x_j e_j.$$

Then we can deduce the following:

$$F(X) = F\left(\sum_{j=1}^{n} x_j e_j\right) = F_0 + \sum_{j=1}^{n} x_j F(e_j) = F_0 + \sum_{j=1}^{n} x_j F_j,$$

which is in the basic form of (2.17). The coefficients x_j in the expansion of X define the decision variables. Note that the number of decision variables n is at most m^2, and it depends on the structure imposed on the matrix variable X.

For example, if the matrix variable X is required to be symmetric, then $\mathbb{X} = \mathbb{S}^m$ and $n = m(m + 1)/2$. Let us also note that a nonstrict LMI is a linear matrix inequality whereby $<$ in (2.17) and (2.18) is replaced by $\leq$. Moreover, matrix inequalities of the type

$$F(x) > 0 \text{ and } F(x) < G(x)$$

with F and G affine functions are obtained as special cases of Definition 2.9, since they can be rewritten as the following LMIs:

$$F(x) > 0 \text{ and } F(x) < G(x) < 0.$$

Let us next introduce a definition of systems of LMIs.

Definition 2.10 (Systems of LMIs). A system of LMIs is a finite set of LMIs

$$F_1(x) < 0, \ldots, F_k(x) < 0. \tag{2.19}$$

As a consequence of the properties of convex sets in a normed vector space, we can infer that the intersection of the feasible sets of each of the inequalities (2.10) is convex. In other words, the set of all variables x that satisfy (2.10) is convex. The question now arises whether this set can be represented as the feasibility set of another LMI. The positive answer to the above question provides an analytic basis for some effective numerical methods associated with the algorithmic solution procedures to Definition 2.9. Indeed,

$$F_1(x) < 0, \ldots, F_k(x) < 0$$

if and only if

$$F(x) := \begin{pmatrix} F_1(x) & 0 & \cdots & 0 \\ 0 & F_2(x) & \cdots & 0 \\ \vdots & & \ddots & \vdots \\ 0 & 0 & \cdots & F_k(x) \end{pmatrix} < 0$$

The last inequality makes sense because $F(x)$ is symmetric (or Hermitian) for every x. Further, since the set of eigenvalues of $F(x)$ is simply the union of the eigenvalues of $F_1(x), \ldots, F_k(x)$, every x that satisfies $F(x) < 0$ also satisfies the system of LMIs (2.19) and conversely. We conclude that multiple LMI constraints can always be converted to a single LMI constraint.

The next important property we consider is related to affine constraints in LMIs. By this, we mean that combined constraints (in the unknown x) of the form

$$\begin{cases} F(x) < 0 \\ Ax = b \end{cases} \quad \text{or} \quad \begin{cases} F(x) < 0 \\ x = Bu + c \quad \text{for some } u \end{cases}$$

where the affine function $F : \mathbb{R}^n \to \mathbb{S}$, matrices A and B, and vectors c and b are given, can be lumped into a single linear matrix inequality

$$\hat{F}(\hat{x}) < 0.$$

More generally, the combined equations

$$\begin{cases} F(x) < 0, \\ x \in \mathcal{M}, \end{cases} \tag{2.20}$$

where $\mathcal{M}$ is an affine set in $\mathbb{R}^n$, can be rewritten in the form of a single linear matrix inequality

$$\hat{F}(\hat{x}) < 0.$$

To do this, let us first recall that affine sets $\mathcal{M}$ can be written as

$$\mathcal{M} = \{x \mid x = x_0 + m, \; m \in \mathcal{M}_0\}$$

with $x_0 \in \mathbb{R}^n$ and $\mathcal{M}_0$ a linear subspace of $\mathbb{R}^n$. Suppose that $\hat{n} = \dim(\mathcal{M}_0)$, and let

$$e_1, \ldots, e_{\hat{n}} \in \mathbb{R}^n$$

be a basis of $\mathcal{M}_0$. Let

$$F(x) = F_0 + T(x)$$

be decomposed as in Remark 2.1. Then (2.20) can be rewritten as

$$0 > F(x) = F_0 + T\left(x_0 + \sum_{j=1}^{\hat{n}} x_j e_j\right) =$$

$$F_0 + T(x_0) + \sum_{j=1}^{\hat{n}} x_j T(e_j) =$$

$$\hat{F}_0 + x_1 \hat{F}_1 + \ldots + x_{\hat{n}} \hat{F}_{\hat{n}} =: \hat{F}(\hat{x})$$

where

$$\hat{F}_0 = F_0 + T(x_0), \ \hat{F}_j = T(e_j)$$

and $\hat{x} = \mathrm{col}(x_1, \ldots, x_{\hat{n}})$ are the coefficients of $x - x_0$ in the basis of $\mathcal{M}_0$. This implies that $x \in \mathbb{R}$ satisfies (2.20) if and only if $\hat{F}(\hat{x}) < 0$. Note that the dimension $\hat{n}$ of $\hat{x}$ is equal to at most the dimension n of x.

We now consider some illustrative examples of LMIs in use.

Example 2.6. Consider the problem of determining the exponential stability of the linear autonomous system

$$\dot{x} = Ax, \tag{2.21}$$

where $A \in \mathcal{R}^{n \times n}$. By this, we mean the problem of deciding whether all functions $x : \mathbb{R} \leftarrow \mathbb{R}^n$ that satisfy (2.21) have the property

$$\lim_{t \to \infty} x(t) = 0.$$

Lyapunov taught us that the system (2.21) is exponentially stable if and only if there exists $X = X^T$ such that

$$X > 0$$

and

$$A^T X + XA < 0.$$

Indeed, in that case, the function

$$V(x) := x^T X x$$

qualifies as a Lyapunov function in that it is positive for all nonzero x and is strictly decaying along solutions x of (2.21). Thus, asymptotic stability of the system (2.21) is equivalent to the feasibility problem of the following LMI:

$$\begin{pmatrix} -X & 0 \\ 0 & A^T X + XA \end{pmatrix} < 0$$

Example 2.7. Consider the linear autonomous system

$$\dot{x} = Ax \tag{2.22}$$

together with an arbitrary (but fixed) initial value $x(0) = x_0$ and the criterion function

$$J := \int_0^\infty x^T(t) Q x(t) dt, \ Q = Q^T \geq 0.$$

Let us assume that the system is asymptotically stable. Then all solutions x of (2.22) are square integrable, so that $J < \infty$. Now consider the nonstrict linear matrix inequality

$$A^T X + XA + Q \leq 0.$$

For every solution $X = X^T$ of this LMI we can differentiate $x^T(t)Xx(t)$ along solutions x of (2.22) to obtain

$$\frac{d}{dt}\left[x^T(t)Xx(t)\right] = x^T(t)\left[A^T X + XA\right]x(t) \leq -x^T(t)Qx(t).$$

We now assume that $X > 0$. Then integrating the latter inequality from $t = 0$ to ∞ yields the upper bound

$$J = \int_0^\infty x^T(t)Qx(t)dt \leq x_0^t X x_0,$$

where we used that

$$\lim_{t \to \infty} x(t) = 0.$$

Moreover, the smallest upper bound of J is obtained by minimizing the objective function

$$f(X) := x_0^T X x_0$$

over all $X = X^T$ that satisfy $X > 0$ y $A^T X + XA + Q \leq 0$. Again, this is an optimization problem with an LMI constraint.

2.3.2   Existence of Solutions of LMIs

Let us finally present some particular existence results for systems of LMIs. The corresponding proofs can be found in Balandin and Kogan (2007). The following supporting lemmas are useful for the corresponding analysis.

Lemma 2.2. *If the matrix equation*

$$AX = C \tag{2.23}$$

(with $A \in \mathbb{R}^{m \times n}$, $C \in \mathbb{R}^{m \times q}$) is resolvable with respect to the unknown matrix $X \in \mathbb{R}^{n \times q}$, then among its solutions there exists a solution $\overset{\circ}{X}$ of minimal rank such that

$$\operatorname{rank}\overset{\circ}{X} = \operatorname{rank}C := r_c, \tag{2.24}$$

which can be represented as

$$\overset{\circ}{X} = VC, \tag{2.25}$$

where $V \in \mathbb{R}^{n \times m}$ is some matrix.

Proof. Without loss of generality, one may suggest that the first r_c columns of the matrix C are linearly independent and the others are linear combinations of these first ones. This means precisely that

$$C = \left[C_1 \vdots C_2 \right], \ C_2 = C_1 D$$

for some matrix $D \in \mathbb{R}^{(m-r_c) \times q}$. Represent X in the form

$$X = \left[X_1 \vdots X_2 \right], \ X_1 \in \mathbb{R}^{n \times r_c}, X_2 \in \mathbb{R}^{(n-r_c) \times q},$$

where X_1 is a solution of (2.23), i.e.,

$$A X_1 = C_1,$$

so that the columns of X_1 are linearly independent. Define

$$\overset{\circ}{X}_2 = X_1 D$$

satisfying

$$A \overset{\circ}{X}_2 = C_2.$$

Then we can say that the matrix $\overset{\circ}{X} = \left[X_1 \vdots \overset{\circ}{X}_2 \right]$ can be taken as a solution of (2.23) with minimal rank. Since we have $A \overset{\circ}{X} = C$, the matrix C is a linear combination of the rows of the matrix $\overset{\circ}{X}$. And conversely, by (2.24), it follows that the rows of $\overset{\circ}{X}$ are a linear combination of the rows of the matrix C, which can be expressed as (2.25). ∎

Lemma 2.3. *The matrix equation*

$$A X B = C \tag{2.26}$$

is feasible (resolvable) with respect to the matrix X if and only if the two matrix equations

$$AY = C, \ ZB = C \tag{2.27}$$

are feasible with respect to the unknown matrices Y and Z.

Proof. (A) *Necessity.* If X is a solution of (2.26), then it is clear that

$$Y = XB \text{ and } Z = AX$$

satisfy (2.27).

(B) *Sufficiency.* Let Y and Z be solutions of (2.27). Then by Lemma 2.2, the first equation in (2.27) has a solution $\overset{\circ}{Y}$ of minimal rank r_c such that it may be represented as $\overset{\circ}{Y} = VC$. Hence,

$$C = A\overset{\circ}{Y} = AVC = AVZB,$$

and as a result, the matrix $X := VZ$ is a solution of (2.26).   ∎

Now let us prove the following theorem on the feasibility of LMIs.

Theorem 2.6 (On the Feasibility of LMIs).   *Let $P \in \mathbb{R}^{l \times n}$ and $Q \in \mathbb{R}^{k \times n}$ be given matrices, and $\Psi \in \mathbb{R}^{n \times n}$ a given symmetric matrix.*

(1) If

$$\operatorname{rank} P = n \text{ and } \operatorname{rank} Q := r_Q < n, \tag{2.28}$$

then the LMI

$$\Psi + P^\mathsf{T} X^\mathsf{T} Q + Q^\mathsf{T} X P < 0 \tag{2.29}$$

has a solution with respect to $X \in \mathbb{R}^{k \times l}$ if and only if

$$W_Q^\mathsf{T} \Psi W_Q < 0, \tag{2.30}$$

where the columns of the matrix W_Q constitute the basis of the kernel

$$\mathcal{N}(Q) = \ker Q := \{x \in \mathbb{R}^n \mid Qx = 0\} \tag{2.31}$$

of the matrix Q, that is, W_Q satisfies the condition

$$QW_Q = 0. \tag{2.32}$$

(2) If

$$\operatorname{rank} P := r_P < n \text{ and } \operatorname{rank} Q := r_Q < n, \tag{2.33}$$

then the LMI (2.29) has a solution with respect to $X \in \mathbb{R}^{k \times l}$ if and only if

$$W_P^\mathsf{T} \Psi W_P < 0 \text{ and } W_Q^\mathsf{T} \Psi W_Q < 0, \tag{2.34}$$

where the columns of the matrix W_P constitute the basis of the kernel

$$\mathcal{N}(P) = \ker P := \{x \in \mathbb{R}^n \mid Px = 0\}$$

of the matrix P, that is, W_P, analogously to (2.32), satisfies the condition

$$P W_P = 0. \tag{2.35}$$

Proof. (1) Consider the case (2.28).

Necessity. Suppose that (2.29) is satisfied. Multiplying (2.29) by W_Q^T from the left and by W_Q^T from the right, we obtain (2.30).

Sufficiency. Suppose that (2.30) is satisfied. Let us represent the space $\mathbb{R}^n$ as

$$\mathbb{R}^n = \mathcal{R}(Q^T) \oplus \mathcal{N}(Q),$$

where

$$\mathcal{R}(Q^\mathsf{T}) = \mathrm{Im}A := \{y \in \mathbb{R}^n : y = Q^\mathsf{T}x, \ x \in \mathbb{R}^k\}$$

is the image (or range) of the matrix Q, and $\mathcal{N}(Q)$ is its kernel (2.31). Select the corresponding basis in $\mathbb{R}^n$ such that the matrix will have the representation

$$Q = \left[Q_1 \vdots 0_{k \times (n-r_Q)} \right],$$

where $Q_1 \in \mathbb{R}^{k \times r_Q}$ has full rank. In this basis, the matrices P and Ψ have the following structure:

$$P = \left[P_1 \vdots P_2 \right], \quad \Psi = \begin{bmatrix} \Psi_{11} & \Psi_{12} \\ \Psi_{12}^\mathsf{T} & \Psi_{22} \end{bmatrix}$$

where

$$P_1 \in \mathbb{R}^{l \times r_Q}, \ P_2 \in \mathbb{R}^{l \times (n-r_Q)}, \ \Psi_{11} \in \mathbb{R}^{r_Q \times r_Q}, \ \Psi_{22} \in \mathbb{R}^{(n-r_Q) \times (n-r_Q)}.$$

Recall that the matrix W_Q has maximal rank r_Q (since it constitutes a basis) and satisfies (2.32). That is why W_Q may be taken as

$$W_Q = \begin{bmatrix} 0_{r_Q \times (n-r_Q)} \\ I_{(n-r_Q) \times (n-r_Q)} \end{bmatrix} \quad \text{or} \quad W_Q^\mathsf{T} = \left[0_{r_Q \times r_Q} \ I_{(n-r_Q) \times (n-r_Q)} \right] \tag{2.36}$$

Then the condition (2.30) may be rewritten as

$$\Psi_{22} < 0,$$

since

$$W_Q^{\mathsf{T}} \Psi W_Q =$$

$$\begin{bmatrix} 0_{r_Q \times r_Q} & I_{(n-r_Q) \times (n-r_Q)} \end{bmatrix} \begin{bmatrix} \Psi_{11} & \Psi_{12} \\ \Psi_{12}^{\mathsf{T}} & \Psi_{22} \end{bmatrix} \begin{bmatrix} 0_{r_Q \times (n-r_Q)} \\ I_{(n-r_Q) \times (n-r_Q)} \end{bmatrix}$$

$$= \begin{bmatrix} 0_{r_Q \times r_Q} & I_{(n-r_Q) \times (n-r_Q)} \end{bmatrix} \begin{bmatrix} \Psi_{12} \\ \Psi_{22} \end{bmatrix} = \Psi_{22}$$

Moreover, the main inequality becomes

$$\begin{bmatrix} \Psi_{11} + Q_1^{\mathsf{T}} X P_1 + P_1^{\mathsf{T}} X Q_1 & \Psi_{12} + Q_1^{\mathsf{T}} X P_2 \\ \Psi_{12}^{\mathsf{T}} + P_2^{\mathsf{T}} X Q_1 & \Psi_{22} \end{bmatrix} < 0. \tag{2.37}$$

According to Lemma 2.3, for a given matrix $K = \begin{bmatrix} K_1 \vdots K_2 \end{bmatrix}, K_1 \in \mathbb{R}^{r_Q \times r_Q}, K_2 \in \mathbb{R}^{r_Q \times (n-r_Q)}$, the matrix equation

$$Q_1^{\mathsf{T}} X P = K \tag{2.38}$$

has a solution with respect to X if and only if the two matrix equations

$$Q_1^{\mathsf{T}} Y = K \text{ and } Z P = K$$

have solutions with respect to $Y \in \mathbb{R}^{k \times n}$ and $Z \in \mathbb{R}^{n \times l}$, respectively. Since $(r_Q \times k)$-matrix Q_1^{T} has rank $r_Q \leq k$ and the $(l \times n)$ matrix P has rank $P = n$, both equations above are solvable with respect to the matrices Y and Z. The LMI (2.37) has a solution with respect to X if the LMI

$$\begin{bmatrix} \Psi_{11} + K_1 + K_1^{\mathsf{T}} & \Psi_{12} + K_2 \\ \Psi_{12}^{\mathsf{T}} + K_2^{\mathsf{T}} & \Psi_{22} \end{bmatrix} < 0$$

has a solution with respect to $K_1 \in \mathbb{R}^{r_Q \times r_Q}, K_2 \in \mathbb{R}^{r_Q \times (n-r_Q)}$. From the inequality $\Psi_{22} < 0$, we see that the last LMI is feasible, for example with $K_2 = -\Psi_{12}$ and $K_1 = -\Psi_{11} - I_{r_Q \times r_Q}$.

(2) Now consider the case (2.33).

Necessity. Suppose that (2.29) is satisfied. Multiplying first, (2.29) by W_P^{T} from the left and by W_P from the right and then, analogously, multiplying (2.29) by W_Q^{T} from the left and by W_Q from the right, we obtain (2.34).

Sufficiency. Suppose now that both inequalities (2.34) are satisfied. Let us represent $\mathbb{R}^n$ as the direct sum:

$$\mathbb{R}^n = (\mathcal{N}(P) \setminus [\mathcal{N}(P) \cap \mathcal{N}(Q)]) \oplus [\mathcal{N}(P) \cap \mathcal{N}(Q)]$$
$$\oplus (\mathcal{N}(Q) \setminus [\mathcal{N}(P) \cap \mathcal{N}(Q)]) \oplus \mathcal{M},$$

where M is the complement of $[\mathcal{N}(P) \cap \mathcal{N}(Q)]$, i.e.,

$$\mathbb{R}^n = [\mathcal{N}(P) \cap \mathcal{N}(Q)] \oplus M$$

Selecting an appropriate basis, we have the following representation:

$$P = \begin{bmatrix} 0 \vdots 0 \vdots P_1 \vdots P_2 \end{bmatrix}, \quad Q = \begin{bmatrix} Q_1 \vdots 0 \vdots 0 \vdots Q_2 \end{bmatrix}$$
$$\Psi = [\Psi_{ij}]_{i,j=1,2,3,4}.$$

Obviously, in such a format, the matrices W_P and W_Q are of the following form:

$$W_P = \begin{bmatrix} I & 0 \\ 0 & I \\ 0 & 0 \\ 0 & 0 \end{bmatrix}, \quad W_Q = \begin{bmatrix} 0 & 0 \\ I & 0 \\ 0 & I \\ 0 & 0 \end{bmatrix},$$

and the inequalities (2.34) become

$$\begin{bmatrix} \Psi_{11} & \Psi_{12} \\ \Psi_{12}^{\mathsf{T}} & \Psi_{22} \end{bmatrix} < 0, \quad \begin{bmatrix} \Psi_{22} & \Psi_{23} \\ \Psi_{23}^{\mathsf{T}} & \Psi_{33} \end{bmatrix} < 0. \tag{2.39}$$

So, now we need to check the feasibility of the following matrix inequality

$$\Psi + P^{\mathsf{T}} X^{\mathsf{T}} Q + Q^{\mathsf{T}} X P =$$
$$\begin{bmatrix} \Psi_{11} & \Psi_{12} & \Psi_{13} + K_{11} & \Psi_{14} + K_{12} \\ \Psi_{12}^{\mathsf{T}} & \Psi_{22} & \Psi_{23} & \Psi_{24} \\ \Psi_{13}^{\mathsf{T}} + K_{11}^{\mathsf{T}} & \Psi_{23}^{\mathsf{T}} & \Psi_{33} & \Psi_{34} + K_{21}^{\mathsf{T}} \\ \Psi_{14}^{\mathsf{T}} + K_{12}^{\mathsf{T}} & \Psi_{24}^{\mathsf{T}} & \Psi_{34}^{\mathsf{T}} + K_{21} & \Psi_{44} + K_{22} + K_{22}^{\mathsf{T}} \end{bmatrix} < 0, \tag{2.40}$$

where

$$K_{ij} = Q_i^{\mathsf{T}} X P_j, \quad i, j = 1, 2.$$

First, we show that the matrix equation

$$\begin{bmatrix} Q_1^{\mathsf{T}} \\ Q_2^{\mathsf{T}} \end{bmatrix} X \begin{bmatrix} P_1 & P_2 \end{bmatrix} = K := \begin{bmatrix} K_{11} & K_{12} \\ K_{21} & K_{22} \end{bmatrix}$$

is resolvable with respect to X for every matrix K of the corresponding size. Indeed, according to Lemma 2.3, to satisfy this demand, it is necessary and sufficient to prove the feasibility of the following two matrix equations:

$$\begin{bmatrix} Q_1^{\mathsf{T}} \\ Q_2^{\mathsf{T}} \end{bmatrix} Y = \begin{bmatrix} K_{11} & K_{12} \\ K_{21} & K_{22} \end{bmatrix}, \quad Z \begin{bmatrix} P_1 & P_2 \end{bmatrix} = \begin{bmatrix} K_{11} & K_{12} \\ K_{21} & K_{22} \end{bmatrix}. \tag{2.41}$$

Note that since $(k \times r_Q)$ - matrix $\left[Q_1 \vdots Q_2 \right]$ has rank $r_Q \leq k$ and the $(l \times r_P)$ matrix $\left[P_1 \vdots P_2 \right]$ has rank $r_P \leq l$, both these equations are solvable. Therefore, for every K_{ij} $(i, j = 1, 2)$, there exists a matrix X such that the relations (2.41) hold.

Define

$$\Phi := \begin{bmatrix} \Psi_{11} & \Psi_{12} & \Psi_{13} + K_{11} \\ \Psi_{12}^{\mathsf{T}} & \Psi_{22} & \Psi_{23} \\ \Psi_{13}^{\mathsf{T}} + K_{11}^{\mathsf{T}} & \Psi_{23}^{\mathsf{T}} & \Psi_{33} \end{bmatrix}.$$

Then by Schur's complement (see Theorem 2.11 below), the relation (2.40) is satisfied if and only if

$$\Phi < 0$$

$$\left(\Psi_{44} + L_{22} + L_{22}^{\mathsf{T}} \right) - \begin{bmatrix} \Psi_{14} + K_{12} \\ \Psi_{24} \\ \Psi_{34} + K_{21}^{\mathsf{T}} \\ \Psi_{44} + K_{22} + K_{22}^{\mathsf{T}} \end{bmatrix}^{\mathsf{T}} \Phi^{-1} \begin{bmatrix} \Psi_{14} + K_{12} \\ \Psi_{24} \\ \Psi_{34} + K_{21}^{\mathsf{T}} \\ \Psi_{44} + K_{22} + K_{22}^{\mathsf{T}} \end{bmatrix} < 0.$$

On can satisfy the second inequality by a suitable selection of K_{22}. So to finish the proof, it is sufficient to demonstrate that $\Phi < 0$. Taking into account that by (2.39), we have $\Psi_{22} < 0$, consider the corresponding quadratic form

$$z^{\mathsf{T}} \Phi z = \begin{bmatrix} z_1 \\ z_2 \\ z_3 \end{bmatrix}^{\mathsf{T}} \begin{bmatrix} \Psi_{11} & \Psi_{12} & \Psi_{13} + K_{11} \\ \Psi_{12}^{\mathsf{T}} & \Psi_{22} & \Psi_{23} \\ \Psi_{13}^{\mathsf{T}} + K_{11}^{\mathsf{T}} & \Psi_{23}^{\mathsf{T}} & \Psi_{33} \end{bmatrix} \begin{bmatrix} z_1 \\ z_2 \\ z_3 \end{bmatrix} =$$

$$\left(z_2 + \Psi_{22}^{-1} \Psi_{12}^{\mathsf{T}} z_1 + \Psi_{22}^{-1} \Psi_{23} z_3 \right)^{\mathsf{T}} \Psi_{22} \left(z_2 + \Psi_{22}^{-1} \Psi_{12}^{\mathsf{T}} z_1 + \Psi_{22}^{-1} \Psi_{23} z_3 \right) +$$

$$\begin{pmatrix} z_1 \\ z_3 \end{pmatrix}^{\mathsf{T}} \begin{bmatrix} \Psi_{11} - \Psi_{12} \Psi_{22}^{-1} \Psi_{12}^{\mathsf{T}} & \Psi_{13} + K_{11} - \Psi_{12} \Psi_{22}^{-1} \Psi_{23} \\ \left[\Psi_{13} + K_{11} - \Psi_{12} \Psi_{22}^{-1} \Psi_{23} \right]^{\mathsf{T}} & \Psi_{33} - \Psi_{23}^{\mathsf{T}} \Psi_{22}^{-1} \Psi_{23} \end{bmatrix} \begin{pmatrix} z_1 \\ z_3 \end{pmatrix}.$$

The first term is negative, since $\Psi_{22} < 0$. Taking K_{11}, for example, as

$$\Psi_{13} + K_{11} - \Psi_{12} \Psi_{22}^{-1} \Psi_{23} = 0,$$

we obtain that the second term is negative, too, since again by Schur's complement, Theorem 2.11, applied to (2.39), it follows that

$$\Psi_{11} - \Psi_{12} \Psi_{22}^{-1} \Psi_{12}^{\mathsf{T}} < 0$$

and

$$\Psi_{33} - \Psi_{23}^{\mathsf{T}} \Psi_{22}^{-1} \Psi_{23} < 0.$$

This completes the proof of the theorem.   ∎

2.3.3 Numerical Approaches to LMIs

This section contains some basic computational approaches to the practical treatment of LMIs. These numerical schemes can be included in a specific LMIs-based (robust) control design procedure. In fact, we follow the modern methodology based on the feasibility and optimization techniques associated with the general types of LMIs.

Let $F : \mathbb{R}^n \to \mathbb{S}$ be affine. Recall that

$$F(x) < 0$$

if and only if

$$\lambda_{\max}(F(x)) < 0.$$

Define, for $x \in \mathbb{R}^n$, the function $f(x) := \lambda_{\max}(F(x))$ and consider the optimal value

$$V_{\text{opt}} := \inf_{x \in \mathcal{R}^n} f(x).$$

Clearly, the LMI $F(x) < 0$ is feasible if and only if

$$V_{\text{opt}} < 0.$$

It is infeasible if and only if

$$V_{\text{opt}} \geq 0.$$

The most powerful family of modern numerical methods for solving the different types of LMIs is known as the family of *interior point methods*. A major breakthrough in convex optimization lies in the introduction of the above-mentioned approaches. These methods were developed in a series of papers and became of true interest in the context of LMI problems in the work of Yurii Nesterov and Arkadii Nemirovskii (see, e.g., Ben-Tal & Nemirovski 2001 and references therein). The main idea is as follows. Let F be an affine function and let

$$\omega := \{x \,|\, F(x) < 0\}$$

be the domain of a convex function $f : \omega \to \mathcal{R}$ that we wish to minimize. That is, we consider the convex optimization problem

$$V_{\text{opt}} = \inf_{x \in \omega} f(x).$$

To solve this problem (that is, to determine optimal or almost optimal solutions), it is first necessary to introduce a barrier function. This is a smooth function ϕ that satisfies the following properties:

(a) It must be strictly convex on the interior of ω.
(b) It must approach $+\infty$ along each sequence of points $\{x_n\}_{n=1}^{\infty}$ in the interior of ω that converges to a boundary point of ω.

Given such a specific barrier function $\phi(\cdot)$, the constrained optimization problem to minimize $f(x)$ over all $x \in \omega$ is replaced by the unconstrained optimization problem to minimize the functional

$$f_t(x) := tf(x) + \phi(x),$$

where $t > 0$ is the *penalty parameter*. Note that f_t is strictly convex on $\mathbb{R}_n$. The main idea is to determine a mapping $t \mapsto x(t)$ that associates with each $t > 0$ a minimizer $x(t)$ of f_t. Subsequently, we consider the behavior of this mapping as the penalty parameter t varies. In almost all interior point methods, the latter unconstrained optimization problem is solved with the classical Newton–Raphson iteration technique Atkinson & Han 2005 to approximate the minimum of f_t. Under mild assumptions and for a suitably defined sequence of penalty parameters t_n with $t_n \to \infty$ as $n \to \infty$, the sequence $x(t_n)$ with $n \in (Z)_+$ will converge to a point x_{opt} that is a solution of the original convex optimization problem. That is, the limit

$$x_{\text{opt}} := \lim_{t \to \infty} x(t)$$

exists and

$$V_{\text{opt}} = f(x_{\text{opt}}).$$

A small modification of this theme is obtained by replacing the original constrained optimization problem by the unconstrained optimization problem to minimize

$$g_t(x) := \phi_0(t - f(x)) + \phi(x),$$

where

$$t > t_0 := V_{\text{opt}},$$

and ϕ_0 is a barrier function for the nonnegative real semiaxis. Again, the idea is to determine for every $t > 0$ a minimizer $x(t)$ of g_t (typically using the classical Newton–Raphson algorithm) and to consider the path $t \mapsto x(t)$ as a function of the penalty parameter t. The curve $t \mapsto x(t)$ with $t > t_0$ is called the path of centers for the optimization problem. Under suitable conditions, the solutions $x(t)$ are analytic and have a limit as $t \downarrow t_0$, say x_{opt}. The point

$$x_{\text{opt}} := \lim_{t \downarrow t_0} x(t)$$

is optimal in the sense that

$$V_{\text{opt}} = f(x_{\text{opt}}),$$

since for $t > t_0$, $x(t)$ is feasible and satisfies $f(x(t)) < t$.

The interior point methods described above can be applied to either of the two LMI problems as defined in the previous section. If we consider the feasibility problem associated with the LMI $F(x) < 0$, then (f does not play a role) one candidate barrier function is the logarithmic function

$$\phi(x) := \begin{cases} \log \det -F(x)^{-1} & \text{if } x \in \omega, \\ \infty & \text{otherwise.} \end{cases}$$

Under the assumption that the feasible set ω is bounded and nonempty, it follows that ϕ is strictly convex, and hence it defines a barrier function for the feasibility set ω. We know that there exists a unique x_{opt} such that $\phi(x_{\text{opt}})$ is the global minimum of ϕ. The point x_{opt} obviously belongs to ω; it is called the analytic center of the feasibility set ω. It is usually obtained in a very efficient way from the classical Newton iteration

$$x_{k+1} = x_k - (\phi''(x_k))^{-1}\phi'(x_k). \tag{2.42}$$

Here ϕ' and ϕ'' denote the gradient and the Hessian of ϕ, respectively. The convergence of this algorithm can be analyzed as follows. Since ϕ is strongly convex and sufficiently smooth, there exist numbers L and M such that for all vectors u with norm $\|u\| = 1$, we have

$$u^T \phi''(x)u \geq M,$$

$$\|\phi''(x)u - \phi''(y)u\| \leq L\|x - y\|.$$

In the case indicated above, we have

$$\|\phi'(x_{k+1})\|^2 \leq \frac{L}{2M^2}\|\phi'(x_k)\|^2,$$

so that whenever the initial value x_0 is such that

$$\frac{L}{2M^2}\|\phi'(x_k)\| < 1,$$

the method is guaranteed to converge quadratically. The idea will be to implement this algorithm in such a way that quadratic convergence can be guaranteed for the largest possible set of initial values x_0. For this reason, the iteration (2.42) is modified as follows:

$$x_{k+1} = x_k - \alpha_k(\lambda(x_k))(\phi''(x_k))^{-1}\phi'(x_k),$$

where

$$\alpha_k(\lambda) := \begin{cases} 1 & \text{if } \lambda < 2 - \sqrt{3}, \\ \dfrac{1}{1 + \lambda} & \text{if } \lambda \geq 2 - \sqrt{3}, \end{cases}$$

and

$$\lambda(x) := \sqrt{\phi'^T \phi''(x) \phi'(x)}$$

is the *Newton decrement* associated with ϕ. It is this damping factor that guarantees that x_k will converge to the analytic center, the unique minimizer of ϕ. It is important to note that the step size is variable in magnitude. The algorithm guarantees that x_k is always feasible in the sense that $x_k \in \omega$ and that x_k converges globally to a minimizer x_{opt} of ϕ. It can be shown that

$$\phi(x_k) - \phi(x_{\text{opt}}) \leq \epsilon$$

whenever

$$k \geq c_1 + c_2 \log \log(1/\epsilon) + c_3(\phi(x_0) - \phi(x_{\text{opt}})),$$

where c_1, c_2, and c_3 are constants. The first and second terms on the right-hand side do not depend on the optimization criterion and the specific LMI constraint. The second term can almost be neglected for small values of ϵ.

2.4 S-Lemma and Some Useful Mathematical Facts

We close our preliminary discussion by consideration of a celebrated linear algebraic result, namely the *S-Lemma* (sometimes called the *S-Procedure*). We follow here the brilliant survey by Polik and Terlaky (2007). Moreover, we also give in that section some additional technical facts that we will use in the course of this book. Recall that the first result related to the *S-Lemma* is known as *Finsler's lemma* (see Yakubovich 1977). Let us give the corresponding formulation.

Theorem 2.7. *Let*

$$A = A^T, \ B = B^T \in \mathbb{R}^{n \times n}$$

be some matrices. Assume that

$$y^T B y = 0, \ (y \in \mathbb{R}^n, \ \neq 0)$$

implies

$$y^T A y > 0.$$

Then there exists a (Lagrange-type multiplier) $\lambda \in \mathbb{R}$ such that $(A + \lambda B)$ is a positive definite matrix.

The next progress in that direction was a result of Yakubovitch (1977).

Theorem 2.8. *Let*

$$A = A^T, \ B = B^T \in \mathbb{R}^{n \times n}$$

be some matrices. Assume that there exists $\bar{y} \in \mathbb{R}^n$ such that

$$\bar{y}^T A \bar{y} > 0.$$

Then the inequality

$$y^T B y \geq 0$$

is a consequence of

$$y^T A y \geq 0$$

if and only if there exists a (Lagrange-type) multiplier $\lambda \geq 0$ such that

$$B \geq \lambda A.$$

Note that the proof of this fact is based on a specific fact from abstract convex analysis, namely on the Dines's lemma (published in 1941). This fundamental fact and related results can be found in Poznyak (2008). Recently, the classical Dines's lemma was generalized by Polyak (2001). An alternative proof of the generic Theorem 2.8 can be given using the so-called relaxed optimization problem

$$\min_{y}\{y^T B y \ | \ y^T A y \geq 0\}.$$

It is easy to see that the versions of the S-lemma given above are closely related to the fundamental *Farkas's lemma* (see, e.g., Ben-Tal & Nemirovski 2001; Polik & Terlaky 2007 for details). Motivated by this fact, let us also present a standard formulation of this classical result in the vector case.

Theorem 2.9. *The inequality $a^T y \geq 0$, where $a, y \in \mathbb{R}^n$, is a consequence of the system of inequalities*

$$\{b_j y \geq 0, \ j = 1, \ldots, M\}$$

if and only if there exists a (Lagrange-type multiplier) $\lambda \in \mathbb{R}^M$, $\lambda_j \geq 0$ such that

$$a = \sum_{j=1}^{M} \lambda_j b_j.$$

Some extended formulations of Farkas's lemma (homogeneous and inhomogeneous cases) can be also found in Ben-Tal and Nemirovski (2001), Rockafellar (1970). Let us also note (by name) some theoretical results related to the conventional S-procedure, namely the Gordon's theorem and Motzkin's theorem (see Polik & Terlaky 2007).

One of the most widely applicable techniques from the family of "S-lemma"-related results is known as the Lagrange duality theorem. We next give a formulation of this famous result. Let $g_j : \mathbb{R}^n \to \mathbb{R}$ be a real-valued function. Consider a system of inequalities

$$g_j(y) \leq 0$$

for $j = 1, \ldots, M$. Recall that the well-known Slater condition associated with the above inequalities is a specific interior point requirement of the following type: there exists at least one point $\bar{y}$ such that

$$g_j(\bar{y}) < 0$$

for all $j = 1, \ldots, M$.

Theorem 2.10. *Assume that $g_j(y) \leq 0$ for $j = 1, \ldots, M$ and that moreover, the Slater condition for the above system of inequalities is satisfied. The inequality*

$$f(y) \geq 0,$$

where $f : \mathbb{R}^n \to \mathbb{R}$ is a real-valued function, is a consequence of the given system of inequalities if there exists a (Lagrange-type multiplier) $\lambda \in \mathbb{R}^n$, $\lambda \geq 0$, such that

$$f(y) + \sum_{j=1}^{M} \lambda_j g_j(y) \geq 0$$

for all $y \in \mathbb{R}^n$.

Among additional useful results similar to the S-lemma, let us mention the so-called inhomogeneous S-lemma (please consult Ben-Tal & Nemirovski 2001; Polik & Terlaky 2007 for the exact formulation and proofs).

We next give an "algorithmic" formulation of a concrete version of the general S-lemma that we use in some technical proofs in our book. Note that recently, many brilliantly written overviews related to this matter have been published.

Lemma 2.4 (S-procedure). *Consider now the numbers $\alpha_0, \alpha_1, \ldots, \alpha_k \in \mathbb{R}$ and the quadratic forms*

$$q_i(x) = x^T Q_i x, \ i = 0, 1 \ldots, k,$$

where $x \in \mathbb{R}^n$, $Q_i \in \mathbb{R}^{n \times n}$, and $Q_i = Q_i^T$. If there exist numbers

$$\tau_1, \ldots, \tau_k \in \mathbb{R}_+ \cup \{0\}$$

such that

$$Q_0 \leq \tau_1 Q_1 + \tau_2 Q_2 + \ldots + \tau_k Q_k,$$

$$\alpha_0 \geq \tau_1 \alpha_1 + \tau_2 \alpha_2 + \ldots \tau_k \alpha_k,$$

(2.43)

then the system of inequalities

$$q_i(x) \leq \alpha_i \ \ for \ i = 1, 2, \ldots, k$$

(2.44)

implies the specific inequality

$$q_0(x) \leq \alpha_0.$$

(2.45)

Conversely, there exist $\tau_i \geq 0$, $i = 1, \ldots, m$, such that the inequalities (2.43) are satisfied if (2.44) implies (2.45) and one of the following conditions holds:

(a) $k = 1$;
(b) $k = 2, n = 3$ and there exist $\mu_1, \mu_2 \in \mathbb{R}$ and $x_0 \in \mathbb{R}^n$ such that

$$\mu_1 Q_1 + \mu_2 Q_2 > 0, \quad q_1(x_0) \leq \alpha_1, \quad q_2(x_0) \leq \alpha_2.$$

The proof of the formal fact we present can be found in Ben-Tal and Nemirovski (2001).

In our book, we frequently use some additional simple mathematical facts. Let us give here a short collection of such results for further reference (see Poznyak 2008; Poznyak et al. 2011).

Lemma 2.5. *Let a function $V : \mathbb{R}_+ \rightarrow \mathbb{R}_+$ satisfy the following differential inequality:*

$$\dot{V}(t) \leq -\alpha V(t) + \beta,$$

where $\alpha > 0$ and $\beta > 0$. Then

$$\overline{\lim}_{t \uparrow \infty} V(t) \leq \beta/\alpha.$$

We next formulate the following useful result (known as *Schur's lemma* Boyd et al. 1994; Poznyak 2008).

Theorem 2.11 (Schur's Complement). *Let $F : \mathbb{X} \rightarrow \mathbb{H}$ be an affine function that is partitioned according to*

$$F(x) = \begin{pmatrix} F_{11}(x) & F_{12}(x) \\ F_{21}(x) & F_{22}(x) \end{pmatrix},$$

where $F_{11}(x)$ and $F_{22}(x)$ are square. Then the following statements are equivalent:

(a)

$$F(x) < 0.$$

(b)

$$\begin{cases} F_{11}(x) < 0, \\ \\ F_{22}(x) - F_{21}(x)\left[F_{11}(x)\right]^{-1} F_{21}(x) < 0. \end{cases}$$

(c)

$$\begin{cases} F_{22}(x) < 0, \\ \\ F_{11}(x) - F_{12}(x)\left[F_{22}(x)\right]^{-1} F_{21}(x) < 0. \end{cases}$$

The proof of this elementary result can be found in Poznyak (2008). The next auxiliary result we present is usually called the Λ-*matrix inequality* (see Poznyak 2008 for details and for the corresponding proof).

Lemma 2.6 (Λ-inequality). *For any matrices $\mathcal{X}, \mathcal{Y} \in \mathbb{R}^{n \times m}$ and any symmetric positive definite matrix $\Lambda \in \mathbb{R}^{n \times n}$, the following holds:*

$$\mathcal{X}^T \mathcal{Y} + \mathcal{Y}^T \mathcal{X} \le \mathcal{X}^T \Lambda \mathcal{X} + \mathcal{Y}^T \Lambda^{-1} \mathcal{Y}.$$

Moreover, we have

$$(\mathcal{X} + \mathcal{Y})^T (\mathcal{X} + \mathcal{Y}) \le \mathcal{X}^T (I + \Lambda)\mathcal{X} + \mathcal{Y}^T (I + \Lambda^{-1})\mathcal{Y}.$$

Note that in the scalar case, we have in Lemma 2.6 an elementary quadratic inequality.

Chapter 3
Robust State Feedback Control

Abstract In this chapter, a particular family of nonlinear affine control systems with a sufficiently general type of uncertainties is considered. Nonlinear uncertain systems, considered here, are governed by vector ordinary differential equations with so-called quasi-Lipschitz right-hand sides admitting a wide class of external and internal uncertainties (including discontinuous nonlinearities such as relay and hysteresis elements, time-delay blocks, and so on). Here, the simplest class of linear state-feedback controllers is analyzed. Sufficient conditions guaranteeing the boundedness of all possible trajectories of controlled systems are presented. Since bounded dynamics can always be imposed on an ellipsoid, it is suggested that the "robust-optimal" gain matrix of the designated linear feedback be selected in such a way that the "size" of this attractive ellipsoid will be minimal. Several numerical and experimental illustrative examples are considered.

Keywords State feedback design • Attractive ellipsoid • Linear matrix inequalities

In this chapter, a particular family of nonlinear affine control systems with a sufficiently general type of uncertainties is considered. Nonlinear uncertain systems, considered here, are governed by vector ordinary differential equations with so-called quasi-Lipschitz right-hand sides admitting a wide class of external and internal uncertainties (including discontinuous nonlinearities such as relay and hysteresis elements, time-delay blocks, and so on). Here, the simplest class of linear state-feedback controllers is analyzed. Sufficient conditions guaranteeing the boundedness of all possible trajectories of controlled systems are presented. Since bounded dynamics can always be imposed on an ellipsoid, it is suggested that the "robust-optimal" gain matrix of the designated linear feedback be selected in such a way that the "size" of this attractive ellipsoid will be minimal. Several numerical and experimental illustrative examples are considered.

© Springer International Publishing Switzerland 2014

A. Poznyak et al., *Attractive Ellipsoids in Robust Control*, Systems & Control:
Foundations & Applications, DOI 10.1007/978-3-319-09210-2_3

3.1 Introduction

In dealing with the control design of a real dynamical system, a researcher tries to satisfy several basic demands to make the control process simpler and more attractive from a practical point of view:

- First, the mathematical model of a plant to be controlled may be known imprecisely or contain some uncertain parameters or structural elements.
- The controlled system should be able to work satisfactorily in the presence of external perturbations (even bounded and not necessarily "smooth").
- The controller should be the simplest structure admitting an easy realization: from this point of view, a linear state-feedback regulator (in spite of the fact that the considered plant is nonlinear) seems to be preferable.

Unsurprisingly, the classical optimal control methods (such as the Pontryagin maximum principle (Pontryagin, Boltyansky, Gamkrelidze, & Mishenko 1969) and Bellman dynamic programming (Bellman 1956)) developed for control design under complete and precise information on the plant turn out to be inapplicable in such uncertain situations. Recent research and practical implementations have shown that the most adequate techniques for control design of different classes of uncertain systems are the following:

- *Robust control theory* (Zhou, Doyle, & Glover 1996) and its various modifications such as robust adaptive (Ioannou & Sun 1996) and adaptive robust controls (Hong 2008).
- Static (Haykin 2009) and dynamic (Poznyak, Sanchez, & Yu 2001) *neural network control*.
- *Sliding mode control* (Utkin 1992).

In this book, a new technique especially designed for the control of uncertain systems is presented. It is referred to as the **attractive ellipsoid method**. It is based on a Lyapunov-like technique with the use of linear (or bilinear) matrix inequalities. For conventional controllable dynamic processes, this Lyapunov-based technique provides a useful theoretical tool not only for the classical stability test but also in connection with several types of robust control design procedures (see Zubov 1962; Isidori et al. 2000; Chen et al. 2001; Lin & Qian 2001; Azhmyakov 2006; Michel et al. 2007; Poznyak 2008). Various feedback control problems for input–output systems with bounded disturbances have been recognized as challenging tasks in modern control engineering (see, e.g., Glover & Schweppe 1971; Kurzhanski & Veliov 1994; Isidori et al. 2000; Lin & Qian 2001; Khalil 2002; Coutinho et al. 2003; Mera et al. 2011).

For example, a newly elaborated methodology based on the optimal rejection of bounded disturbances was examined by Yakubovich (1976), Barabanov and Granichin (1984), and Dahleh, Pearson, and Boyd (1988). If we view a control system with a class of restricted additive uncertainties as a nonlinearly affine model, it is quite intuitive to exploit Lyapunov-type tools in the design of stabilizing controls. In this direction, one of the main tools for generating a practically stable trajectory is the invariant (or attractive) ellipsoid method. Recall that a set in the state space is said to be positively invariant for a given dynamical system if every trajectory that begins in this set remains inside the set at all future times. We refer to Kurzhanski and Veliov (1994), Kurzhanski and Varaiya (2006), Polyak et al. (2004), Polyak and Topunov (2008), Mera, Poznyak, and Azhmyakov (2011), and Azhmyakov (2011) for some basic theoretical details and also for further references.

In this chapter, we investigate a particular family of nonlinear affine control systems with the above-mentioned type of uncertainties. The nonlinear uncertainties in our contribution are modeled by the state equations with *quasi-Lipschitz* right-hand sides. From the computational point of view, we are interested in creating effective control algorithms constituting a corresponding numerical extension of the proposed robust control design schemes. Generally speaking, we restrict our consideration to a class of *linear state-feedback control laws*. In other chapters, more complex feedback is considered. An abstract existence question of an invariant set for a general dynamical system constitutes a very sophisticated mathematical question. Under some technical assumptions related to the structure of the closed-loop system, it is possible to characterize an attractive set constructively. In modern control engineering, this set is usually chosen in the form of an ellipsoid in the state space of the given system. Here we construct an attractive ellipsoid for some specific control systems with an affine structure with respect to control actions. The resulting ellipsoidal invariant attractive set possesses some optimal properties and is used constructively in the main design procedure of a robust feedback strategy. It is not surprising that a robust synthesis problem associated with a linear system that possesses some size minimality properties of the obtained invariant attractive set can usually be reduced to an auxiliary linear matrix inequality-constrained optimization problem. We refer to Boyd et al. (1994), Polyak et al. (2004), Polyak and Topunov (2008), Blanchini and Miami (2008), Mera et al. (2011), and Gonzalez-Garcia, Polyakov, and Poznyak (2011) for some related ideas and technical results.

3.2 Proportional Feedback Design

3.2.1 Model Description

Here we consider quasi-Lipschitz affine controlled systems of the form

$$\dot{x}(t) = Ax(t) + Bu(t) + Df(t, x, u), \tag{3.1}$$

where

- $x(t) \in \mathbb{R}^n$ is the state vector at time $t \in \mathbb{R}_+ := \{t : t \geq 0\}$,
- $A \in \mathbb{R}^{n \times n}$ is the constant system matrix,
- $u(t) \in \mathbb{R}^m$ is the vector of control inputs,

- $B \in \mathbb{R}^{n \times m}$ is the constant matrix of control gains,
- $D \in \mathbb{R}^{n \times k}$ is the gain constant matrix of uncertain inputs,
- $f : \mathbb{R} \times \mathbb{R}^n \times \mathbb{R}^m \to \mathbb{R}^k$ is the uncertain vector function, assumed to be bounded as

$$f^T (t, x, u) \, Q_f f (t, x, u) \leq f_0 + x^T Q_x x + u^T Q_u u, \tag{3.2}$$

where the number f_0 is positive, and the positive definite quadratic matrices Q_f, Q_x, Q_u are assumed to be known.

We assume in this section that the control function u has a form of linear feedback

$$u(t) = K x(t) \tag{3.3}$$

with a matrix $K \in \mathbb{R}^{m \times n}$, referred to below as a gain matrix, which should be designed to obtain a desired behavior of the closed-loop system. With this controller, the uncertainty $f(t, x, u)$ 3.2 satisfies the inequality

$$f^T (t, x, u) \, Q_f f (t, x, u) \leq f_0 + x^T Q_x x + x^T K^T Q_u K x.$$

3.2.2 *Problem Formulation*

Problem 3.1. *The problem now is to present a stabilizing control design scheme, that is, to find a gain matrix K that allows us to guarantee the boundedness of all possible trajectories $\{x(t)\}_{t \geq 0}$ of the closed-loop system (3.1)–(3.3) and to estimate, adjust, and minimize the "attractive ellipsoid" containing these bounded trajectories asymptotically.*

3.3 S-Procedure-Based Approach

Consider the quadratic function

$$V(x) = x^T P^{-1} x, \quad P > 0, \tag{3.4}$$

referred to below as the *storage* (or energetic) function, and find its total derivative along the trajectories of the system (3.1)–(3.3):

$$\dot{V}(x) = \begin{pmatrix} x \\ f \end{pmatrix}^T \begin{pmatrix} P^{-1}(A + BK) + (A + BK)^T P^{-1} & P^{-1} D \\ D^T P^{-1} & 0 \end{pmatrix} \begin{pmatrix} x \\ f \end{pmatrix}$$

Definition 3.1. The ellipsoid

$$\mathcal{E}(P) := \{ x \in \mathbb{R}^n : x^T P^{-1} x < 1 \} \tag{3.5}$$

with the center in the origin and the ellipsoidal matrix $P = P^\mathsf{T} > 0$ is said to be **invariant** for the system (3.1) with uncertainties (3.2) and the control 3.3 if the storage function $V(x)$ 3.4 does not increase outside (including the boundary) of this ellipsoid, that is, if

$$\dot{V}(x(t)) \leq 0 \quad \text{for all } x(t) \text{ such that } V(x(t)) \geq 1$$

or equivalently, if the following two inequalities

$$\begin{pmatrix} x \\ f \end{pmatrix}^T W_1 \begin{pmatrix} x \\ f \end{pmatrix} \leq \alpha_1 := -1, \ W_1 = \begin{bmatrix} -P^{-1} & 0 \\ 0 & 0 \end{bmatrix}$$

$$\begin{pmatrix} x \\ f \end{pmatrix}^T W_2 \begin{pmatrix} x \\ f \end{pmatrix} \leq \alpha_2 := f_0, \ W_2 = \begin{bmatrix} -Q_x - K^T Q_u K & 0 \\ 0 & Q_f \end{bmatrix}$$

(3.6)

imply

$$\begin{pmatrix} x \\ f \end{pmatrix}^T W_0 \begin{pmatrix} x \\ f \end{pmatrix} \leq \alpha_0 := 0$$

(3.7)

$$W_0 = \begin{bmatrix} P^{-1}(A + BK) + (A + BK)^T P^{-1} & P^{-1}D \\ D^T P^{-1} & 0 \end{bmatrix}$$

According to *S*-procedure (see, for example, Poznyak 2008 or Lemma 2.4 of this book for the details) the required implication holds if and only if there exist the nonnegative numbers $\tau_1, \tau_2 \in \mathbb{R}$, $\tau_1 \geq 0$, $\tau_2 \geq 0$ such that

$$W_0 \leq \tau_1 W_1 + \tau_2 W_2 \text{ and } \alpha_0 \geq \tau_1 \alpha_1 + \tau_2 \alpha_2$$

or equivalently,

$$W = \begin{bmatrix} P^{-1}(A + BK) + (A + BK)^T P^{-1} + & P^{-1}D \\ \tau_1 P^{-1} + \tau_2 Q_x + \tau_2 K^T Q_u K & \\ D^T P^{-1} & -\tau_2 Q_f \end{bmatrix} \leq 0 \qquad (3.8)$$

with

$$\tau_1, \tau_2 \in \mathbb{R}, \tau_1 \geq 0, \tau_2 \geq 0, \tau_1 \geq f_0 \tau_2$$

Notice that this matrix inequality is nonlinear with respect to scalar parameter τ_1 and two unknown matrices P^{-1} and K which we are interested in. Since the matrix inequality

$$W \leq 0$$

implies that for any non-singular matrix T (of the same size as W) it follows

$$W_T := T^\mathsf{T} W T \le 0$$

and inverse, we will try to find a transformation matrix T which simplifies the obtained matrix W_T making it at least linear with respect to some new matrix variables affirmatively defining the initially interested matrices P and K. To do this let us apply the equivalent nonsingular transformation

$$T = \begin{bmatrix} P & 0 \\ 0 & I_{n \times n} \end{bmatrix}$$

to the matrix inequality (3.8) which gives

$$\begin{bmatrix} AP + BKP + PA^T + PK^T B^T + \\ \tau_1 P + \tau_2 PQ_x P + \tau_2 PK^T Q_u KP & D \\ D^T & -\tau_2 Q_f \end{bmatrix} \le 0$$

It can be easily shown that the positiveness of the number τ_2 is required for the feasibility of the last inequality with D of the full rank. So, using the Schur complement (see Poznyak 2008 or Theorem 2.11 of this book) we obtain the following equivalent inequalities:

$$AP + BKP + PA^T + PK^T B^T + \tau_1 P +$$

$$\tau_2 PQ_x P + \tau_2 PK^T Q_u KP + \frac{1}{\tau_2} DQ_f^{-1} D^T \le 0$$

and

$$\begin{bmatrix} AP + BKP + PA^T + PK^T B^T + \\ \tau_1 P + \frac{1}{\tau_2} DQ_f^{-1} D^T & P & PK^T \\ P & -\frac{1}{\tau_2} Q_x^{-1} & 0 \\ KP & 0 & -\frac{1}{\tau_2} Q_u^{-1} \end{bmatrix} \le 0$$

Finally, introducing the notations

$$Y := KP \text{ and } \bar{\tau}_2 := \frac{1}{\tau_2}$$

we can formulate the following result.

Lemma 3.1. *If the matrices* $P \in \mathbb{R}^{n \times n}, Y \in \mathbb{R}^{m \times n}$ *and the nonnegative numbers* $\tau_1, \overline{\tau}_2 \in \mathbb{R}$ *satisfy the system of inequalities*

$$
\begin{bmatrix}
\begin{array}{c} AP + BY + PA^T + Y^T B^T + \\ \tau_1 P + \overline{\tau}_2 DQ_f^{-1} D^T \end{array} & P & Y^T \\
P & -\overline{\tau}_2 Q_x^{-1} & 0 \\
Y & 0 & -\overline{\tau}_2 Q_u^{-1}
\end{bmatrix} \le 0
$$

$$
P > 0, \ \tau_1 > 0, \ \overline{\tau}_2 > 0, \ \tau_1 \overline{\tau}_2 \ge f_0
$$

(3.9)

then $\mathcal{E}(P)$ *(3.5) is the **invariant** ellipsoid for the closed-loop system (3.1)–(3.3) with the stabilizing linear feedback gain matrix*

$$
K = YP^{-1}
$$

(3.10)

Remark 3.1. This control design scheme is rather classical and well-known for disturbed linear control systems (i.e., $f(t,x,u) := f(t)$, $Q_x = 0$ and $Q_u = 0$) (see Polyak and Topunov 2008; Blanchini & Miami 2008). In the linear case the obtained invariant set is also "attractive". For **quasi-Lipschitz system,** satisfying (3.2) this fact is not obvious but expectable, so the attractivity property of the ellipsoid $\mathcal{E}(P)$ satisfying (3.9) must be carefully studied.

3.4 Storage Function Method

Consider again the quadratic storage function (3.4) and calculate its total derivative along the trajectories of the system (3.1)–(3.3)

$$
\dot{V} = \begin{pmatrix} x \\ f \end{pmatrix}^T \begin{bmatrix} P^{-1}(A + BK) + (A + BK)^I P^{-1} & P^{-1}D \\ D^T P^{-1} & 0 \end{bmatrix} \begin{pmatrix} x \\ f \end{pmatrix} =
$$

$$
\begin{pmatrix} x \\ f \end{pmatrix}^T \begin{bmatrix} P^{-1}(A + BK) + (A + BK)^T P^{-1} + \tau_1 P^{-1} & D \\ D^T P^{-1} & -\tau_2 Q_f \end{bmatrix} \begin{pmatrix} x \\ f \end{pmatrix}
$$

$$
+ \tau_2 f^T Q_f f - \tau_1 V,
$$

where $\tau_1 > 0$ and $\tau_2 > 0$ are some positive numbers. Taking into account the quasi-Lipschitz constraint (3.2), we derive

$$
\dot{V} \le -\tau_1 V + f_0 \tau_2 +
$$

$$
\begin{pmatrix} x \\ f \end{pmatrix}^T \begin{bmatrix} \begin{array}{c} P^{-1}(A + BK) + (A + BK)^T P^{-1} + \\ \tau_1 P^{-1} + \tau_2 Q_x + \tau_2 K^T Q_u K \end{array} & P^{-1}D \\ D^T P^{-1} & -\tau_2 Q_f \end{bmatrix} \begin{pmatrix} x \\ f \end{pmatrix}.
$$

Hence, the feasibility of the matrix inequality (3.8) for $\tau_1, \tau_2 > 0$ implies that the corresponding storage function $V(x) = x^T P^{-1} x$ satisfies the following inequality:

$$\dot{V}(x(t)) \leq -\tau_1 V(x(t)) + f_0 \tau_2. \qquad (3.11)$$

In studying (3.11), we obtain

$$V(x(t)) \leq \frac{f_0 \tau_2}{\tau_1} + \left(V(x(0)) - \frac{f_0 \tau_2}{\tau_1} \right) e^{-\tau_1 t} \underset{t \to \infty}{\longrightarrow} \frac{f_0 \tau_2}{\tau_1},$$

and for $\dfrac{f_0 \tau_2}{\tau_1} \leq 1$, the inequality (3.8) guarantees that $\mathcal{E}(P)$ is the attractive ellipsoid of the closed-loop system (3.1)–(3.3). This yields the following expectable result.

Lemma 3.2. *If the matrices $P \in \mathbb{R}^{n \times n}, Y \in \mathbb{R}^{m \times n}$ and the numbers $\tau_1, \overline{\tau}_2 \in \mathbb{R}$ satisfy the system of inequalities (3.9), then the ellipsoid $\mathcal{E}(P)$ (3.5) is the* **exponentially attractive ellipsoid** *(with rate of attraction τ_1) of the closed-loop system (3.1)–(3.3) with* **stabilizing linear feedback** *gain matrix $K = Y P^{-1}$.*

Summary 3.1.

- *Based on the results of two sections presented above, one may conclude that* **both approaches** *(based on the S-procedure and on the storage function method)* **are equivalent** *in the sense that both provide sufficient conditions guaranteeing the boundedness of every trajectory of the controlled system with any admissible uncertainty or external perturbation.*
- *The invariant ellipsoidal set of the quasi-Lipschitz control system (3.1)–(3.3), (3.10) obtained wit the S-procedure approach is also globally attractive.*

Therefore, throughout this book, we will effectively use *both introduced approaches* for an attractive ellipsoid description of closed-loop quasi-Lipschitz systems.

Based on the results obtained above, control design procedures for some different control problems can now be presented:

- One is *minimization of an attractive ellipsoid* by the corresponding selection of the feedback gain matrix K.
- Another is *practical stabilization* , consisting in the design of a proportional linear feedback that guarantees the convergence of all trajectories of the corresponding closed-loop system (3.1)–(3.3) in the predetermined ellipsoid $\mathcal{E}(P_{\text{ref}})$ with some given positive definite matrix $P_{\text{ref}} > 0, P_{\text{ref}} \in \mathbb{R}^{n \times n}$.

3.5   Minimization of the Attractive Ellipsoid

To find the "optimal" linear feedback minimizing the attractive ellipsoid of the closed-loop system (3.1)–(3.3), we will consider the following optimization problem corresponding to the minimization of the "size" of the ellipsoid:

$$\min_{P,Y,\tau_1,\bar\tau_2} \ \mathrm{tr}(P) \tag{3.12}$$

subject to (3.9).

The trace of the matrix P defines the sum of the squares of the ellipsoid's semiaxes.

Remark 3.2. When we speak about the "size" of an ellipsoid with a matrix P, we do not mean its volume. The volume of an ellipsoid (or equivalently, its determinant) is, in fact, a bad function for the characterization of its "size" for the following two reasons:

– Since

$$\det P = \prod_{i=1}^{N} \lambda_i(P) \ \text{and} \ \lambda_i(P) = 1/\lambda_i(P^{-1}) = \rho_i^2(P),$$

where $\lambda_i(P), i = 1,\ldots,N)$, are the eigenvalues of the ellipsoid matrix P and $\rho_i(P)$ is the distance from the center to each semiaxis of the ellipsoid, we may conclude that minimization of $\det(P)$ is equivalent to minimization of

$$\prod_{i=1}^{n} \rho_i(P) = \mathrm{vol}(P),$$

that is, the minimization of its volume. But the product $\prod_{i=1}^{N} \rho_i(P)$ admits a very large value of one semiaxis, for example, $\rho_{i_0}(P)$, and all others may be very small. This means precisely that the designed controller guarantees a very good quality of control in practically all directions except one, where it works with very bad quality. That is why the criterion $\mathrm{tr}(P)$ is preferable, since

$$\mathrm{tr}\,P = \sum_{i=1}^{N} \lambda_i(P) \geq \max_{i=1,\ldots,N} \lambda_i(P) = \lambda_{\max}(P),$$

and the minimization of $\mathrm{tr}\,P$ guarantees at least the minimization of its maximum eigenvalue, and hence, this guarantees the minimization of the corresponding maximal semiaxis $\rho_{\max}(P) = \sqrt{\lambda_{\max}(P)}$ of the given ellipsoid.
– The second reason, important from the numerical-computational point of view, is that $\mathrm{tr}\,P$ is a linear function of the matrix P, and $\det(P)$ is not!

If the tuple $(P^{opt}, Y^{opt}, \tau_1^{opt}, \bar\tau_2^{opt})$ is a solution of (3.12), then the gain matrix of the corresponding "optimal" linear feedback can be calculated as

$$K^{opt} = Y^{opt}\left[P^{opt}\right]^{-1}. \tag{3.13}$$

The problem (3.12) consists in the minimization of a linear criterion under bilinear constraints. The resolution of such optimization problems is not a trivial task in general (Henrion, Loefberg, Kocvara, & Stingl 2006),

(Boyd & Vandenberghe 1997). Fortunately, here all bilinear terms in (3.9) contain the same scalar variable τ_1. In this case, the following algorithm can be used for solving the problem 3.12.

Algorithm 3.1. *1. For fixed τ_1, the constraints (3.9) can be put in LMI format, and the corresponding optimization problem can be effectively solved using appropriate mathematical software such as MATLAB with any SDP solver such as SEDUMI or SDPT3. Let us denote by $g(\tau_1) = \min\limits_{P,Y,\tau_1,\bar{\tau}_2} \mathrm{tr}(P)$ the corresponding minimal value. 2. The optimization of the function $g(\tau_1)$ with respect to the parameter τ_1 can be realized locally based on some derivative-free method, for example using the MATLAB function fminsearch.*

Remark 3.3. The trace-constrained optimization problem may be replaced by some alternative constrained optimization problem such as $\det(P)$ characterizing the ellipsoid's volume or $\lambda_{\max}(P)$ defining the square of the maximal semiaxis of the ellipsoid $\mathcal{E}(P)$.

If the system (3.1) has some controlled output

$$z = Gx, \quad z \in \mathbb{R}^r, \quad G \in \mathbb{R}^{r \times n},$$

then the problem of minimization of the attractive ellipsoid of the output z has the form

$$\min_{P,Y,\tau_1,\bar{\tau}_2} \mathrm{tr}(GPG^T)$$
$$\text{subject to (3.9).}$$

3.6   Practical Stabilization

The *practical stabilization problem* (Lakshmikantham, Leela, & Martynyuk 1990) also requires the consideration of an optimization procedure slightly different from the previous one. This problem consists in designing a proportional linear feedback that guarantees convergence of all trajectories of the closed-loop system (3.1)–(3.3) to the predetermined ellipsoid $\mathcal{E}(P_{\mathrm{ref}})$, where $P_{\mathrm{ref}} > 0$ is a given positive definite matrix. Such a requirement can be easily taken into account by incorporating the linear matrix inequality

$$P \leq P_{\mathrm{ref}} \tag{3.14}$$

into the system of inequalities (3.9).

Practical implementations always restrict the maximum admissible control magnitude, for example by the following inequality:

$$\|u\|^2 = x^T K^T K x \leq \mu \quad \text{for} \quad \forall x \in \mathbb{R}^n, \tag{3.15}$$

where μ is a given positive number. In this case, the maximization of the attraction region for the closed-loop system (3.1)–(3.3), (3.15) should be resolved. Denote this domain by Ω. Notice that in the general case, the set Ω may not be of elliptic form. So using the technique developed in this book, we may maximize only an *internal ellipsoid* contained in Ω.

Lemma 3.3. *If in addition to (3.9) and (3.14), the inequalities*

$$\begin{bmatrix} P & Y^T \\ Y & \delta\mu I_m \end{bmatrix} \geq 0,$$

$$\delta P_{\mathrm{ref}} \leq P, \quad 0 < \delta < 1, \tag{3.16}$$

hold, then $\mathcal{E}(P_{\mathrm{ref}})$ is an attractive set of the closed-loop system (3.1)–(3.3), (3.15), and

$$\mathcal{E}(P) \subseteq \mathcal{E}(P_{\mathrm{ref}}) \subset \mathcal{E}(\delta^{-1}P) \subseteq \Omega,$$

where Ω is the region of attraction.

Proof. According to Lemmas 3.1 and 3.2 given above, the ellipsoid $\mathcal{E}(P)$ is attractive and invariant for the closed-loop system (3.1)–(3.3) (without consideration of the constraint (3.15)). So the attractivity property of $\mathcal{E}(P_{\mathrm{ref}})$ follows from the implication

$$P \leq P_{\mathrm{ref}} \;\Rightarrow\; \mathcal{E}(P) \subseteq \mathcal{E}(P_{\mathrm{ref}}).$$

Analogously, the inequality $\delta P_{\mathrm{ref}} \leq P$ implies

$$\mathcal{E}(P_{\mathrm{ref}}) \subseteq \mathcal{E}(\delta^{-1}P).$$

Since the quadratic function $V(x) = x^T P^{-1} x$ satisfies the inequality (3.11) and $x(0) \in \mathcal{E}(\delta^{-1}P)$ implies $V(x(0)) \leq \delta^{-1}$, then

$$V(x(t)) \leq (\delta^{-1} - \tau_2 f_0/\tau_1)e^{-\tau_1 t} + \tau_2 f_0/\tau_1.$$

Hence, due to the relations

$$\tau_2 f_0/\tau_1 \leq 1, \; \delta^{-1} > 1,$$

we obtain $V(x(t)) \leq \delta^{-1}$, or equivalently, $x(t) \in \mathcal{E}(\delta^{-1})$ for all $t > 0$. Therefore, for every $\delta : 0 < \delta < 1$, the ellipsoid $\mathcal{E}(\delta^{-1}P)$ is an invariant set for the closed-loop system (3.1)–(3.3). Rewriting the constraint (3.15) in the form

$$\mu^{-1}x^T K^T K x \leq 1,$$

we obviously obtain that the matrix inequality

$$\mu^{-1} K^T K \le \delta P^{-1}$$

implies that $\mathcal{E}(\delta^{-1} P)$ belongs to the region of attraction of Ω. Multiplying the obtained matrix inequality by P from both sides and applying again the Schur complement, we derive the first LMI from (3.16). This completes the proof of the lemma.   ■

The procedure of attraction domain maximization using the attractive ellipsoids method can be formulated in the form of the following optimization problem:

$$\min_{P,Y,\tau_1,\bar{\tau}_2,\delta} \delta \tag{3.17}$$
$$\text{subject to} \quad (3.9), (3.14), (3.16).$$

3.7   Other Restrictions on Control and Uncertainties

I. If for control purposes, we need to restrict the control magnitude inside the attractive ellipsoid as

$$\|u\|^2 = x^T K^T K x \le \alpha \quad \text{for} \quad \forall x \in \mathcal{E}(P),$$

then it is sufficient to incorporate the matrix inequality

$$\frac{1}{\alpha} K^T K \le P^{-1} \tag{3.18}$$

into the optimization problem (3.12) or (3.17). Multiplying the last inequality (3.18) by P from both sides, we obtain the equivalent matrix inequality

$$\frac{1}{\alpha} Y^T Y \le P \tag{3.19}$$

written in terms of the matrix variables Y and P. Finally, applying the Schur complement to the last constraint (3.19), one can represent it in the following LMI form:

$$\begin{bmatrix} P & Y^T \\ Y & \alpha I_m \end{bmatrix} \ge 0. \tag{3.20}$$

II. From the formula (3.11), it follows that the rate of exponential convergence of the trajectories of the closed-loop system (3.1)–(3.3) is defined by the parameter τ_1. So the introduction of additional restrictions such as

$$\tau_1^{\min} \le \tau_1 \le \tau_1^{\max}$$

allows us to adjust the convergence rate.

III. To avoid a case of high-gain control, the matrix inequality

$$K^T K \leq M_u, \qquad \text{where } M_u > 0, \tag{3.21}$$

is required. The positive definite matrix M_u restricts the maximal values of the control gains matrix K. In particular, the constraint

$$\|K\| \leq \beta$$

for

$$\|K\| := \sqrt{\lambda_{\max}(K^T K)}$$

is equivalent to (3.21) with

$$M_u := \beta^2 I_n.$$

Similarly to (3.18), the inequality (3.21) can be transformed into the matrix inequality

$$\begin{bmatrix} PM_u P & Y^T \\ Y & I_m \end{bmatrix} \geq 0, \tag{3.22}$$

which is not linear, but quadratic.

So the optimization problems (3.12) and (3.16) considered with additional constraint (3.22) require a modified resolving procedure based on, for example, the PENBMI package of MATLAB. This package allows one to resolve the optimization problems subject to bilinear and quadratic constraints.

Proposition 3.1. *If the matrices* $P = P^T \in \mathbb{R}^{n \times n}$, $Y \in \mathbb{R}^{m \times n}$, *satisfy the system of LMI*

$$\begin{bmatrix} P & Y^T \\ Y & \theta I_m \end{bmatrix} \geq 0, \quad P \geq \theta M_u^{-1}, \quad \theta > 0 \quad \theta \in \mathbb{R}, \tag{3.23}$$

then they satisfy the inequality (3.22).

Proof. The inequality $P \geq \theta M_u^{-1}$ is equivalent to

$$M_u \geq \theta P^{-1} \quad \text{and} \quad PM_u P \geq \theta P.$$

By Schur's complement, the first inequality of (3.23) is equivalent to

$$\theta P \geq Y^T Y.$$

Hence we obtain the matrix inequality $PK_u P \geq Y^T Y$, which is equivalent to (3.22). ∎

The given proposition presents the LMI approximation of the quadratic constraint (3.22). Based on this approximation, suboptimal solutions of the optimization problems (3.12) and (3.17) can be effectively found using semidefinite programming techniques:

IV. In the general case, the estimate of the quasi-Lipschitz function $f(t, x, u)$ may not depend on some components of x or u. In this case, the matrices Q_x and Q_u are positive semidefinite, but not invertible. For instance, let us assume

$$Q_x = V_x^T \tilde{Q}_x V_x \quad \text{and} \quad Q_u = V_u^T \tilde{Q}_u V_u,$$

where

$$\tilde{Q}_x \in \mathbb{R}^{\tilde{n} \times \tilde{n}} \quad \text{and} \quad V_x \in \mathbb{R}^{\tilde{n} \times n}$$

$$\tilde{Q}_u \in \mathbb{R}^{\tilde{m} \times \tilde{m}} \quad \text{and} \quad V_u \in \mathbb{R}^{\tilde{m} \times m}$$

are some constant matrices such that $\tilde{Q}_x > 0$ and $\tilde{Q}_u > 0$ with $\tilde{n} < n$ and $\tilde{m} < m$.

In this case, the matrix inequality (3.9) assumes the form

$$\begin{bmatrix} AP + BY + PA^T + Y^T B^T + \\ \tau_1 P + \bar{\tau}_2 DQ_f^{-1} D^T & PV_x^T & Y^T V_u^T \\ V_x P & -\bar{\tau}_2 \tilde{Q}_x^{-1} & 0 \\ Y V_u^T & 0 & -\bar{\tau}_2 \tilde{Q}_u^{-1} \end{bmatrix} \leq 0. \qquad (3.24)$$

3.8 Illustrative Example

In this section, we consider the inverted pendulum depicted in Fig. 3.1.

The control to be designed is intended to maintain (stabilize) the pendulum in the vertical position.

Dynamic Model

The mathematical model of the disturbed inverted pendulum can be presented as

$$ml^2 \ddot{q} + mgl \sin(q) = \tilde{b}u + p(t), \qquad (3.25)$$

Fig. 3.1 Inverted pendulum

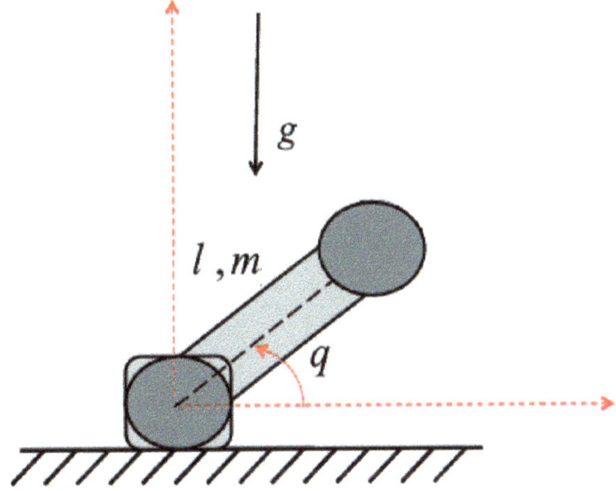

where

- the position coordinates $q \in \mathbb{R}$ with associated velocities $\dot{q}$ and accelerations $\ddot{q}$ are controlled by the driving force $u \in \mathbb{R}$ with gain $\tilde{b} \in \mathbb{R}$,
- m is the mass of the pendulum,
- l is the distance from the pivot point to the center of mass,
- the function $p : \mathbb{R} \to \mathbb{R}$ describes the bounded exogenous disturbances

$$|p(t)| < p_0 \quad \forall t > 0.$$

The parameters of the system (3.25) are assumed to be calculated with some errors

$$m = m_0(1 + \delta m), \quad l = l_0(1 + \delta l), \quad \tilde{b} = b_0(1 + \delta b),$$

where m_0, l_0, b_0 are given values, $\delta m, \delta l, \delta b$ are small modeling errors.

Let us make the change of time $\tau = t l_0 \sqrt{m_0}$. In this case, the original system can be rewritten as

$$\frac{d}{d\tau} x = Ax + bu + df(\tau, x, u),$$

where $x = (x_1, x_2)^\top = \left(q, \frac{dq}{d\tau} \right)^\top \in \mathbb{R}^2$,

$$A = \begin{pmatrix} 0 & 1 \\ -m_0 g l_0 & 0 \end{pmatrix}, \quad b = \begin{pmatrix} 0 \\ b_0 \end{pmatrix}, \quad d = \begin{pmatrix} 0 \\ 1 \end{pmatrix},$$

$$f(\tau, x, u) = \frac{p\left(\tau/(\sqrt{ml})\right)}{(1+\delta m)(1+\delta l)^2} + m_0 g l_0 \left[x_1 - \frac{1}{1+\delta l} \sin(x_1) \right]$$

$$+ b_0 \left(\frac{1+\delta b}{(1+\delta m)(1+\delta l)^2} - 1 \right) u,$$

and f satisfies the quasi-Lipschitz condition

$$f^2(t, x, u) \le f_0 + x^T V_x^T \tilde{Q}_x V_x x + Q_u u^2$$

with

$$\tilde{Q}_x = 3 m_0^2 g^2 l_0^2 \left(1 + \tfrac{1}{4(1+\delta l)} \right)^2 \quad \text{and} \quad V_x = \begin{pmatrix} 1 & 0 \end{pmatrix},$$

$$f_0 = \frac{3 p_0^2}{(1+\delta m)^2 (1+\delta l)^4} \quad \text{and} \quad Q_u = 3 b_0^2 \left(\frac{1+\delta b}{(1+\delta m)(1+\delta l)^2 - 1} \right)^2.$$

Let us consider the following parameters of the model:

$$m_0 = 0.075 \text{ kg}, \quad \delta m = -0.02, \quad l_0 = 0.3 \text{ m}, \quad \delta l = -0.01,$$

$$b_0 = 1, \quad \delta b = 0.05, \quad g = 9.81 \text{ m/sec}^2, \quad p_0 = 0.02,$$

and restrict the maximum value of the control input inside the attractive ellipsoid by $\alpha = 0.3$ (see the inequality (3.20) for details).

Numerical Simulations Results

Taking into account (3.24) and applying Algorithm 3.1, we obtain

$$\tau_1^{opt} = 1.202 \quad \text{and} \quad \tau_2^{opt} = 0.0055,$$

$$P^{opt} = \begin{bmatrix} 0.0119 & 0.0128 \\ -0.0128 & -0.0199 \end{bmatrix},$$

$$K^{opt} = \begin{bmatrix} -4.2606 & -3.8566 \end{bmatrix}.$$

The dynamics were simulated using the Runge–Kutta method (ODE45) within MATLAB Simulink 7.8 with

$$x(0) = \begin{bmatrix} \frac{\pi}{2} \\ 0 \end{bmatrix}.$$

Figure 3.2 shows the evolution of the system state (angle and angular velocity) with the designated control law. Figure 3.3 depicts the obtained attractive ellipsoid.

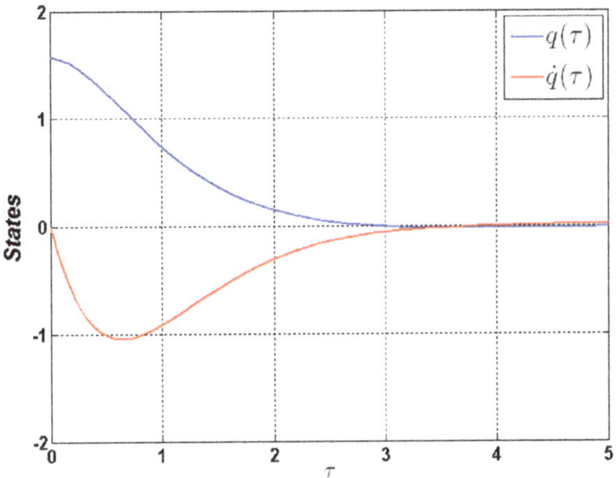

Fig. 3.2 Trajectory of the controlled inverted pendulum

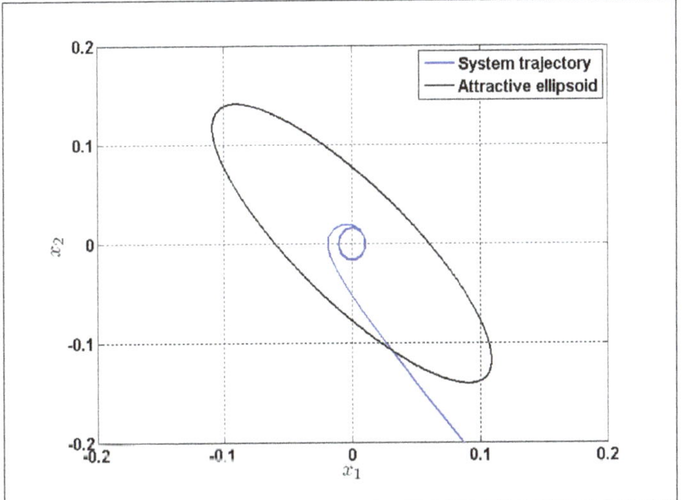

Fig. 3.3 Attractive ellipsoid

3.9   What to Do If We Don't Know the Matrix A?

3.9.1   Description of the Dynamic Model in This Case

Consider again the dynamic plant given by

$$\dot{x}(t) = g(t, x(t), u(x(t))), \quad t \geq 0,$$

$$x(0) = x_0 \in \mathbb{R}^n,$$

as in (2.2). It can be rewritten in the following quasi-linear format (3.1):

$$\dot{x}(t) = Ax(t) + Bu(t) + f(t, x(t), u(t)),\tag{3.26}$$

where $A \in \mathbb{R}^{n \times n}$ is a constant matrix to be selected, $B \in \mathbb{R}^{n \times m}$ is a known constant matrix of control gains, and

$$f(t, x, u) := g(t, x, u) - Ax - Bu$$

is the unmodeled dynamic part, satisfying (3.2), namely

$$f^T(t, x, u)\, Q_f f(t, x, u) \leq f_0 + x^T Q_x x + u^T Q_u u.$$

Below, for simplicity, we will consider

$$g(t, x, u) := g_0(t, x) + Bu,$$

which implies

$$f(t, x, u) := g_0(t, x) - Ax.$$

In the case $g_0(t, x) \in \mathcal{C}(A, \delta_1, \delta_2)$, it is a *quasi-Lipschitz* function satisfying

$$\|g_0(x) - Ax\|^2 \leq \delta_1 + \delta_2\|x\|^2$$

for all $x \in \mathbb{R}$ and some nonnegative constants δ_1 and δ_2.

If we do not know the matrix A characterizing the class $\mathcal{C}(A, \delta_1, \delta_2)$, we suggest considering that

$$g_0(t, x) \in \mathcal{C}(0, \delta_1, \delta_2),$$

satisfying

$$\|g_0(x)\|^2 \leq \delta_1 + \delta_2\|x\|^2.\tag{3.27}$$

In this situation, we have

$$f^T(t, x, u)\, Q_f f(t, x, u) = (g_0(t, x) - Ax)^{\mathsf{T}} Q_f (g_0(t, x) - Ax) \leq$$

$$2g_0(t, x)^{\mathsf{T}} Q_f g_0(t, x) + 2x^{\mathsf{T}} A^{\mathsf{T}} Q_f Ax \leq$$

$$2\lambda_{\max}(Q_f)(\delta_1 + \delta_2\|x\|^2) + 2x^{\mathsf{T}} A^{\mathsf{T}} Q_f Ax =$$

$$f_0 + x^T Q_x x + u^T Q_u u$$

with

$$f_0 = 2\lambda_{\max}\left(Q_f\right)\delta_1,$$

$$Q_x = 2\lambda_{\max}\left(Q_f\right)\delta_2 I_{n\times n} + 2A^\mathsf{T}Q_f A, \tag{3.28}$$

$$Q_u = 0.$$

If we select the control in the form of linear feedback (3.3),

$$u\left(t\right) = Kx\left(t\right),$$

the model (3.26) becomes

$$\dot{x}\left(t\right) = (A + BK)\,x\left(t\right) + f(t, x\left(t\right), u\left(t\right)), \tag{3.29}$$

where $f(t, x\left(t\right), u\left(t\right))$ satisfies the constraint

$$f^T\left(t, x, u\right)Q_f f\left(t, x, u\right) \le f_0 + x^T Q_x x \tag{3.30}$$

for all u with f_0 and Q_x defined by (3.28).

In this section, we have tried to find the gain feedback matrix K to guarantee that all trajectories of the system (3.29)–(3.28) finally enter an ellipsoid $\mathcal{E}\left(P\right)$ of minimal size. Now we will try to find two matrices A and K to solve the same problem.

Proposition 3.2. *Let us try to find the matrix A in the following form:*

$$A = A_0 + \Delta A, \tag{3.31}$$

where A_0 is a preselected matrix that creates the controllable pair with B, that is, the pair (A, B) is controllable, and ΔA is the correction term in which we are interested.

Problem 3.2. *The problem now is to present a stabilizing control design scheme, that is, to find a matrix ΔA (characterized by (3.31), the "best linear approximator" A, and a gain matrix K that guarantee the boundedness of all possible trajectories $\{x\left(t\right)\}_{t\geq 0}$ of the closed-loop system (3.29)–(3.28) and to estimate, adjust, and minimize the "attractive ellipsoid" containing these bounded trajectories asymptotically.*

3.9.2   Sufficient Conditions of Attractiveness

Repeating all considerations described above, we may again conclude that for the storage function (3.4), we have

$$V(x) = x^T P^{-1}x, \quad P > 0.$$

Its derivative on the trajectories of (3.29)–(3.28) is as follows:

$$
\dot{V}(x) = \begin{pmatrix} x \\ f \end{pmatrix}^{T} \begin{pmatrix} P^{-1}(A+BK)+(A+BK)^{T}P^{-1} & P^{-1} \\ P^{-1} & 0 \end{pmatrix} \begin{pmatrix} x \\ f \end{pmatrix} =
$$

$$
\begin{pmatrix} x \\ f \end{pmatrix}^{T} \begin{pmatrix} P^{-1}(A+BK)+(A+BK)^{T}P^{-1} & P^{-1} \\ P^{-1} & -\varepsilon Q_f \end{pmatrix} \begin{pmatrix} x \\ f \end{pmatrix} + \varepsilon f^{T} Q_f f \le
$$

$$
\begin{pmatrix} x \\ f \end{pmatrix}^{T} W_0 \begin{pmatrix} x \\ f \end{pmatrix} + \varepsilon f_0
$$

with

$$
W_0 = \begin{pmatrix} P^{-1}(A+BK)+(A+BK)^{T}P^{-1}+ & P^{-1} \\ 2\varepsilon\lambda_{\max}(Q_f)\,\delta_2\,I_{n\times n}+2\varepsilon A^{\mathsf{T}}Q_f A & \\ P^{-1} & -\varepsilon Q_f \end{pmatrix}.
$$

To guarantee the *asymptotic attractivity property*

$$
\dot{V}(x(t)) \le 0 \quad \text{for all } x(t) \text{ such that } V(x(t)) \ge 1,
$$

it is necessary and sufficient to satisfy the following inequalities:

$$
\begin{pmatrix} x \\ f \end{pmatrix}^{T} W_0 \begin{pmatrix} x \\ f \end{pmatrix} \le -f_0 := \alpha_0,
$$

$$
\begin{pmatrix} x \\ f \end{pmatrix}^{T} \begin{matrix} -P^{-1} & 0 \\ 0 & 0 \end{matrix} W_1 \begin{pmatrix} x \\ f \end{pmatrix} \le -1 := \alpha_1, \ P > 0, \ W_1 := \begin{bmatrix} -P^{-1} & 0 \\ 0 & 0 \end{bmatrix},
$$

or by the *S*-procedure (see Lemma 2.4), equivalently

$$
W_0 \le \tau_1 W_1 \ \text{ and } \ \alpha_0 \ge \tau_1 \alpha_1
$$

for some $\tau_1 \ge 0$. The last matrix inequality can be equivalently represented as

$$
W := \begin{pmatrix} P^{-1}(A+BK)+(A+BK)^{T}P^{-1}+ & P^{-1} \\ \tau_1 P^{-1}+2\varepsilon\lambda_{\max}(Q_f)\,\delta_2\,I_{n\times n}+2\varepsilon A^{\mathsf{T}}Q_f A & \\ P^{-1} & -\varepsilon Q_f \end{pmatrix} \le 0
$$

$$
\tag{3.32}
$$

with

$$
\tau_1 \in \mathbb{R}, \ \tau_1 \ge 0, \ \tau_1 \ge f_0.
$$

Applying to W the nonsingular transformation $T = \begin{bmatrix} P & 0 \\ 0 & I \end{bmatrix}$, we may conclude that (3.32) is equivalent to the following one:

$$W := \begin{pmatrix} (A + BK)P + (A + BK)^T P^\mathsf{T} + \\ \tau_1 P + 2\varepsilon\lambda_{\max}(Q_f)\delta_2 P^2 + 2\varepsilon PA^\mathsf{T} Q_f AP & I \\ I & -\varepsilon Q_f \end{pmatrix} \le 0. \tag{3.33}$$

Introducing the notation

$$Y := KP, \quad Z := \Delta AP$$

and the estimates

$$2\,PA^\mathsf{T}Q_f AP = 2\,(A_0 P + Z)^\mathsf{T} Q_f (A_0 P + Z) \le Q_z,$$
$$2\lambda_{\max}(Q_f)\delta_2 P^2 \le Q_P,$$

which is equivalent (by Schur's complement) to

$$\begin{bmatrix} Q_z & A_0 P + Z \\ PA_0^\mathsf{T} + Z^\mathsf{T} & \frac{1}{2}Q_f^{-1} \end{bmatrix} \ge 0,$$

$$\begin{bmatrix} Q_P & P \\ P & \dfrac{1}{2\lambda_{\max}(Q_f)\delta_2} I_{n\times n} \end{bmatrix} \ge 0,$$

the matrix inequality (3.33) can be finally represented as three LMIs:

$$W(Z, Y, P, Q_z, Q_P, \varepsilon, \tau_1) :=$$
$$\begin{pmatrix} A_0 P + PA_0^\mathsf{T} + Z + Z^\mathsf{T} + \\ BY + Y^\mathsf{T} B^\mathsf{T} + \tau_1 P + \varepsilon Q_z + \varepsilon Q_P & I \\ I & -\varepsilon Q_f \end{pmatrix} \le 0,$$

$$\begin{bmatrix} -Q_z & -A_0 P - Z \\ -PA_0^\mathsf{T} - Z^\mathsf{T} & -\frac{1}{2}Q_f^{-1} \end{bmatrix} \le 0, \tag{3.34}$$

$$\begin{bmatrix} -Q_P & -P \\ -P & -\dfrac{1}{2\lambda_{\max}(Q_f)\delta_2} I_{n\times n} \end{bmatrix} \le 0.$$

Now we are ready to formulate the main result of this subsection.

Lemma 3.4. *If the matrices* $0 < Q_Z, P \in \mathbb{R}^{n \times n}, Y \in \mathbb{R}^{m \times n}, Z \in \mathbb{R}^{n \times n}$ *and the nonnegative numbers* $\varepsilon, \tau_1 \in \mathbb{R}$ *satisfy the LMIs (3.34) and also satisfy* $\tau_1 \geq f_0$, *then* $\mathcal{E}(P)$ *(3.5) is the **invariant** ellipsoid for the closed-loop system (3.29)–(3.28) with the stabilizing linear feedback gain matrix*

$$K = Y P^{-1}, \tag{3.35}$$

and the matrix A is equal to

$$A = A_0 + Z P^{-1}. \tag{3.36}$$

3.9.3   Optimal Robust Linear Feedback as a Solution of an Optimization Problem with LMI Constraints

The optimal robust linear feedback K^* can be found as

$$K^* = Y^* \left(P^*\right)^{-1}, \tag{3.37}$$

minimizing the "size" of the attractive ellipsoid $\mathcal{E}(P)$, where Y^* and P^* are the solutions of the following optimization problem:

$$\mathrm{tr}\{P\} \to \min_{Z,Y,P>0,Q_Z>0,Q_P>0,\varepsilon>0,\tau_1\geq f_0} \tag{3.38}$$

subject to the LMI constraints (3.34).

Remark 3.4. If we are interested in practical stability with prescribed level P_{pract}, we should solve the following optimization problem:

$$\left.\begin{array}{c} \mathrm{tr}\{P\} \to \min\limits_{Z,Y,P>P_{pract},Q_Z>0,Q_P>0,\varepsilon>0,\tau_1\geq f_0} \\[2ex] \text{subject to (3.34).} \end{array}\right\} \tag{3.39}$$

3.10   Conclusions

In this chapter, we have accomplished the following:

- We have shown how it possible to design a linear feedback controller for a class of quasi-Lipschitz nonlinear systems satisfying (3.2) without knowledge of the exact mathematical model of the system to be controlled.

- We have suggested characterizing the set of all stabilizing gain-feedback matrices by a system of corresponding LMIs, providing the boundedness of all possible trajectories of every controlled plant from the considered class (3.2) of uncertain systems.
- Since every set of bounded trajectories can be imposed on a multidimensional ellipsoid, called attractive, we have suggest selecting the optimal feedback gain matrix from the described class of stabilizing feedbacks as the one that minimizes the "size" of this attractive ellipsoid (in fact, maximizing the trace of the ellipsoidal matrix).
- We introduced the corresponding numerical procedure for designing the optimal feedback gain matrices.
- We have shown that practical stabilization and additional restrictions for control actions can be considered by introducing some additional constraints of LMI type.

Chapter 4
Robust Output Feedback Control

Abstract In this chapter, we consider three types of possible linear feedbacks using only the current output information: static feedback proportional to the output measurable signal, observer-based feedback proportional to the state estimation vector, and full-order linear dynamic controllers. For each type of possible linear feedback, we suggest that one characterize the set of all stabilizing gain-feedback matrices by a system of the corresponding linear matrix inequalities, providing the boundedness of all possible trajectories of every controlled plant from the considered class of uncertain systems. We also suggest selecting the optimal feedback gain matrix from the described class of stabilizing feedbacks as the one that minimizes the "size" of the attractive ellipsoid containing all possible bounded dynamic trajectories. The corresponding numerical procedures for designing the best feedback gain matrices are introduced and discussed for each type of considered feedback. Several illustrative examples clearly show the effectiveness of the suggested technique.

Keywords Static output feedback • Observer-based output feedback • Full-order dynamic control

In this chapter, we consider three types of possible linear feedbacks using only the current output information: static feedback proportional to the output measurable signal, observer-based feedback proportional to the state estimation vector, and full-order linear dynamic controllers. For each type of possible linear feedback, we suggest that one characterize the set of all stabilizing gain-feedback matrices by a system of the corresponding linear matrix inequalities (LMIs), providing the boundedness of all possible trajectories of every controlled plant from the considered class of uncertain systems. We also suggest selecting the optimal feedback gain matrix from the described class of stabilizing feedbacks as the one that minimizes the "size" of the attractive ellipsoid containing all possible bounded dynamic trajectories. The corresponding numerical procedures for designing the best feedback gain matrices are introduced and discussed for each type of considered feedback. Several illustrative examples clearly show the effectiveness of the suggested technique.

© Springer International Publishing Switzerland 2014
A. Poznyak et al., *Attractive Ellipsoids in Robust Control*, Systems & Control:
Foundations & Applications, DOI 10.1007/978-3-319-09210-2_4

4.1 Static Feedback Control

4.1.1 System Description and Problem Statement

Consider here the quasi-Lipschitz affine control system with quasi-Lipschitz state-output mapping given by

$$\dot{x}(t) = Ax(t) + Bu(t) + Df(t, x),$$

$$y(t) = Cx(t) + Eg(t, x),$$

$$(4.1)$$

where

- $x(t) \in \mathbb{R}^n$ is the state vector at time $t \geq 0$,
- $u(t) \in \mathbb{R}^m$ is the control input,
- $y(t) \in \mathbb{R}^k$ is the system output,
- $f : \mathbb{R} \times \mathbb{R}^n \to \mathbb{R}^r$ and $g : \mathbb{R} \times \mathbb{R}^n \to \mathbb{R}^s$ are unknown nonlinear functions, and
- $A \in \mathbb{R}^{n \times n}, B \in \mathbb{R}^{n \times m}, C \in \mathbb{R}^{k \times n}, D \in \mathbb{R}^{n \times r}, E \in \mathbb{R}^{k \times s}$ are the system matrices.

The following assumptions will be in force throughout:

(A1) The nonlinear function $f(x, t)$ is *quasi-Lipschitz*, namely, it belongs to the class of functions satisfying

$$f^T Q_f f \leq f_0 + x^T Q_x x, \qquad (4.2)$$

where

$$0 < Q_f \in \mathbb{R}^{n \times n}, \quad 0 < Q_x \in \mathbb{R}^{n \times n}$$

are known matrices and $f_0 \geq 0$ is a given constant.

(A2) The unknown nonlinear function $g(t, x)$ in the output satisfies the similar quasi-Lipschitz constraint

$$g^T Q_g g \leq g_0 + x^T Q'_x x, \qquad (4.3)$$

where

$$0 < Q_g \in \mathbb{R}^{s \times s}, \quad 0 < Q'_x \in \mathbb{R}^{n \times n}$$

are given positive definite matrixes, $g_0 \geq 0$ is a known number.

(A3) The matrices B and C have *full rank*, that is,

$$B^T B > 0 \quad \text{and} \quad CC^T > 0. \qquad (4.4)$$

In this section, we consider *static linear output feedback* of the form

$$u = Ky, \tag{4.5}$$

where $K \in \mathbb{R}^{m \times k}$ is the control gain matrix to be designed.

4.1.2   Application of the Attractive Ellipsoids Method

We introduce the orthogonal matrices

$$G_B \in \mathbb{R}^{n \times n}, \quad G_B = G_B^T = I_n \ : \ G_B B = \begin{bmatrix} 0 \\ \tilde{B} \end{bmatrix}, \tag{4.6}$$

$$\tilde{B} \in \mathbb{R}^{m \times m}, \quad \det(\tilde{B}) \neq 0.$$

The corresponding matrices can be easily found in MATLAB using the function `null`. For given matrix M, the function `null(M)` returns the matrix whose columns are an orthonormal basis of the null space of the matrix M. In this case,

$$G_B = \begin{pmatrix} B^{\perp} \\ B' \end{pmatrix}, \text{ where } B^{\perp} = (\text{null}(B^T))^T \text{ and } B' = \left(\text{null}\left(B^{\perp}\right)\right)^T.$$

Now we are ready to describe the class of feedback matrices K providing the boundedness of all possible trajectories of the quasi-Lipschitz class of nonlinear systems (4.1), closed by the linear feedback (4.5).

Theorem 4.1. *If the matrices*

$$P_1 \in \mathbb{R}^{(n-m) \times (n-m)}, \ P_2 \in \mathbb{R}^{m \times m}, \ Y \in \mathbb{R}^{m \times k}$$

and the numbers $\tau_i > 0, i = 1, 2, 3$ *satisfy the matrix inequalities*

$$\begin{bmatrix} \begin{matrix} P\tilde{A} + \tilde{A}^T P + \tau_1 P + \tau_2 \tilde{Q}_x + \tau_3 \tilde{Q}'_x \\ + \begin{bmatrix} 0 \\ Y \end{bmatrix} C G_B^T + G_B C^T \begin{bmatrix} 0 & Y^T \end{bmatrix} \end{matrix} & P G_B D & G_B^T \begin{bmatrix} 0 \\ Y \end{bmatrix} \\ (P G_B D)^{\mathsf{T}} & -\tau_2 Q_f & 0 \\ \begin{bmatrix} 0 & Y^T \end{bmatrix} G_B & 0 & -\tau_3 Q_g \end{bmatrix} \leq 0, \tag{4.7}$$

$$P := \begin{bmatrix} P_1 & 0 \\ 0 & P_2 \end{bmatrix}, \ \tilde{A} := G_B A G_B^T, \ \tilde{Q}_x := G_B Q_x G_B^T, \ \tilde{Q}'_x := G_B Q'_x G_B^T,$$

and

$$\tau_1 \geq \tau_2 f_0 + \tau_3 g_0, P_1 > 0, P_2 > 0, \tag{4.8}$$

then the ellipsoid

$$\varepsilon(\tilde{P}^{-1}) := \{x \in \mathbb{R}^n : x^T \tilde{P} x \le 1\}, \quad \tilde{P} := G_B^T \begin{bmatrix} P_1 & 0 \\ 0 & P_2 \end{bmatrix} G_B,$$

is an **attractive ellipsoid** *of the system* (4.1)–(4.5) *with output control gain matrix*

$$K := \tilde{B}^{-1} P_2^{-1} Y.$$

Proof. Consider a quadratic storage function of the form

$$V(x) = x^T \tilde{P} x, \quad \tilde{P} = G_B^T \begin{bmatrix} P_1 & 0 \\ 0 & P_2 \end{bmatrix} G_B,$$

$$P_1 \in \mathbb{R}^{(n-m)\times(n-m)}, \, P_2 \in \mathbb{R}^{m\times m}, \, P_1 > 0, \, P_2 > 0,$$

and calculate its time derivative along the trajectories of the system (4.1)–(4.5):

$$\dot{V}(x) = 2x^T \tilde{P} \dot{x} = 2x^T \tilde{P} \left[(A + BKC) x + Df(x(t), t) \right] =$$

$$x^T \left[\tilde{P}(A + BKC) + (A + BKC)^T \tilde{P} \right] x + 2x^T \tilde{P} BKEg + 2x^T \tilde{P} Df(x(t), t) =$$

$$\begin{pmatrix} x \\ f \\ g \end{pmatrix}^T \begin{bmatrix} \tilde{P}(A + BKC) + (A + BKC)^T \tilde{P} & \tilde{P} D & \tilde{P} BK \\ (\tilde{P} D)^T & 0 & 0 \\ (\tilde{P} BK)^T & 0 & 0 \end{bmatrix} \begin{pmatrix} x \\ f \\ g \end{pmatrix} =$$

$$\begin{pmatrix} x \\ f \\ g \end{pmatrix}^T \begin{bmatrix} \begin{array}{c} \tilde{P}(A + BKC) + \\ (A + BKC)^T \tilde{P} + \tau_1 \tilde{P} \end{array} & \tilde{P} D & \tilde{P} BK \\ (\tilde{P} D)^T & -\tau_2 Q_f & 0 \\ (\tilde{P} BK)^T & 0 & -\tau_3 Q_g \end{bmatrix} \begin{pmatrix} x \\ f \\ g \end{pmatrix}$$

$$-\alpha V + \tau_2 \|f\|_{Q_f}^2 + \tau_3 \|g\|_{Q_g}^2 \le \begin{pmatrix} x \\ f \\ g \end{pmatrix}^T W \begin{pmatrix} x \\ f \\ g \end{pmatrix} - \tau_1 V + \tau_2 f_0 + \tau_3 g_0,$$

where $\tau_1, \tau_2, \tau_3 > 0$, and

$$W :=$$
$$\begin{bmatrix} \tilde{P}(A + BKC) + (A + BKC)^T \tilde{P} + \alpha \tilde{P} + \tau_2 Q_x + \tau_3 Q'_x & \tilde{P} D & \tilde{P} BK \\ (\tilde{P} D)^T & -\tau_2 Q_f & 0 \\ (\tilde{P} BK)^T & 0 & -\tau_3 Q_g \end{bmatrix}.$$

If $W \leq 0$ and $\tau_1 \geq \tau_2 g_0 + \tau_2 f_0$, then the ellipsoid $\varepsilon(\tilde{P}^{-1})$ is an attractive ellipsoid for the system (4.1)–(4.5).

Applying the equivalent transformation $T = \mathrm{diag}(G_B^T, I_r, I_s)$ to the matrix W, setting

$$Y := P_2 \tilde{B} K,$$

and taking into account the sequence of equalities

$$\tilde{P} BK = G_B^T \begin{bmatrix} P_1 & 0 \\ 0 & P_2 \end{bmatrix} G_B BK =$$

$$G_B^T \begin{bmatrix} P_1 & 0 \\ 0 & P_2 \end{bmatrix} \begin{bmatrix} 0 \\ \tilde{B} \end{bmatrix} K = G_B^T \begin{bmatrix} 0 \\ Y \end{bmatrix},$$

we obtain that the matrix inequality $T^T W T \leq 0$ has the form (4.7).   ∎

We need to minimize the trace of the matrix $\tilde{P}^{-1}$ under constraints (4.7), (4.8) in order to optimize an attractive ellipsoid. In view of the identity

$$\mathrm{tr}(\tilde{P}^{-1}) = \mathrm{tr}\left(G_B^T \begin{bmatrix} P_1^{-1} & 0 \\ 0 & P_2^{-1} \end{bmatrix} G_B \right) =$$

$$\mathrm{tr}(G_B^T G_B \begin{bmatrix} P_1^{-1} & 0 \\ 0 & P_2^{-1} \end{bmatrix}) = \mathrm{tr}(P_1^{-1}) + \mathrm{tr}(P_2^{-1}),$$

we can rewrite the optimization problem in the form

$$\min_{H_1,H_2,P_1,P_2,Y,\tau_1,\tau_2,\tau_3} [\mathrm{tr}(H_1) + \mathrm{tr}(H_2)]$$

subject to (4.7), (4.8), and

$$\begin{bmatrix} H_1 & I_{n-m} \\ I_{n-m} & P_1 \end{bmatrix} \geq 0 \quad \text{and} \quad \begin{bmatrix} H_2 & I_m \\ I_m & P_2 \end{bmatrix} \geq 0, \tag{4.9}$$

where

$$H_1 \in \mathbb{R}^{(n-m)\times(n-m)}, \quad H_2 \in \mathbb{R}^{m\times m}.$$

The constraints (4.9) appear here by the Schur complement implementation for the matrix inequalities

$$H_1 \geq P_1^{-1}$$

and

$$H_2 \geq P_2^{-1},$$

which are introduced in order to avoid a nonlinear representation of the optimizing
functional.

4.1.3   Example: Stabilization of a Discontinuous System

Consider the model given by the strongly nonlinear differential equations

$$\begin{cases} \dot{x}_1 = -x_1 + x_2 + 0.1x_1\text{sign}[x_2], \\ \dot{x}_2 = -x_1 + 0.2\text{sign}[x_1] + u, \qquad x_1, x_2 \in \mathbb{R}, \qquad (4.10) \\ y = x_1 + x_2 + g(t), \end{cases}$$

where $|g(t)|^2 \leq g_0 = 0.01$. The given system can be represented in the form

$$\dot{x}(t) = Ax(t) + Bu(t) + f(t, x(t))$$

with the correspondingly defined system matrices

$$A = \begin{bmatrix} -1 & 1 \\ -1 & 0 \end{bmatrix}, \quad B = \begin{bmatrix} 0 \\ 1 \end{bmatrix}, \quad C = \begin{bmatrix} 1 & 1 \end{bmatrix},$$

and the nonlinear function

$$f(t, x) = \begin{pmatrix} 0.1\text{sign}[x_1]\sin(x_2) \\ 0.2\text{sign}[x_1] \end{pmatrix}$$

satisfying the quasi-Lipschitz condition

$$f^T f \leq f_0 + x^T Q_x x,$$

where

$$f_0 = 0.04, \quad Q_x = 0.01 \begin{bmatrix} 1 & 0 \\ 0 & 0 \end{bmatrix}.$$

Solving the optimization problem (4.1.2), we obtain

$$\tilde{P} = \begin{pmatrix} 247.9034 & 0 \\ 0 & 75.3833 \end{pmatrix}; \quad K = -2.2886.$$

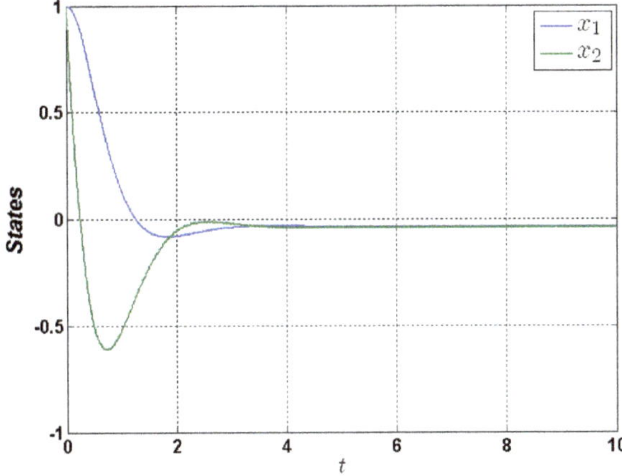

Fig. 4.1 Evolution of a discontinuous system

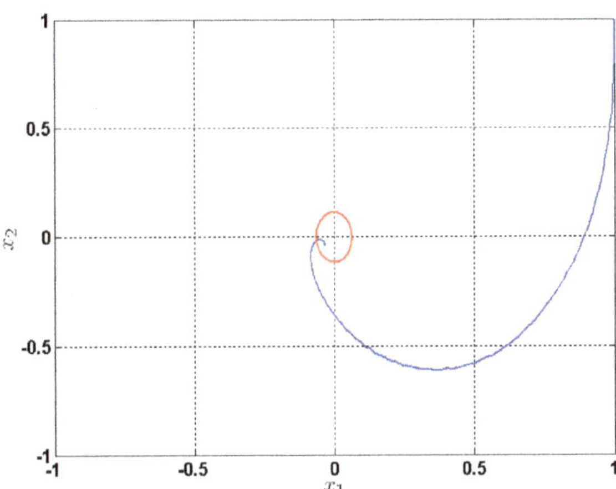

Fig. 4.2 Attractive ellipsoid and the discontinuous system

Figures 4.1 and 4.2 present the results of a numerical simulation of the closed-loop system (4.10) with static output feedback $u = Ky$.

The function

$$g(t) = 0.01 \cos(100t)$$

was selected in order to simulate the deterministic noise of measurements. The initial state was defined as follows:

$$x_0 = (1, \ 1)^T.$$

This nonlinear example illustrates the effectiveness and potential applicability of the proposed theoretical robust control design techniques for the considered class of uncertain systems.

4.2 Observer-Based Feedback Design

4.2.1 State Observer and the Extended Dynamic Model

We consider here the system (4.1)–(4.3) with the linear feedback control

$$u(t) = K\hat{x}(t), \quad K \in \mathbb{R}^{m \times n} \tag{4.11}$$

with respect to the observer state $\hat{x} \in \mathbb{R}^n$, referred to below as the *state estimate*, which is generated by the *classical Luenberger observer* having the structure

$$\dot{\hat{x}}(t) = A\hat{x}(t) + Bu(t) + F(y(t) - C\hat{x}(t)), \quad F \in \mathbb{R}^{n \times k}. \tag{4.12}$$

The robust stabilization of the system (4.1) will be realized using the attractive ellipsoid method (AEM).

Problem 4.1. *The main problem here is to design an observer-based linear feedback control providing the boundedness of every trajectory of the system (4.1), (4.11), (4.12) within an attractive ellipsoid of "minimal size."*

We define the state estimation error as

$$e(t) := x(t) - \hat{x}(t). \tag{4.13}$$

Then its time derivative satisfies

$$\dot{e}(t) = (A - FC)e(t) + Df(t, x) - FEg(t, x).$$

We next introduce the extended vector $z := \begin{pmatrix} \hat{x} \\ e \end{pmatrix} \in \mathbb{R}^{2n}$. Obviously, it is governed by the following ordinary differential equation (ODE):

$$\dot{z}(t) = \hat{A}z(t) + \hat{F}w(t, x), \tag{4.14}$$

where

$$\hat{A} := \begin{bmatrix} A + BK & FC \\ 0 & A - FC \end{bmatrix},$$

$$\hat{F} := \begin{bmatrix} 0 & FE \\ D & -FE \end{bmatrix}, \quad w(t, x) := \begin{pmatrix} f(t, x) \\ g(t, x) \end{pmatrix}.$$

4.2.2   Stabilizing Feedback Gains K and F

Our aim here is to find the control gain matrix K and the observer gain matrix F providing a stabilization (boundedness) of the state dynamics $x(t)$ as well as state estimation $\hat{x}(t)$ of the system (4.14) such that the corresponding attractive ellipsoid in z-space will be "minimal."

The following theorem gives a solution to this problem.

Theorem 4.2. *If the matrices*

$$P_1, P_2, R_1 \in \mathbb{R}^{n \times n}, \quad Y_1 \in \mathbb{R}^{m \times n}, \quad Y_2 \in \mathbb{R}^{n \times k}$$

and the numbers $\tau_1, \tau_2, \tau_3 \in \mathbb{R}$ satisfy the system of matrix inequalities

$$\begin{bmatrix} -R_1 & Y_2 C & 0 & Y_2 E & P_2 P_1 & P_2 P_1 \\ C^T Y_2^T & \begin{matrix} A^T P_2 + P_2 A \\ -Y_2 C - C^T Y_2^T + \tau_1 P_2 \end{matrix} & P_2 D & -Y_2 E & I & I \\ 0 & D^T P_2 & -\tau_2 Q_f & 0 & 0 & 0 \\ E^T Y_2^T & -E^T Y_2^T & 0 & -\tau_3 Q_g & 0 & 0 \\ P_1 P_2 & I & 0 & 0 & -\frac{1}{\tau_2} Q_x^{-1} & 0 \\ P_1 P_2 & I & 0 & 0 & 0 & -\frac{1}{\tau_3} Q_x^{-1} \end{bmatrix} \le 0,$$

$$(4.15)$$

and

$$\begin{bmatrix} R_1 - 2P_2 & I \\ I & P_1 A^T + A P_1 + Y_1^T B^T + B Y_1 + \tau_1 P_1 \end{bmatrix} \le 0$$

$$(4.16)$$

$$\tau_1 \ge f_0 \tau_2 + g_0 \tau_3, \ \tau_2 \ge 0, \tau_3 \ge 0,$$

$$P_1 > 0, P_2 > 0, R_1 > 0,$$

then the ellipsoid

$$\varepsilon(P) := \left\{ z \in \mathbb{R}^{2n} : z^T P^{-1} z \le 1 \right\}$$

with the matrix

$$P = \begin{bmatrix} P_1 & 0 \\ 0 & P_2^{-1} \end{bmatrix} \qquad (4.17)$$

*is the **attractive ellipsoid** of the system (4.14), (4.2), (4.3), (4.11) with the feedback control gain matrix*

$$K = Y_1 P_1^{-1} \qquad (4.18)$$

and the observer gain matrix

$$F = P_2^{-1} Y_2. \tag{4.19}$$

Proof. Define the Lyapunov function as

$$V(z) := z^T P^{-1} z, \quad P^{-1} = \begin{bmatrix} P_1^{-1} & 0 \\ 0 & P_2 \end{bmatrix} := \mathrm{diag}(P_1^{-1}, P_2),$$

where $P > 0$ is the matrix of an attractive ellipsoid to be minimized. Then

$$\dot{V} = \begin{pmatrix} z \\ w \end{pmatrix}^T \begin{bmatrix} \hat{A}^T P^{-1} + P^{-1}\hat{A} & P^{-1}\hat{F} \\ \hat{F}^T P^{-1} & 0 \end{bmatrix} \begin{pmatrix} z \\ w \end{pmatrix} =$$

$$\begin{pmatrix} z \\ f \\ g \end{pmatrix}^T \begin{bmatrix} \hat{A}^T P^{-1} + P^{-1}\hat{A} \\ +\tau_1 P^{-1} & P^{-1}\hat{F} \\ \hat{F}^T P^{-1} & -\begin{bmatrix} \tau_2 Q_f & 0 \\ 0 & \tau_3 Q_g \end{bmatrix} \end{bmatrix} \begin{pmatrix} z \\ f \\ g \end{pmatrix} - \tau_1 V$$

$$+ \tau_2 f^T Q_f f + \tau_3 g^T Q_g g \leq$$

$$\begin{pmatrix} z \\ f \\ g \end{pmatrix}^T W \begin{pmatrix} z \\ f \\ g \end{pmatrix} - \tau_1 V + f_0 \tau_2 + g_0 \tau_3,$$

where $\tau_1 > 0$, $\tau_2 > 0$,and

$$W := \begin{bmatrix} \hat{A}^T P^{-1} + P^{-1}\hat{A} + \tau_1 P^{-1} + \tau_2 Q_z + \tau_3 Q'_z & P^{-1}\hat{F} \\ \hat{F}^T P^{-1} & -\begin{bmatrix} \tau_2 Q_f & 0 \\ 0 & \tau_3 Q_g \end{bmatrix} \end{bmatrix},$$

$$Q_z = \begin{bmatrix} Q_x & Q_x \\ Q_x & Q_x \end{bmatrix}, \quad Q'_z = \begin{bmatrix} Q'_x & Q'_x \\ Q'_x & Q'_x \end{bmatrix},$$

in view of the representations

$$x^T Q_x x = z^T Q_z z, \quad x^T Q'_x x = z^T Q'_z z.$$

Hence the ellipsoid

$$\varepsilon(P) = \{z \in \mathbb{R}^{2n} : z^T P^{-1} z < 1\}$$

will be attractive for the system (4.14) if

$$W \leq 0 \quad \text{and} \quad \tau_1 \geq f_0 \tau_2 + g_0 \tau_3, \tag{4.20}$$

where $\tau_i \geq 0$, $i = 1, 2, 3$. Define

$$A_K := A + BK$$

and

$$A_F = A - FC.$$

Then we obtain

$$W =$$

$$\begin{bmatrix}
A_K^T P_1^{-1} + P_1^{-1} A_K \\ + \tau_2 Q_x + \tau_3 Q_x' & P_1^{-1} FC + \tau_2 Q_x + \tau_3 Q_x' & 0 & P_1^{-1} FE \\
C^T F^T P_1^{-1} + \tau_2 Q_x + \tau_3 Q_x' & A_F^T P_2 + P_2 A_F \\ + \tau_2 Q_x + \tau_3 Q_x' & P_2 D & -P_2 FE \\
0 & D^T P_2 & -\tau_2 Q_f & 0 \\
E^T F^T P_1^{-1} & -E^T F^T P_2 & 0 & -\tau_3 Q_g
\end{bmatrix} \leq 0.$$

Applying the quadratic nonsingular transformation

$$T_1 = \text{diag}[P_1, I_{n\times n}, I_{r\times r}, I_{s\times s}]$$

to the matrix W, we get

$$W_1 = T_1 W T_1^T =$$

$$\begin{pmatrix}
\Psi_1 & FC + \tau_2 P_1 Q_x + \tau_3 P_1 Q_x' & 0 & FE \\
C^T F^T + \tau_2 Q_x P_1 + \tau_3 Q_x' P_1 & \Psi_2 & P_2 D & -P_2 FE \\
0 & D^T P_2 & -\tau_2 Q_f & 0 \\
E^T F^T & -E^T F^T P_2 & 0 & -\tau_3 Q_g
\end{pmatrix} \leq 0,$$

where

$$\Psi_1 := P_1 A_K^T + A_K P_1 + \tau_1 P_1 + \tau_2 P_1 Q_x P_1 + \tau_3 P_1 Q_x' P_1,$$

$$\Psi_2 := A_F^T P_2 + P_2 A_F + \tau_1 P_2 + \tau_2 Q_x + \tau_3 Q_x'.$$

Obviously,

$$W_1 = \tilde{W} + \tau_2 \begin{bmatrix} P_1 \\ I \\ 0 \\ 0 \end{bmatrix} Q_x \begin{bmatrix} P_1 & I & 0 & 0 \end{bmatrix} +$$

$$\tau_3 \begin{bmatrix} P_1 \\ I \\ 0 \\ 0 \end{bmatrix} Q_x' \begin{bmatrix} P_1 & I & 0 & 0 \end{bmatrix} \leq 0,$$

$$(4.21)$$

where

$$\tilde{W} = \begin{bmatrix} P_1 A_K^T + A_K P_1 + \tau_1 P_1 & FC & 0 & FE \\ C^T F^T & A_F^T P_2 + P_2 A_F + \tau_1 P_2 & P_2 D & -P_2 FE \\ 0 & D^T P_2 & -\tau_2 Q_f & 0 \\ E^T F^T & -E^T F^T P_2 & 0 & -\tau_3 Q_g \end{bmatrix}.$$

Using the Schur complement to (4.21) twice, we obtain that the inequality

$$W_1 \leq 0$$

is equivalent to

$$W_2 \leq 0,$$

where

$$W_2 =$$

$$\begin{bmatrix} P_{11}^{(2)} & FC & 0 & FE & P_1 & P_1 \\ C^T F^T & & P_2 & -P_2 FE & I & I \\ 0 & P_2 & -\tau_2 Q_f & 0 & 0 & 0 \\ E^T F^T & -E^T F^T P_2 & 0 & -\tau_3 Q_g & 0 & 0 \\ P_1 & I & 0 & 0 & -\frac{1}{\tau_2} Q_x^{-1} & 0 \\ P_1 & I & 0 & 0 & 0 & -\frac{1}{\tau_3} (Q_x')^{-1} \end{bmatrix}$$

and

$$P_{11}^{(2)} = P_1 A_K^T + A_K P_1 + \tau_1 P_1,$$

$$P_{22}^{(2)} = A_F^T P_2 + P_2 A_F + \tau_1 P_2.$$

Analogously, in applying the equivalent transformation

$$T_2 = \text{diag}[P_2, I_{n \times n}, I_{r \times r}, I_{s \times s}, I_{n \times n}, I_{n \times n}]$$

to the matrix W_2, we obtain

$$W_3 = T_2 W_2 T_2^T =$$

$$\begin{bmatrix} P_{11}^{(3)} & P_2 FC & 0 & P_2 FE & P_2 P_1 & P_2 P_1 \\ C^T F^T P_2 & P_{22}^{(3)} & D P_2 & -P_2 FE & I & I \\ 0 & D^T P_2 & -\tau_2 Q_f & 0 & 0 & 0 \\ E^T F^T P_2 & -E^T F^T P_2 & 0 & -\tau_3 Q_g & 0 & 0 \\ P_1 P_2 & I & 0 & 0 & -\frac{1}{\tau_2} Q_x^{-1} & 0 \\ P_1 P_2 & I & 0 & 0 & 0 & -\frac{1}{\tau_3} (Q_x')^{-1} \end{bmatrix} \leq 0$$

with

$$P_{11}^{(3)} = P_2 \left(P_1 A_K^T + A_K P_1 + \tau_1 P_1 P_2 \right),$$

$$P_{22}^{(3)} = A_F^T P_2 + P_2 A_F + \tau_1 P_2.$$

From the Λ-inequality (see Poznyak 2008)

$$XY^T + YX^T \le X\Lambda X^T + Y\Lambda^{-1}Y^T,$$

valid for every $X \in R^{n\times k}$, $Y \in R^{n\times k}$, and $0 < \Lambda = \Lambda^T \in R^{k\times k}$, applied with $Y = I$, it follows that

$$X + X^T \le X\Lambda X^T + \Lambda^{-1},$$

which for

$$\Lambda := -(P_1 A_K^T + A_K P_1 + \tau_1 P_1)$$

implies

$$P_2(P_1 A_K^T + A_K P_1 + \tau_1 P_1)P_2 \le$$
$$-P_2 - P_2 - (P_1 A_K^T + A_K P_1 + \tau_1 P_1)^{-1}. \tag{4.22}$$

Let $R_1 \in \mathbb{R}^{n\times n}$ be the new matrix variable such that $R_1 > 0$. Applying the Schur complement to the matrix inequality

$$-2P_2 - (P_1^{-1} A_K^T + A_K P_1^{-1} + \tau_1 P_1^{-1})^{-1} \le -R_1, \tag{4.23}$$

we get

$$\begin{bmatrix} R_1 - 2P_2 & I \\ I & P_1^{-1} A_K^T + A_K P_1^{-1} + \tau_1 P_1^{-1} \end{bmatrix} \le 0. \tag{4.24}$$

Defining

$$Y_1 := KP_1, Y_2 := P_2 F$$

and using (4.22), (4.23), (4.24) together with the matrix inequality

$$\begin{bmatrix} X & Y^T \\ Y & Z \end{bmatrix} \le \begin{bmatrix} X' & Y^T \\ Y & Z \end{bmatrix},$$

valid for every $X = X^T$, $Z = Z^T$, and $X' = X'^T \geq X$, we may conclude that the matrix inequalities (4.15), (4.16) imply (4.20). ∎

To minimize the attractive ellipsoid of the system (4.14), we need to resolve the optimization problem

$$\min_{P1,P2,H,Y1,Y2,\tau_1,\tau_2,\tau_3} [\operatorname{tr}(P_1) + \operatorname{tr}(H)] \qquad (4.25)$$

subject to (4.15), (4.16), and

$$\begin{bmatrix} H & I \\ I & P_2 \end{bmatrix} \geq 0, \quad H > 0. \qquad (4.26)$$

The additional constraint (4.26) is incorporated into the optimization problem in order to avoid a nonlinear term in the optimizing functional. Indeed, due to the Schur complement, the matrix inequality (4.26) is equivalent to

$$H \geq P_2^{-1}.$$

So the minimization of the functional

$$\operatorname{tr}(P_1) + \operatorname{tr}(H)$$

is equivalent to the minimization of the functional

$$\operatorname{tr}(P_1) + \operatorname{tr}(P_2^{-1}),$$

which is the trace of the attractive ellipsoid.

Remark 4.1. Since the proof of Theorem 4.2 is based on a Λ-inequality, which gives only upper estimates for the matrix inequalities, we can guarantee that the solution of the optimization problem (4.25) provides only a **quasiminimal** attractive ellipsoid.

Remark 4.2. If $Q_x = 0$ and $Q'_x = 0$, then the constraint (4.15) converts to LMI form:

$$\begin{bmatrix} -R_1 & Y_2^T C & 0 & Y_2^T E \\ C^T Y_2 & A^T P_2 + P_2 A - Y_2^T C - C^T Y_2 + \tau_1 P_2 & P_2 D & -Y_2^T E \\ 0 & D^T P_2 & -\tau_2 Q_f & 0 \\ E^T Y_2 & -E^T Y_2 & 0 & -\tau_3 Q_f \end{bmatrix} \leq 0.$$

4.2.3 *Numerical Aspects*

The problem (4.25) is, in fact, a bilinear optimization problem because of the term $P_1 P_2$ participating in the matrix constraint (4.15). The feasible set for a BMI is nonconvex in general. Semidefinite relaxations (as in Boyd & Vandenberghe 1997) and solutions through nonlinear programming methods (e.g., "branch-bound" algorithms) can be considered two alternatives for solving a given BMI. The efficiency of the MATLAB Toolbox PENBMI is critical to initial point selection, which one would like to be very feasible and close to a solution (Henrion, Loefberg, Kocvara, & Stingl 2006). It is based on the penalty function approach. The following lemma simplifies the problem of finding a feasible starting point.

Lemma 4.1. *The feasibility set of the BMIs (4.15) contains a set of variables satisfying the following system of matrix inequalities:*

$$\begin{bmatrix} -R_1 & Y_2 C & 0 & Y_2 E & 0 & 0 \\ C^T Y_2^T & \begin{matrix} A^T P_2 + P_2 A - \\ Y_2 C - C^T Y_2^T + \tau_1 P_2 \end{matrix} & P_2 D & -Y_2 E & I & I \\ 0 & D^T P_2 & -\tau_2 Q_f & 0 & 0 & 0 \\ E^T Y_2^T & -E^T Y_2^T & 0 & -\tau_3 Q_g & 0 & 0 \\ 0 & I & 0 & 0 & -R_2 & 0 \\ 0 & I & 0 & 0 & 0 & -R_3 \end{bmatrix} \le 0,$$

$$\begin{pmatrix} R_1 - 2P_2 & I \\ I & \begin{matrix} P_1 A^T + AP_1 + Y_1^T B^T + BY_1 \\ +\tau_1 P_1 + \Lambda_1 + \Lambda_2 \end{matrix} \end{pmatrix} \le 0, \tag{4.27}$$

$$\begin{bmatrix} R_2 - \frac{1}{\tau_2} Q_x^{-1} & P_1 \\ P_1 & -\Lambda_1 \end{bmatrix} \le 0, \quad \begin{bmatrix} R_3 - \frac{1}{\tau_3} (Q_x')^{-1} & P_1 \\ P_1 & -\Lambda_2 \end{bmatrix} \le 0,$$

$$P_1 > 0, P_2 > 0, \Lambda_1 > 0, \Lambda_2 > 0, R_1 > 0, R_2 > 0, R_3 > 0,$$

$$\tau_1 \ge f_0 \tau_2 + g_0 \tau_3, \ \tau_i \ge 0, \ i = 1, 2, 3,$$

$$P_1, P_2, R_1 R_2, R_3, \Lambda_1, \Lambda_2 \in \mathbb{R}^{n \times n}, \quad Y_1 \in \mathbb{R}^{m \times n}, \quad Y_2 \in \mathbb{R}^{n \times k}.$$

Proof. By the Λ-inequality (see Poznyak 2008),

$$XY^T + YX^T \le X\Lambda_1^{-1} X^T + Y\Lambda_1 Y^T,$$

valid for every $X, Y \in \mathbb{R}^{n \times m}$ and $0 < \Lambda_1 = \Lambda_1^T \in \mathbb{R}^{n \times m}$ with

$$X^T := \begin{pmatrix} 0 & 0 & 0 & 0 & P_1 & 0 \end{pmatrix} \text{ and } Y := \begin{pmatrix} P_2 & 0 & 0 & 0 & 0 & 0 \end{pmatrix},$$

the matrix W_3 can be estimated as

$$W_3 \le W_3' \le 0$$

where

$$W_3' :=$$

$$\begin{bmatrix} P_{11}' & P_2 FC & 0 & P_2\, FE & 0 & P_2 P_1 \\ C^T F^T P_2 & P_{22}' & P_2 & -P_2\, FE & I & I \\ 0 & P_2 & -\tau_2\, Q_f & 0 & 0 & 0 \\ E^T F^T P_2 & -E^T F^T P_2 & 0 & -\tau_3 Q_g & 0 & 0 \\ 0 & I & 0 & 0 & P_{55}' & 0 \\ P_1 P_2 & I & 0 & 0 & 0 & -\frac{1}{\tau_3}(Q_x')^{-1} \end{bmatrix} \le 0 \qquad (4.28)$$

and

$$P_{11}' = P_2\left(P_1 A_K^T + A_K P_1 + \tau_1 P_1 + \Lambda_1\right) P_2,$$

$$P_{22}' = A_F^T P_2 + P_2 A_F + \tau_1 P_2,$$

$$P_{55}' = -\tfrac{1}{\tau_2} Q_x^{-1} + P_1 \Lambda_1^{-1} P_1.$$

The term

$$-\frac{1}{\tau_2} Q_x^{-1} + P_1 \Lambda_1^{-1} P_1$$

in (4.28) can be bounded by $(-R_2)$ as

$$P_1 \Lambda_1^{-1} P_1 - \frac{1}{\tau_2} Q_x^{-1} \le -R_2. \qquad (4.29)$$

Applying again the Schur complement, we can express (4.29) as

$$\begin{bmatrix} R_2 - \frac{1}{\tau_2} Q_x^{-1} & P_1 \\ P_1 & -\Lambda_1 \end{bmatrix} \le 0.$$

A similar consideration for the Λ-inequality with

$$X^T := \begin{bmatrix} 0 & 0 & 0 & 0 & 0 & P_1 \end{bmatrix} \text{ and } Y := \begin{bmatrix} P_2 & 0 & 0 & 0 & 0 & 0 \end{bmatrix}$$

helps to eliminate the remaining bilinear terms $P_1 P_2$. This completes the proof of the lemma. ∎

Remark 4.3. Note that for the fixed scalar parameters τ_1, τ_2, and τ_3, the matrix inequalities (4.27) become LMIs. They can be solved using the MATLAB toolboxes LMI-toolbox, SeDuMi, and Yalmip.

4.2.4 Example: Robust Stabilization of a Spacecraft

Consider a simplified model of a spacecraft (Yefremov, Polyakov, & Strygin 2006), (Gonzalez-Garcia, Polyakov, & Poznyak 2011) with two dynamic elastic elements (rods having dissipative properties, as shown in Fig. 4.3.

We are interested only in the controlled motion around the longitudinal axis of the spacecraft. Suppose the rods execute antisymmetric oscillations. The position of the system $O_{x_1 y_1 z_1}$ is defined by the rotation angle $\gamma(t)$ of the spacecraft. The deflection of the rod from the O_{x_1}-axis will be denoted by $y(x,t)$. If control is achieved by means of a torque $u(t)$ applied to this body, the rods possess dissipative properties. In this case, the dynamic equations of the system take the following form (Yefremov et al. 2006):

$$J\ddot{\gamma}(t) + 2\int_0^l m(x+r)\frac{\partial^2 y(x,t)}{\partial t^2}dx = u(t) + \bar{f}_0(t,x), \tag{4.30}$$

$$m(x+r)\ddot{\gamma} + m\frac{\partial^2 y(x,t)}{\partial t^2} + EI\frac{\partial^4 y(x,t)}{\partial x^4} + EI\chi\frac{\partial^5 y(x,t)}{\partial t\,\partial x^4} = \bar{f}(t,x), \tag{4.31}$$

with boundary conditions

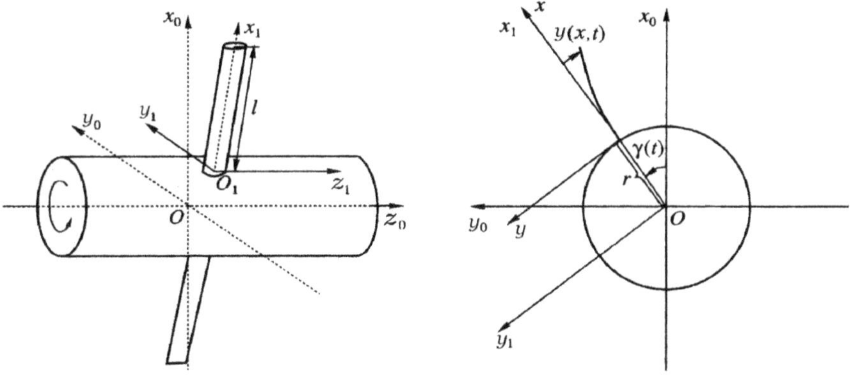

Fig. 4.3 Model of a spacecraft

$$y(0,t) = \left.\frac{\partial y(x,t)}{\partial x}\right|_{x=0} = \left.\frac{\partial^2 y(x,t)}{\partial x^2}\right|_{x=l} = \left.\frac{\partial^3 y(x,t)}{\partial x^3}\right|_{x=l} = 0,$$

where

- r is the distance from the longitudinal axis to the point where the rod is fastened,
- l is the rod length,
- EI is the flexural stiffness of the rod,
- χ is the coefficient of internal viscous friction,
- m is the mass per unit length of the rod,
- J_0 is the moment of inertia of the spacecraft with respect to the OZ-axis,
- u is the control moment applied to the spacecraft,
- the functions $\bar{f}_0(t,x)$, $\bar{f}(t,x)$ describe system uncertainties and disturbances acting on the rigid body and the elastic elements, respectively.

The moment of inertia of the whole system J is given by

$$J = J_0 + 2\int_0^l m(x+r)^2 dx.$$

Using Galerkin's method (see, for example, Sect. 18.6.2 in Poznyak 2008), we can suppose approximately that

$$y(x,t) \simeq \sum_{i=1}^{k} q_i(t)\Phi_i(x), \tag{4.32}$$

where $\Phi_i(x)$ is the eigenfunction corresponding to the positive eigenvalue λ_i of the positive self-conjugate operator

$$L\Phi(x) = \frac{d^4\Phi(x)}{dx^4}, \quad \Phi(0) = \Phi'(0) = \Phi''(l) = \Phi'''(l) = 0,$$

where

$$\frac{d^4\Phi_i(x)}{dx^4} \equiv \lambda_i\Phi_i(x), \quad 0 \le x \le l.$$

After substituting (4.32) into (4.30), (4.31), multiplication of (4.31) by $\Phi_i(x)$, and an integration on the interval $[0,l]$, we get

$$J\ddot{y} + 2\sum_{i=1}^{k} p_i\ddot{q}_i = u(t) + f_0, \tag{4.33}$$

$$p_i\ddot{y} + a_i\ddot{q}_i + b_i\dot{q}_i + c_iq_i = f_i, \quad i = 1,2,\ldots,k, \tag{4.34}$$

with

$$p_i = \int\limits_0^l m(x + r)\Phi_i(x)dx,$$

$$a_i = \int\limits_0^l m\Phi_i^2(x)dx, \quad b_i = \lambda_i EI\chi \int\limits_0^l \Phi_i^2(x)dx,$$

$$c_i = \lambda_i EI \int\limits_0^l \Phi_i^2(x)dx, \quad f_i = \int\limits_0^l \bar{f}(t,x)\,\Phi_i(x)dx.$$

Here the functions f_i describe bounded disturbances, uncertainties, and unmodeled dynamics of the rigid body.

Consider the case of one frequency of oscillations ($k = 1$) in (4.33) and in (4.34). Introduce the state space variables

$$x_1 = \gamma, \ x_2 = \dot{\gamma}, \ x_3 = q_1 \text{ and } x_4 = \dot{q}_1.$$

For the spacecraft with parameters

$$J_0 = 150, \ l = 7.5, \ m = 0.53, \ r = 3, \ EI = 20, \ \chi = 0.1,$$

the state space representation of the dynamic model is

$$\begin{pmatrix} \dot{x}_1 \\ \dot{x}_2 \\ \dot{x}_3 \\ \dot{x}_4 \end{pmatrix} = \begin{bmatrix} 0 & 1 & 0 & 0 \\ 0 & 0 & 0.0028 & 0.0142 \\ 0 & 0 & 0 & 1 \\ 0 & 0 & -0.0825 & -0.4126 \end{bmatrix} \begin{pmatrix} x_1 \\ x_2 \\ x_3 \\ x_4 \end{pmatrix}$$

$$+ \begin{pmatrix} 0 \\ 0.0076 \\ 0 \\ -0.1676 \end{pmatrix} u + w_x,$$

where

$$w_x = \begin{bmatrix} 0 & 0 \\ 0.0076 & 0.0037 \\ 0 & 0 \\ -0.1676 & -0.0979 \end{bmatrix} \begin{pmatrix} f_0 \\ f_1 \end{pmatrix}.$$

Supposing that only the angle γ and the angular velocity $\dot{\gamma}$ are measurable, the state-output mapping is

$$y = \begin{bmatrix} 1 & 0 & 0 & 0 \\ 0 & 1 & 0 & 0 \end{bmatrix} \begin{pmatrix} x_1 \\ x_2 \\ x_3 \\ x_4 \end{pmatrix} + w_y,$$

where $w_y \in \mathbb{R}^2$ is an output noise that can be estimated as follows:

$$w_y^T Q_g w_y \leq 1, \quad Q_g = \begin{bmatrix} 530 & 25 \\ 25 & 1960 \end{bmatrix}.$$

Assume also that system disturbances are bounded as

$$|f_0|^2 \leq 0.05|\gamma|^2 + 0.1|\dot{\gamma}|^2 + 0.06|q|^2 + 0.11|\dot{q}|^2$$

and

$$|f_1| \leq 0.1.$$

One can rearrange them in the form (4.2) with

$$Q_f = \begin{bmatrix} 1 & 0 \\ 0 & 1 \end{bmatrix}, \quad Q_x = \begin{bmatrix} 0.05 & 0 & 0 & 0 \\ 0 & 0.1 & 0 & 0 \\ 0 & 0 & 0.06 & 0 \\ 0 & 0 & 0 & 0.11 \end{bmatrix}.$$

We consider here that the bounding matrix for the constraint control (3.21) is defined as follows:

$$M_u = I_4.$$

The parameters K and F realizing the robust output linear controller are obtained by solving the optimization problem (4.25):

$$K = \begin{pmatrix} -128.8577 & -5509.47037 & 84.5673 & -88.4066 \end{pmatrix},$$

$$F = \begin{bmatrix} 0.0955 & 0.0170 \\ 0.0034 & 0.0014 \\ -0.0502 & -0.0268 \\ 0.0029 & -0.0099 \end{bmatrix}.$$

Figures 4.4–4.6 show the angular position of the spacecraft $\gamma(t)$, the control input $u(t)$, and the projection of the invariant ellipsoid to the phase plane Oe_1e_2 for the observation error $e = x - \hat{x}$.

Fig. 4.4 Angular dynamic of
the spacecraft

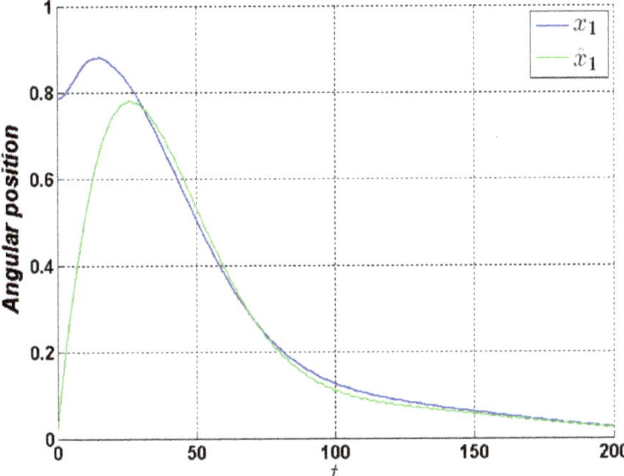

Fig. 4.5 The control law for
the spacecraft

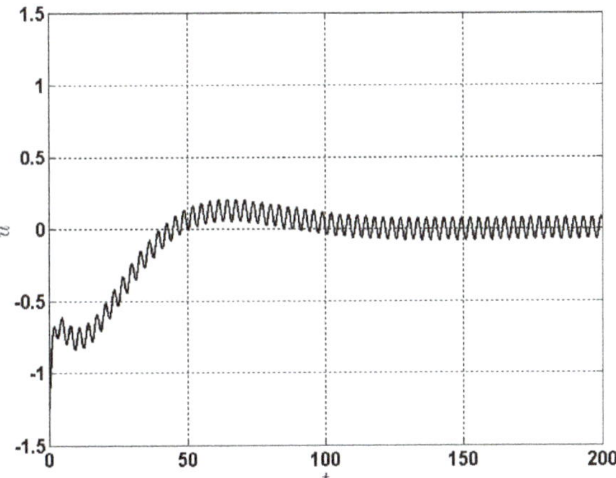

Fig. 4.6 Projection of the
attractive ellipsoid and
estimation error

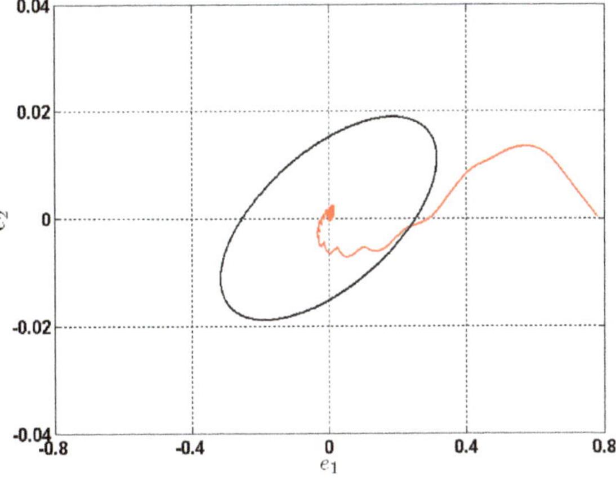

4.3 Dynamic Regulator

4.3.1 Full-Order Linear Dynamic Controllers

The proposed minimization approach to the class of quasi-Lipschitz nonlinear output control systems is used here in combination with the classical full-order linear dynamic output controllers. Roughly speaking, we choose the necessary controller parameters (gain matrices) that minimize the size of an attractive ellipsoid related to the resulting closed-loop control system.

Consider the quasi-Lipschitz nonlinear output control system (4.1), (4.2), (4.3). The admissible feedback control strategies for this system are chosen from the class of *full-order linear dynamic controllers* of the following structure:

$$u(t) = C_r x_r(t) + D_r y(t),$$

$$\dot{x}_r(t) = A_r x_r(t) + B_r y(t), \tag{4.35}$$

$$x_r(0) = x_0^r,$$

where

$$x_r \in \mathbb{R}^n, \ A_r \in \mathbb{R}^{n \times n}, \ B_r \in \mathbb{R}^{n \times k}, \ D_r \in \mathbb{R}^{m \times k}, \ C_r \in \mathbb{R}^{m \times n}.$$

The control design associated with (4.35) is completely determined by the selection of the matrix

$$\Theta := \begin{bmatrix} D_r & C_r \\ B_r & A_r \end{bmatrix} \in \mathbb{R}^{(n+m) \times (n+k)}. \tag{4.36}$$

We call Θ the dynamic controller matrix. The closed-loop realization of (4.1)–(4.3) under (4.35) is given by

$$\dot{z}(t) = (A_0 + B_0 \Theta C_0)z(t) + D_0 f(t, x(t)) + B_0 \Theta E_0 g(t, x(t)) \quad \forall t \geq 0,$$

$$z(0) = (x_0, x_0^r),$$

$$\tag{4.37}$$

where

$$z(t) := \left[x^T(t), x_r^T(t) \right]^T$$

and

$$A_0 := \begin{bmatrix} A & 0 \\ 0 & 0 \end{bmatrix}, \quad B_0 := \begin{bmatrix} B & 0 \\ 0 & I_n \end{bmatrix},$$

$$C_0 := \begin{bmatrix} C & 0 \\ 0 & I_n \end{bmatrix}, \quad D_0 := \begin{bmatrix} D \\ 0 \end{bmatrix}, \quad E_0 := \begin{bmatrix} E \\ 0 \end{bmatrix}.$$

Let the matrix $G_B \in \mathbb{R}^{n \times n}$ be defined by (4.6). In this case, the matrix

$$\tilde{G}_B = \begin{bmatrix} G_B & 0 \\ 0 & I_n \end{bmatrix}$$

satisfies

$$\tilde{G}_B = \tilde{G}_B^T = I_{2n} \quad \text{and} \quad \tilde{G}_B B_0 = \begin{bmatrix} 0 \\ I_{n+m} \end{bmatrix}.$$

Furthermore, define

$$\tilde{Q}_x = \tilde{G}_B \begin{bmatrix} Q_x & 0 \\ 0 & 0 \end{bmatrix} \tilde{G}_B^T \quad \text{and} \quad \tilde{Q}_x' = \tilde{G}_B \begin{bmatrix} Q_x' & 0 \\ 0 & 0 \end{bmatrix} \tilde{G}_B^T,$$

where

$$\tilde{Q}_x, \tilde{Q}_x' \in \mathbb{R}^{2n \times 2n}.$$

4.3.2   Main Result on the Attractive Ellipsoid for a Dynamic Controller

Now we are ready to formulate our principal result concerning the attractive ellipsoid guaranteed by the dynamic controller (4.35).

Theorem 4.3. *If the matrices*

$$P_1 \in \mathbb{R}^{(n-m) \times (n-m)}, \quad P_2 \in \mathbb{R}^{(n+m) \times (n+m)}, \quad Y \in \mathbb{R}^{(n+m) \times (n+k)}$$

and the numbers

$$\tau_1, \tau_2, \tau_3 \geq 0$$

satisfy the system of matrix inequalities

$$
\begin{bmatrix}
\tilde{P}_{11} & \begin{bmatrix} P_1 & 0 \\ 0 & P_2 \end{bmatrix} \tilde{G}_B D_0 & \begin{bmatrix} 0 \\ Y E_0 \end{bmatrix} \\
D_0^T \tilde{G}_B^T \begin{bmatrix} P_1 & 0 \\ 0 & P_2 \end{bmatrix} & -\tau_2 Q_f & 0 \\
\begin{bmatrix} 0 & E_0^T Y^T \end{bmatrix} & 0 & -\tau_3 Q_g
\end{bmatrix} \le 0,
\tag{4.38}
$$

$$
\tau_1 \ge f_0 \tau_2 + g_0 \tau_3, \quad P_1 > 0, \quad P_2 > 0,
\tag{4.39}
$$

where

$$
\tilde{P}_{11} = \begin{bmatrix} P_1 & 0 \\ 0 & P_2 \end{bmatrix} \tilde{G}_B A_0 \tilde{G}_B^T + \tilde{G}_B A_0^T \tilde{G}_B^T \begin{bmatrix} P_1 & 0 \\ 0 & P_2 \end{bmatrix} +
$$
$$
\tau_1 \begin{bmatrix} P_1 & 0 \\ 0 & P_2 \end{bmatrix} + \tau_2 \tilde{Q}_x + \tau_3 \tilde{Q}_x' + \begin{bmatrix} 0 \\ Y \end{bmatrix} C_0 \tilde{G}_B^T + \tilde{G}_B C_0^T \begin{bmatrix} 0 & Y^T \end{bmatrix},
$$

then the ellipsoid

$$
\varepsilon(\tilde{P}^{-1}) := \{ z \in \mathbb{R}^{2n} : z^T \tilde{P} z < 1 \}, \quad \tilde{P} = \left(\tilde{G}_B^T \begin{bmatrix} P_1 & 0 \\ 0 & P_2 \end{bmatrix} \tilde{G}_B \right),
$$

is attractive for the system (4.37) with

$$
\Theta = P_2^{-1} Y.
$$

Proof. Consider a quadratic function of the form

$$
V(z(t)) = z^T(t) \tilde{P} z, \quad P_1 > 0, \ P_2 > 0,
$$

and calculate the total derivative of $V(z)$ along the trajectories of the system (4.37):

$$
\dot{V}(z(t)) = \begin{pmatrix} z \\ f \\ g \end{pmatrix}^T \begin{bmatrix} \bar{P}_{11} & \tilde{P} D_0 & \tilde{P} B_0 \Theta E_0 \\ D_0^T \tilde{P} & -\tau_2 Q_f & 0 \\ C_0^T \Theta^T B_0^T \tilde{P} & 0 & -\tau_3 Q_g \end{bmatrix} \begin{pmatrix} z \\ f \\ g \end{pmatrix}
$$

$$
-\tau_1 V(z) + \tau_2 f^T Q_f f + \tau_3 g^T Q_g g \le
$$
$$
\begin{pmatrix} z \\ f \\ g \end{pmatrix}^T W \begin{pmatrix} z \\ f \\ g \end{pmatrix} - \tau_1 V(z) + \tau_2 f_0 + \tau_3 g_0,
$$

where

$$
\bar{P}_{11} = \tilde{P} A_0 + A_0^T \tilde{P} + \tau_1 \tilde{P} + \tilde{P} B_0 \Theta C_0 + C_0^T \Theta^T B_0^T \tilde{P}
$$

and

$$
W := \begin{bmatrix} W_{11} & \tilde{P}D_0 & \tilde{P}B_0\Theta E_0 \\ D_0^T\tilde{P} & -\tau_2 Q_f & 0 \\ C_0^T\Theta^T B_0^T\tilde{P} & 0 & -\tau_3 Q_g \end{bmatrix},
$$

$$
W_{11} = \tilde{P}A_0 + A_0^T\tilde{P} + \tau_1\tilde{P} + \tau_2\tilde{Q}_x + \tau_3\tilde{Q}_x' \\ + \tilde{P}B_0\Theta C_0 + C_0^T\Theta^T B_0^T\tilde{P}.
$$

Hence we conclude that the ellipsoid $\varepsilon(\tilde{P}^{-1})$ is attractive if

$$
W \le 0
$$

and

$$
\tau_1 \ge f_0\tau_2 + g_0\tau_3.
$$

Taking into account the equalities

$$
\tilde{P}B_0\Theta = \tilde{G}_B^T \begin{bmatrix} P_1 & 0 \\ 0 & P_2 \end{bmatrix} \tilde{G}_B B_0\Theta =
$$

$$
\tilde{G}_B^T \begin{bmatrix} P_1 & 0 \\ 0 & P_2 \end{bmatrix} \begin{bmatrix} 0 \\ I_{n+m} \end{bmatrix} \Theta = \tilde{G}_B^T \begin{bmatrix} 0 \\ Y \end{bmatrix}
$$

and applying the equivalent transformation $T = \mathrm{diag}(\tilde{G}_B^T, I_r, I_s)$, we finally obtain that $T^T W T \le 0$ has the form (4.38). ∎

The optimization procedure for finding the best parameters Θ for the dynamic controller, those that provide the minimal size of the attractive ellipsoid, can be formulated now similarly to the problem from Sect. 4.1, namely,

$$
\min_{H_1,H_2,P_1,P_2,Y,\tau_1,\tau_2,\tau_3} [\mathrm{tr}(H_1) + \mathrm{tr}(H_2)],
$$

subject to (4.38), (4.39), and additionally

$$
\begin{bmatrix} H_1 & I_{n-m} \\ I_{n-m} & P_1 \end{bmatrix} \ge 0 \quad \text{and} \quad \begin{bmatrix} H_2 & I_{n+m} \\ I_{n+m} & P_2 \end{bmatrix} \ge 0, \tag{4.40}
$$

where

$$
H_1 \in \mathbb{R}^{(n-m)\times(n-m)}, H_2 \in \mathbb{R}^{(n+m)\times(n+m)}.
$$

4.4 Conclusions

In this chapter, we have done the following:

- We considered three types of possible linear feedbacks using only the current output information:

  - the direct feedback proportional to the output measurable signal;
  - the observer-based feedback proportional to the state estimation vector;
  - the full-order linear dynamic controller.

- For each type of possible linear feedback, we suggested that one characterize the set of all stabilizing gain-feedback matrices by a system of the corresponding LMIs, providing the boundedness of all possible trajectories of every controlled plant from the considered class of uncertain systems.
- We also suggested that one select the optimal feedback gain matrix from the described class of stabilizing feedbacks as the one that minimizes the "size" of the attractive ellipsoid containing all possible bounded dynamic trajectories.
- The corresponding numerical procedures for designing the best feedback gain matrices were introduced and discussed for each type of feedback considered.
- Several illustrative examples clearly showed the effectiveness of the suggested technique.

Chapter 5
Control with Sample-Data Measurements

Abstract In this chapter, we formulate our main problem and discuss some necessary mathematical concepts related to feedback control design for nonlinear systems under sample-data output measurements. Then we present a theoretical analysis of an extended version of the invariant ellipsoid method. Then two feedbacks are analyzed:

- a linear feedback proportional to the current state estimate obtained by a Luenberger-type estimator; and
- a full-order linear dynamic controller governed by a linear ordinary differential equation with available sample data as input.

Then we construct a minimal attractive ellipsoid that guarantees stability of the system in a practical sense by varying all parameters of the suggested feedbacks. The associated numerical techniques are also presented. An implementable algorithm for the constructive treatment of the robust control design problem is proposed.

Keywords Sampled-data systems • Output-based control • Robust stabilization

In this chapter, we formulate our main problem and discuss some necessary mathematical concepts related to feedback control design for nonlinear systems under sample-data output measurements. Then we present a theoretical analysis of an extended version of the invariant ellipsoid method. Then two feedbacks are analyzed:

- a linear feedback proportional to the current state estimate obtained by a Luenberger-type estimator; and
- a full-order linear dynamic controller governed by a linear ordinary differential equation with available sample data as input.

Then we construct a minimal attractive ellipsoid that guarantees stability of the system in a practical sense by varying all parameters of the suggested feedbacks. The associated numerical techniques are also presented. An implementable algorithm for the constructive treatment of the robust control design problem is proposed.

© Springer International Publishing Switzerland 2014

A. Poznyak et al., *Attractive Ellipsoids in Robust Control*, Systems & Control: Foundations & Applications, DOI 10.1007/978-3-319-09210-2_5

5.1 Introduction and Motivation

Nonlinear dynamical systems with sample-data output are mathematical models of various modern control systems consisting of a part governed by differential equations and discrete (or stepwise) outputs. These models can represent an extremely wide range of systems of practical interest (see, e.g., Sontag 1998; Poznyak 2008). Sample-data outputs can be also interpreted in some practical systems as a result of application of a quantified procedure to an original continuous model. In this chapter, we study a specific family of dynamical systems with quasi-Lipschitz uncertainties. Roughly speaking, nonlinear uncertainty effects in the given dynamic models are modeled by the quasi-Lipschitz right-hand sides of the corresponding state equations. We are particularly interested in effective algorithms for an appropriate robust control design that guarantees the "BIBO (bounded input–bounded output) stability" of the resulting closed-loop system.

Recently, robust control methodologies have attracted considerable attention, and both theoretical results and practical applications have been developed (see, e.g., Duncan & Schweppe 1971; Fridman 2010; Fridman & Shaked 2002; Gu, Kharitonov, & Chen (2003); Kurzhanski & Veliov 1994; Polyak, Nazin, Durieu, & Walter (2004); Polyak and Topunov 2008). Recall that in H^∞-theory (see, e.g., Basar & Olsted 1999) and robust dynamic games (Zhou, Doyle, & Glover 1996), external perturbations are assumed to be from a class of square integrable functions, which means that they tend to zero over time. Here we assume only the boundedness of perturbations, which makes H^∞-theory inapplicable in our case. The robust control approach to be discussed in our contribution is based on two fundamental ideas, namely,

– the well-known *invariant ellipsoid* approach;
– certain advanced Lyapunov techniques, known as the *descriptor method*.

We refer to Fridman (2006) for the corresponding theoretical and computational details.

Recall that a set in state space is said to be *positively invariant* (for a given dynamical system) if every trajectory initiated in that set remains inside the set at all future times (Kurzhanski & Veliov 1994; Michel, Hou, & Liu 2007). The theoretical questions related to the existence and possible constructive characterizations of an invariant set are very sophisticated mathematical questions. Under some structural assumptions, it is possible to apply the celebrated invariant ellipsoid method and to determine an invariant set constructively (see, e.g., Polyak et al. 2004; Polyak and Topunov 2008). This set will usually be chosen in the form of an ellipsoid in the given state space of the system. Note that the main theoretical tools in the usual "practical stability" analysis are the celebrated Lyapunov and Lyapunov–Krasovskii approaches (see, e.g., Yakubovich 1976; Poznyak 2008). The question of existence and a constructive characterization of an invariant set for a given control system is ordinarily a sophisticated mathematical problem. From the practical applications point of view, this problem can be replaced by a relaxed concept of the above invariant property, namely, by the attractivity property. Under some weak assumptions related to the structure of the examined dynamics, it is possible to specify an attractive set constructively. This set will also be chosen in the form of an (attractive) ellipsoid. This chapter deals with a generalization of the conventional ellipsoid schemes in the sense of the above-mentioned attractivity set. We create an attractive ellipsoid that possesses some minimal properties and use it in the design of the stabilizing feedback strategy. From the numerical point of view, the control synthesis problem is reduced to an auxiliary linear matrix inequality (LMI)-constrained optimization problem (Polyak et al. 2004; Polyak and Topunov 2008; Polyakov & Poznyak 2009). The last constitutes an analytical consequence of the advanced Lyapunov-based techniques used in our manuscript (see Fridman & Shaked 2002).

5.2   Problem Formulation and Some Preliminaries

The basic inspiration for studying continuous-time control systems with sample-data output is the dynamical model of the form

$$\left.\begin{aligned}
\dot{x}(t) &= f(x(t)) + Bu(t) + \upsilon_x(x(t),t) \\[2mm]
x(0) &= x_0 \\[2mm]
y(t) &= Cx(t) + \omega_y(x(t),t) \\[2mm]
\bar{y}(t) &= y(t_k)\chi_{[t_k,t_{k+1})}(t)
\end{aligned}\right\} \qquad (5.1)$$

where

- $x_0, x(t) \in R^n$, $t \in \mathbb{R}_+$ is the state vector,
- $u(t) \in R^m$ is the control input,
- $y(t) \in R^q$ describes the system output for every $t \in \mathbb{R}_+$,
- $\upsilon_x(x(t),t) \in \mathbb{R}^n$ and $\omega_y(x(t),t) \in R^q$ are bounded perturbations associated with the inputs $x(t)$ and outputs $y(t)$, respectively,
- $f : R^n \to R^n$ is a nonlinear function,
- $B \in R^{n\times m}, C \in R^{q\times n}$ are given matrices,

$\bar{y}(t)$ describes the real available sample-data outputs of the system for every $t \in \mathbb{R}_+$.
By

$$\chi_{[t_k,t_{k+1})}(t) := \begin{cases} 1 & \text{if } t \in [t_k, t_{k+1}), \\ 0 & \text{otherwise,} \end{cases}$$

we denote in (5.1) the *characteristic function of the time interval* $[t_k, t_{k+1})$. Note that in contrast to $y(t)$, the stepwise value $\bar{y}(t)$ is a given "measurable" output of the system under consideration. For a feasible control function $u(\cdot)$, the dynamical system (7.1) is considered on the positive half-line $\mathbb{R}_+$.

We also introduce our standing assumptions, namely, the following hypothesis (**A**):

-

$$h := t_{k+1} - t_k = \text{const } \forall k = 0, 1, \dots ;$$

-

$$\|\upsilon_x(x(t), t)\|^2_{Q^1_\eta} + \|\omega_y(x(t), t)\|^2_{Q^2_\eta} \leq 1 \text{ for every } t \in \mathbb{R}_+;$$

- there exist a matrix $A \in \mathbb{R}^{n \times n}$, nonnegative definite matrices Q_f, Q_x, and a constant $\delta \geq 0$ such that

$$\|f(x) - Ax\|^2_{Q_f} \leq \delta + \|x\|^2_{Q_x} \tag{5.2}$$

for every $x \in \mathbb{R}^n$;
- the pair (A, B) is controllable and (A, C) is observable (see, e.g., Pytlak 1999).

Here

$$\| \cdot \|_{Q^1_\eta}, \ \| \cdot \|_{Q^2_\eta}, \ \| \cdot \|_{Q_f}, \ \| \cdot \|_{Q_x}$$

are weighted (by some given nonnegative matrices Q^1_η, Q^2_η, Q_f and Q_x) *Euclidean norms* of the corresponding vectors.

Remark 5.1. Here we consider a class of so-called quasi-Lipschitz uncertain systems (5.1), which is quite general (see Definition 2.1), since it includes nonlinearities of both types, namely bounded discontinuities (as, for example, in the sliding mode control) containing some sign-terms and those of Lipschitz type as well. This class is very general and includes practically all nonlinear models considered in engineering applications.

A control system of type (5.1) is usually associated with a set $\mathcal{U}$ of feasible control functions $u(\cdot)$. Here we deal with a class of linear feedback control strategies proportional to a state-estimate vector. Let us also assume that for every admissible control input $u(\cdot)$, the closed-loop variant of (5.1) has an absolutely continuous solution $x(\cdot)$. Using the quasi-Lipschitz property of $f(\cdot)$ (see (**A**)), we define the auxiliary function

$$\omega_x(x(t), t) := \upsilon_x(x(t), t) + f(x(t)) - Ax(t)$$

and rewrite (5.1) as follows:

$$\left.\begin{aligned}
\dot{x}(t) &= Ax(t) + Bu(t) + \omega_x(x(t),t) \\[4pt]
y(t) &= Cx(t) + \omega_y(x(t),t) \\[4pt]
\bar{y}(t) &= y(t_k)\chi_{[t_k,t_{k+1})}(t)
\end{aligned}\right\} \qquad (5.3)$$

5.3   Linear Feedback Proportional to a State Estimate Vector

5.3.1   Description in Extended Form

The presence of uncertainties in the state and output dynamics motivates an additional formal step in the design procedure of an appropriate control law. This necessary step is related to a suitable observation scheme for the given initial system (5.1). Here we use the standard *Luenberger observer* for this purpose:

$$\begin{aligned}
\dot{\hat{x}}(t) &= A\hat{x}(t) + Bu(t) + L(\bar{y}(t) - C\hat{x}(t)) \quad \forall t \geq 0, \\[4pt]
\hat{x}(0) &= \hat{x}_0,
\end{aligned} \qquad (5.4)$$

where L is an $n \times q$ matrix. Application of (5.4) to the initial system (5.1) gives rise to the explicit definition of the set $\mathcal{U}$ as the set of all feedback control functions of the type

$$u(t) = K\hat{x}(t). \qquad (5.5)$$

Evidently, a control of type (5.5) is characterized by am $(m \times n)$ gain matrix K.
We now introduce some additional auxiliary variables, namely

$$\Delta y(t) := \bar{y}(t) - y(t),$$

and also consider the error vector

$$e(t) := x(t) - \hat{x}(t).$$

Certainly, the estimated error $e(\cdot)$ satisfies the differential equation

$$\dot{e}(t) = (A - LC)e(t) + \omega_x(t) - L(\omega_y(t) + \Delta y(t)) \quad \forall t \geq 0 \qquad (5.6)$$

with an initial condition $e(0) = e_0$. Combining (5.4)–(5.6), we now can rewrite the resulting closed-loop variant of system (5.3) in the following extended compact form:

$$\left.\begin{aligned} \dot{z}(t) &= \tilde{A}z(t) + F\omega(t) + \psi(t) \ \ \forall t > 0 \\[2mm] z(0) &= (\hat{x}_0, e_0) \end{aligned}\right\} \tag{5.7}$$

where

$$z(t) := (\hat{x}^{\mathsf{T}}(t),\, e^{\mathsf{T}}(t))^{T},$$

$$\omega(t) := (\omega_x(t), \omega_y(t)),$$

for all $t \in \mathbb{R}_+$ and

$$\tilde{A} := \begin{bmatrix} A + BK & LC \\ 0 & A - LC \end{bmatrix},$$

$$F := \begin{bmatrix} 0 & L \\ I & -L \end{bmatrix}, \ \psi(t) := \begin{bmatrix} L\Delta y(t) \\ -L\Delta y(t) \end{bmatrix}.$$

Note that the bounded variable $\omega(t)$ introduced above can also be interpreted as a bounded uncertain perturbation or as *unmodeled dynamics* for the given system (5.7). From our basic assumptions, we easily deduce that

$$||\omega(t)||_Q^2 = ||\omega_x(t)||_{Q_f}^2 + ||\omega_y(t)||_{Q_\eta}^2 \le \delta + ||x(t)||_{Q_x}^2 + 1 \tag{5.8}$$

for every $t \in \mathbb{R}_+$. Here $Q := (Q_f, Q_\eta)$.

Let us also introduce the transformation block matrix

$$M := [I_{n\times n},\, I_{n\times n}],$$

where $I_{n\times n}$ denotes the $n \times n$ unit matrix, in order to obtain the relation

$$x(t) = Mz(t).$$

We use this representation of the vector $x(t)$ in the next subsections.

5.3.2 Lyapunov-Like Analysis

The conventional ellipsoid method is usually based on an adequate Lyapunov analysis of the attracting sets. We refer to Zubov (1962) for some classical attraction Lyapunov theorems.

Let us now introduce the *Lyapunov–Krasovskii-type* functional associated with trajectories $z(\cdot)$ and the corresponding derivatives $\dot{z}(\cdot)$ of (5.7) (see Fridman 2006):

$$V(z(\cdot),\dot{z}(\cdot))(t) = z^T(t)P^{-1}z(t) + \int_{t-h}^{t} e^{a(s-t)}z^T(s)Sz(s)ds$$

$$+h\int_{\theta=-h}^{0}\int_{t+\theta}^{t} e^{a(s-t)}\dot{z}^T(s)R\dot{z}(s)dsd\theta, \tag{5.9}$$

where S and R are positive definite symmetric matrices,

$$h := \max_{k}|t_{k+1}-t_k|,$$

and $a > 0$. This analytical construction will be used in the further analysis of the attractive/invariance properties of $\mathcal{E}$. Let us first compute the Lie derivative (the derivative calculated on trajectories of the considered system) of the function $V(z(\cdot),\dot{z}(\cdot))(t)$ (5.9) introduced above.

Lemma 5.1. *The Lie derivative of $V(\dot{z}(\cdot),z(\cdot))$ along the trajectory $x(\cdot)$ (and its derivative $\dot{z}(\cdot)$) of (5.7) is given by the following expression:*

$$\left.\begin{aligned}
&\frac{d}{dt}V(\dot{z}(\cdot),z(\cdot)) = \\[2mm]
&2z^T(t)P^{-1}\dot{z}(t) \quad a\int_{t-h}^{t} e^{a(s-t)}z^T(s)Sz(s)ds + z^T(t)Sz(t)- \\[2mm]
&e^{-ah}z(t-h)Sz(t-h) - ah\int_{-h}^{0}\int_{t+\theta}^{t} e^{a(s-t)}\dot{z}^T(s)R\dot{z}(s)dsd\theta- \\[2mm]
&h\int_{t-h}^{t} e^{a(s-t)}\dot{z}^T(s)R\dot{z}(s)ds + h^2\dot{z}^T(t)R\dot{z}(t)
\end{aligned}\right\} \tag{5.10}$$

Proof. Evidently, the derivative of the first term of $V(\cdot,\cdot)$ is equal to $2z^T(t)P^{-1}\dot{z}(t)$. The derivative of the second term of this function can be evaluated by a direct computation. Application of the well-known differentiation formula for integrals with variable limits (see, e.g., Poznyak 2008) to parts of V implies the resulting relation for $\frac{dV}{dt}(z(\cdot),\dot{z}(\cdot))(t)$. This completes the proof. ∎

We are now able to formulate our first important result of this chapter.

Theorem 5.1. *Let all conditions (A) from Sect. 5.2 be satisfied. Let in Lemma 2.5,*

$$\alpha := a, \ \beta\,(b,\lambda_1,\lambda_2) := b(1+\delta) + (\lambda_1+\lambda_2)\|L\|^2_{Q_\eta},$$

where λ_1 and λ_2 are the maximal eigenvalues of some positive definite $(2n \times 2n)$ matrices Λ_1 and Λ_2, respectively. Assume also that the nonlinear constraint minimization problem

$$\left[\frac{\beta\,(b,\lambda_1,\lambda_2)}{a}\mathrm{tr}\,\{P\}\right] \to \min_{P,a,b,\lambda_1,\lambda_2}$$
$$\text{subject to}$$
$$P > 0, \ P^T = P; \ (K,L) \in \Gamma_2; \ a,b,\lambda_1,\lambda_2 > 0$$

has an optimal solution $(\hat{P}, \hat{K}, \hat{L}, \hat{a}, \hat{b}, \hat{\lambda}_1, \hat{\lambda}_2)$. Then the following property holds:

$$\left.\begin{array}{ll} z^T(t)P^{-1}z(t) \le 1 & \text{if} \quad z(0) \in \mathcal{E} \\[2mm] \dfrac{dV}{dt}(z(\cdot),\dot{z}(\cdot))(t) \le -\alpha V(z(\cdot),\dot{z}(\cdot))(t) + \beta\,(b,\lambda_1,\lambda_2) & \text{if } z(0) \in \mathbb{R}^{2n} \setminus \mathcal{E} \end{array}\right\}$$

where $z(\cdot)$ is a solution of the closed-loop system (5.7), or in other words, the ellipsoid $\mathcal{E}(0, P)$ determined by the matrix

$$P^{-1} = \hat{\alpha}\,\hat{P}^{-1}/\hat{\beta}$$

with $\hat{\alpha} = a$, $\hat{\beta} = \beta\left(\hat{b}, \hat{\lambda}_1, \hat{\lambda}_2\right)$ is a minimal attractive ellipsoid for (5.7).

Proof. Let us consider the two families of constraints in the minimization problem, namely the cases

$$z(0) \in E$$

and

$$z(0) \in R^{2n} \setminus E.$$

Our aim is to show that the composite function $V(t) = V(\dot{z}(\cdot), z(\cdot))(t)$ satisfies the inequality from Lemma 2.5. Let us estimate some terms of the derivative of $V(z(\cdot),\dot{z}(\cdot))(t)$ in (5.10):

(1)

$$-h\int_{t-h}^{t} e^{a(s-t)}\dot{z}^T(s)R\dot{z}(s)ds \le -he^{-ah}\int_{t-h}^{t}\dot{z}^T(s)R\dot{z}(s)ds.$$

Applying the celebrated Jensen's inequality (see Poznyak 2008) to the above integrals, we get

$$he^{-ah}\int_{t-h}^{t}\dot{z}^{T}(s)R\dot{z}(s)ds \geq he^{-ah}\left[\int_{t-h}^{t}\dot{z}^{T}(s)ds\right]R\left[\int_{t-h}^{t}\dot{z}^{T}(s)ds\right].$$

Now we have

$$\frac{dV}{dt}(z(\cdot),\dot{z}(\cdot))(t)\leq 2z^{T}(t)P^{-1}\dot{z}(t)-a\int_{t-h}^{t}e^{a(s-t)}z^{T}(s)Sz(s)ds+z^{T}(t)Sz(t)-$$

$$e^{-ah}z^{T}(t-h)Sz(t-h)-ah\int_{-h}^{0}\int_{t+\theta}^{t}e^{a(s-t)}\dot{z}^{T}(s)R\dot{z}(s)dsd\theta-$$

$$he^{-ah}\int_{t-h}^{t}\dot{z}^{T}(s)dsR\int_{t-h}^{t}\dot{z}^{T}(s)ds+h^{2}\dot{z}^{T}(t)R\dot{z}(t).$$

Adding

$$\alpha V((z(\cdot),\dot{z}(\cdot))(t)-b||\omega(t)||_{Q}^{2}$$

with $b>0$ to both sides of the last inequality, we next deduce

$$\frac{dV}{dt}(z(\cdot),\dot{z}(\cdot))(t)+\alpha V(z(\cdot),\dot{z}(\cdot))(t)-b||\omega(t)||_{Q}^{2}\leq 2z^{T}(t)P^{-1}\dot{z}(t)-$$

$$a\int_{t-h}^{t}e^{a(s-t)}z^{T}(s)Sz(s)ds+z^{T}(t)Sz(t)-e^{-ah}z^{T}(t-h)Sz(t-h)-$$

$$ah\int_{-h}^{0}\int_{t+\theta}^{t}e^{a(s-t)}\dot{z}^{T}(s)R\dot{z}(s)dsd\theta-$$

$$he^{ah}\int_{t-h}^{t}\dot{z}^{T}(s)dsR\int_{t-h}^{t}\dot{z}^{T}(s)ds+h^{2}\dot{z}^{T}(t)R\dot{z}(t)+$$

$$\alpha z^{T}(t)P^{-1}z(t)+\alpha\int_{t-h}^{t}e^{a(s-t)}z^{T}(s)Sz(s)ds+$$

$$\alpha h\int_{-h}^{0}\int_{t+\theta}^{t}e^{a(s-t)}\dot{z}^{T}(s)R\dot{z}(s)dsd\theta-b||\omega(t)||_{Q}^{2}.$$

Let $\alpha:=a$. Since

$$b||\omega(t)||_{Q}^{2}\leq b(\delta+1)+b||Mz(t)||_{Q}^{2},$$

we modify the right-hand side of the last inequality and obtain the following estimate:

$$2z^{T}(t)P^{-1}\dot{z}(t)+z^{T}(t)Sz(t)-e^{-ah}z^{T}(t-h)Sz(t-h)-$$

$$he^{ah}\int_{t-h}^{t}\dot{z}^{T}(s)dsR\int_{t-h}^{t}\dot{z}^{T}(s)ds+h^{2}\dot{z}^{T}(t)R\dot{z}(t)+$$

$$az^{T}(t)P^{-1}z(t)+z^{T}(s)Sz(s)ds-b\omega^{T}(t)Q\omega(t)+bz^{T}(t)M^{T}QMz(t).$$

(2) Let Π_b and Π_c be some matrices of suitable dimension. Following the idea of the "descriptor method" developed in Fridman (2006) and Fridman and Orlov (2007), we now consider the following

$$2(z^T(t)\Pi_b + \dot{z}^T(t)\Pi_c)\left(\tilde{A}z(t) + F\omega(t) + \psi(t) - \dot{z}(t)\right).$$

It is evident that this relation defined on admissible trajectories (solutions) $z(\cdot)$ of the main system (5.7) is equal to zero. Using Lemma 2.6, we can compute the upper bounds for the terms involving $\Delta\omega_y(t)$:

$$2z^T(t)\Pi_b\psi^T(t) \leq z^T(t)\Pi_b\Lambda_1^{-1}\Pi_b z(t) + \psi^T(t)\Lambda_1\psi(t),$$

$$2\dot{z}^T(t)\Pi_c\psi^T(t) \leq \dot{z}^T(t)\Pi_c\Lambda_2^{-1}\Pi_c\dot{z}(t) + \psi^T(t)\Lambda_2\psi(t).$$

We now rewrite $\Delta y(t)$ as

$$\Delta y(t) = CM\Delta z(t) + \Delta\omega_y(t)$$

and compute the quotient

$$\Delta z(t) = \int_{t-h}^t \dot{z}(s)ds.$$

We use the additional notation

$$\Delta\omega_y(t) := \omega_y(t) - \omega_{\bar{y}}(t)$$

and finally obtain

$$\psi^T(t)(\Lambda_1^{-1} + \Lambda_2^{-1})\psi(t) \leq (\lambda_1 + \lambda_2)\|L\|^2_{Q_y}\|CM\|^2\int_{t-h}^t \|\dot{z}(s)\|ds.$$

Here Λ_1 and Λ_2 are $2n \times 2n$ symmetric positive definite matrices, and λ_1 and λ_2 are the maximal eigenvalues, respectively. Consider the extended vector

$$\eta(t) := (z(t),\ \dot{z}(t),\ \int_{t-h}^t \dot{z}(s)ds,\ z(t-h),\ \omega(t))^T \tag{5.11}$$

and compute the estimate

$$\frac{dV}{dt}(z(\cdot),\dot{z}(\cdot))(t) + \alpha V(z(\cdot),\dot{z}(\cdot))(t) - b\|\omega(t)\|^2_Q \leq$$

$$\frac{dV}{dt}(z(\cdot),\dot{z}(\cdot))(t) + a V(z(\cdot),\dot{z}(\cdot))(t) - \tag{5.12}$$

$$(\tilde{b} + (\lambda_1 + \lambda_2)\|L\|^2_{Q_\eta} \leq \eta^T W\eta,$$

where $\tilde{b} := b(1 + \delta)$ and

$$
W := \begin{bmatrix}
w_{11} & w_{12} & w_{13} & 0 & w_{15} \\
w_{12} & w_{22} & w_{23} & 0 & w_{25} \\
w_{13} & w_{23} & w_{33} & 0 & 0 \\
0 & 0 & 0 & w_{44} & 0 \\
w_{15} & w_{25} & 0 & 0 & w_{55}
\end{bmatrix} .
$$

The elements of W are as follows:

$$
\begin{aligned}
w_{11} &:= ((2 + a)I + M^T Q_\eta M)P^{-1} + S + 2\Pi_b \tilde{A} + \Pi_b \Lambda_1^{-1} \Pi_b, \\
w_{12} &:= P^{-1} - \Pi_b + \Pi_c \tilde{A}, \quad w_{13} := \mathcal{M}\Pi_b, \quad w_{15} := \Pi_b F, \\
w_{22} &:= h^2 R - 2\Pi_c + \Pi_c \Lambda_2^{-1} \Pi_c, \\
w_{23} &:= \mathcal{M}\Pi_c, \quad w_{25} := \Pi_c F, \quad w_{31} := \Pi_b \mathcal{M}^T, \\
w_{32} &:= \Pi_c \mathcal{M}^T, \quad w_{33} := -Re^{-ah}, \quad w_{44} := -Se^{-ah}, \\
w_{51} &:= F^T \Pi_b, \quad w_{52} := F^T \Pi_c, \quad w_{55} := -bQ_\eta,
\end{aligned}
$$

where $M := (0, \ LCM)^T$. Additionally, define

$$
\beta := \tilde{b} + (\lambda_1 + \lambda_2)\|L\|_{Q_\eta}^2
$$

and select some matrices Π_b, Π_c, the matrices P, K, L, and scalars $a, b, \lambda_1, \lambda_2 > 0$ such that

$$
W \le 0. \tag{5.13}
$$

Then we conclude that the function $V(t)$ defined as $V(z(\cdot), \dot{z}(\cdot))(t)$ satisfies the conditions of Lemma 2.5. From this fact, it follows that the ellipsoid given by $P^{-1} - \hat{\alpha}\hat{P}^{-1}/\hat{\beta}$, where $\hat{P}$ has the minimality property in the sense of the above optimization problem, is a minimal attractive ellipsoid of the closed-loop system (5.7). This completes the proof.   ∎

Evidently, a set of admissible matrices P, K, L and scalars $a, b, \lambda_1, \lambda_2$ such that $W \le 0$ is an unspecified set in the corresponding finite-dimensional Euclidean space. The initial minimization problem from Theorem 5.1 is a strongly nonlinear problem of mathematical programming with constraints given in the form of nonlinear matrix inequalities. The general solution procedures for these optimization problems are usually given by highly sophisticated algorithms. The same is also true with respect to possible effective numerical approximations. Therefore, we replace this initial nonlinear optimization problem by an adequate linear problem. Our aim is to relax the given nonlinear matrix constraint $W \le 0$ by a suitable system of LMIs under fixed scalar variables. Moreover, the cost functional will also be relaxed to a linear functional that approximates the initial costs.

We next use the following additional notation:

$$P^{-1} := \operatorname{diag}\{P_1, P_2\} \quad R := \operatorname{diag}\{R_1, R_2\}, \quad S := \operatorname{diag}\{S_1, S_2\},$$

$$\Lambda_1^{-1} = \Lambda_2^{-1} := \operatorname{diag}\{P_1^{-1}, P_2^{-1}\}, \quad Q_\eta := \operatorname{diag}\{Q_\eta^1, Q_\eta^2\}.$$

Here $\{S_1, S_2\}$, $\{R_1, R_2\}$ and $\{P_1, P_2\}$ are $n \times n$ symmetric positive definite block matrices that constitute the diagonal-type factorization of S, R, and P, respectively. Evidently, these matrices are also symmetric and positive definite. Moreover, define

$$X_1 = P_1, \ X_2^{-1} = P_2, \ Y_1 = X_1 K, \ Y_2 = L X_2.$$

We now rewrite the matrix P as

$$P^{-1} = \operatorname{diag}\{X_1, X_2^{-1}\}$$

and introduce the auxiliary block matrices

$$\mathcal{W} := \begin{bmatrix} \mathcal{W}_{11} & \mathcal{W}_{12} \\ \mathcal{W}_{12}^T & \mathcal{W}_{22} \end{bmatrix},$$

where

$$\mathcal{W}_{11} := \begin{bmatrix} G_1 & Y_1^T C & G_2 & 0 & 0 & 0 \\ \cdot & w & C^T Y_2 & A X_2 - C^T Y_2 & -2 Y_2^T C & -2 Y_2^T C \\ \cdot & \cdot & G_3 & 0 & 0 & 0 \\ \cdot & \cdot & \cdot & h^2 R_2 - X_2 & -2 Y_2^T C & -2 Y_2^T C \\ \cdot & \cdot & \cdot & \cdot & -R_1 e^{-ah} & 0 \\ \cdot & \cdot & \cdot & \cdot & \cdot & -R_2 e^{-ah} \end{bmatrix}$$

is a symmetric matrix with the second block diagonal matrix

$$w := (1 + a) X_2 + A^T X_2 + X_2 A - C^T Y_2 - C^T Y_2 + S_2.$$

We use here the additional matrix notation

$$\mathcal{W}_{12} := \begin{bmatrix} 0 & 0 & 0 & Y_2^T & 0 & 0 \\ 0 & 0 & X_2 & -Y_2^T & 0 & 0 \\ 0 & 0 & 0 & Y_2^T & 0 & 0 \\ 0 & 0 & X_2 & -Y_2^T & 0 & 0 \\ 0 & 0 & 0 & 0 & 0 & 0 \\ 0 & 0 & 0 & 0 & 0 & 0 \end{bmatrix}$$

and

$$W_{22} := \mathrm{diag}\left\{-S_1 e^{-ah}, \; -S_2 e^{-ah}, \; -bQ_\eta^1, \; -bQ_\eta^2, \; G_4, \; G_5\right\}.$$

Consider some symmetric $n \times n$ positive definite matrices Λ_3, Λ_4 satisfying the conditions of the auxiliary Lemma 2.6 and consider also the block matrices

$$\mathcal{G}^1 := \begin{bmatrix} \mathcal{G}_{11}^1 & \mathcal{G}_{12}^1 \\ \mathcal{G}_{21}^1 & \mathcal{G}_{22}^1 \end{bmatrix}, \quad \mathcal{G}^2 := \begin{bmatrix} -G_2 - 2X_2 & I \\ I & X_1 A^T + Y_1 B^T \end{bmatrix},$$

$$\mathcal{G}^3 := \begin{bmatrix} -G_3 - 2X_2 & I \\ I & X_1 + \Lambda_4 \end{bmatrix}, \quad \mathcal{G}^4 := \begin{bmatrix} -G_4 - S_1 & X_1 \\ X_1 & -\Lambda_3^{-1} \end{bmatrix},$$

$$\mathcal{G}^5 := \begin{bmatrix} -G_5 - h^2 R_1 & X_1 \\ X_1 & -\Lambda_4^{-1} \end{bmatrix}, \quad \mathcal{G}^6 := -\begin{bmatrix} H & I \\ I & X_2 \end{bmatrix},$$

where

$$G_{11}^1 := -G_1 - 2X_2, \quad G_{21}^1 = G_{12}^1 := I$$

and

$$\mathcal{G}_{22}^1 := (1+a)X_1 + X_1 A^T + A X_1 + Y_1 B^T + B Y_1^T + \Lambda_3.$$

We now introduce the family of the associated matrix parameters

$$\Upsilon := \{X_1, X_2, Y_1, Y_2, R, S, G_1, G_2, G_3, G_4, G_5, H\}$$

of suitable dimension. We are now able to formulate an overrelaxation to the initial minimization problem from Theorem 5.1. The solution set to this relaxed auxiliary problem contains the solution set of the above-mentioned initial problem.

Theorem 5.2. *Assume that the auxiliary optimization problem*

$$\min_{\Upsilon;a,b,\lambda_1,\lambda_2>0} [\mathrm{tr}\,\{X_1\} + \mathrm{tr}\,\{H\}]$$

$$subject\ to$$

$$\mathcal{W} \leq 0, \; \mathcal{G}^1 \leq 0, \; \mathcal{G}^2 \leq 0, \; \mathcal{G}^3 \leq 0,$$

$$\mathcal{G}^4 \leq 0, \; \mathcal{G}^5 \leq 0, \mathcal{G}^6 \leq 0, \; X_1 \geq 0, \; X_2 \geq 0, \; H \geq 0$$

$$(5.14)$$

has a solution

$$\hat{\Upsilon} := \left\{\hat{X}_1, \hat{X}_2, \hat{Y}_1, \hat{Y}_2, \hat{R}, \hat{S}, \hat{G}_1, \hat{G}_2, \hat{G}_3, \hat{G}_4, \hat{G}_5, \hat{H}\right\}$$
$$and\ \hat{a}, \hat{b}, \hat{\lambda}_1, \hat{\lambda}_2 > 0.$$

Then the solution set of (5.14) contains the solution set of the initial minimization problem from Theorem 5.1. The ellipsoid $\mathcal{E}$ determined by the matrix $P^{-1} = \dfrac{\hat{\alpha}}{\hat{\beta}}\mathrm{diag}\left\{\hat{X}_1, \hat{X}_2^{-1}\right\}$ is an overapproximation of the minimal attractive ellipsoid associated with (5.7). Moreover, the corresponding system and observer matrices are given by the following expressions:

$$\hat{K} = \hat{X}_1^{-1}\hat{Y}_1, \quad L = \hat{Y}_2\hat{X}_2^{-1}. \tag{5.15}$$

The proof of this result is based on a polyhedral overapproximation (relaxation) of the set of admissible solutions in (5.13). Evidently, the resulting optimization problem (5.14) is a linear LMI-constrained minimization problem. The above linear-type relaxation of the initial nonlinear minimization problem provides a basis for numerical approaches to practically or BIBO stable feedback control design. The corresponding design procedure is now characterized by the optimal system/observer matrices (5.15). In that case, the concrete overapproximation of the minimal attractive ellipsoid for the system (5.7) is given by the matrix P from Theorem 5.2.

5.3.3 Numerical Aspects

In this subsection we discuss numerical extensions of the theoretical techniques developed in the previous sections. Our aim is to illustrate the implementability of the proposed robust control design and to test the corresponding computational aspects by means of simple examples. Observe that in general, the relaxed minimization problem (5.14) may have an optimal solution that does not satisfy Theorem 5.1. So for practical applications, the alternative numerical procedure presented below is preferable. It is based on the following restriction of the class of the quadratic matrices P:

$$\tilde{P} = G_E^T \begin{pmatrix} P_1 & 0 \\ 0 & P_2 \end{pmatrix} G_E, \tag{5.16}$$

where

$$P_1 \in R^{n-m \times n-m}, \quad P_2 \in R^{n+m \times n+m}$$

are positive definite matrices and G_E is an orthogonal matrix of the form

$$G_E \in \mathbb{R}^{2n \times 2n}, \quad G_E G_E^T = I_{2n} \; : \; G_E E_1 = \begin{bmatrix} 0 \\ \tilde{E}_1 \end{bmatrix},$$
$$\tilde{E}_1 \in \mathbb{R}^{n+m \times n+m}, \quad \det(\tilde{E}_1) \neq 0. \tag{5.17}$$

Since the matrix $B^T B$ is positive, the orthogonal transformation G_E always exists and can be easily calculated in MATLAB, for example using the function `null`. For a given matrix M, the function `null(M)` returns the matrix whose columns are an orthonormal basis of the null space of the matrix M. So the following representation of G_E holds:

$$G_E = \begin{pmatrix} E_1^{\perp} \\ E_1' \end{pmatrix}, \text{ where } E_1^{\perp} = (\text{null}(E_1^T))^T \text{ and } E_1' = (\text{null}(E_1^{\perp}))^T.$$

In that case, the elements of the matrix $\tilde{W}$ take the form

$$w_{11} = A_0^T \tilde{P}^{-1} + G_1^T \begin{pmatrix} 0 \\ P_2^{-1} \tilde{E}_1 \Theta \end{pmatrix}^T G_E + \tilde{P}^{-1} A_0 +$$
$$G_E^T \begin{pmatrix} 0 \\ P_2^{-1} \tilde{E}_1 \Theta \end{pmatrix} G_1 + (1+a)\tilde{P}^{-1} + S,$$

$$w_{12} = w_{13} = w_{14} = A_0^T \tilde{P}^{-1} + G_1^T \begin{pmatrix} 0 \\ P_2^{-1} \tilde{E}_1 \Theta \end{pmatrix}^T G_E, \tag{5.18}$$

$$w_{15} = w_{25} = w_{35} = w_{45} = \tilde{P}^{-1} F_0 + G_E^T \begin{pmatrix} 0 \\ P_2^{-1} \tilde{E}_1 \Theta \end{pmatrix} E_2,$$

$$w_{22} = h^2 R - \tilde{P}^{-1}, \quad w_{23} = w_{24} = -\tilde{P}^{-1}, \quad w_{33} = -e^{-ah} S + \tilde{P}^{-1},$$

$$w_{44} = he^{-ah} R + \|CM\|^2 + \tilde{P}^{-1}, \quad w_{55} = -bQ_{\eta}.$$

We set $Y = P_2^{-1} \tilde{E}_1 \Theta$ and obtain the following minimization problem:

$$\left. \begin{array}{c} \min\limits_{Y; a,b,\lambda_1,\lambda_2 > 0} \operatorname{tr}\left\{ \dfrac{\alpha}{\beta} \tilde{P} \right\} \\ \text{subject to} \\ \tilde{P}, S, R > 0, \quad \tilde{P}^T = \tilde{P}, \quad S^T = S, \quad R^T = R \\ \tilde{W}(\tilde{P}, S, R, Y; a, b, \lambda_1, \lambda_2) \leq 0 \end{array} \right\} \tag{5.19}$$

where the elements of the matrix $\tilde{W}$ are defined by (5.18). If the tuple $(\tilde{P}^{\text{opt}}, S^{\text{opt}}, R^{\text{opt}}, Y^{\text{opt}})$ denotes an optimal solution of (5.19), then the control gain matrix can be found in the form

$$\Theta^{\text{opt}} = \tilde{E}_1^{-1} P_2 Y_{\text{opt}},$$

where P_2 is the corresponding block of $\tilde{P}$ (see (5.16)), and E_1 is defined by (5.17). The obtained control gain matrix Θ^{opt} will always satisfy the initial constraints in (5.15). However, the restriction on the form of the matrix $\tilde{P}$ (see (5.16)) determines only a suboptimal solution.

Example 5.1. Consider a dynamic model of a separately excited DC motor (see Leonhard 1966),

$$\left.\begin{aligned}
J\frac{d\Omega}{dt} &= c\Phi_s I_r - B\Omega - \eta_\omega \\
L_r\frac{dI_r}{dt} &= U_r - R_r I_r - c\Phi_s \Omega \\
\frac{d\Phi_s}{dt} &= U_s - R_s \Phi
\end{aligned}\right\}
\tag{5.20}$$

and select the concrete matrices for the transformed system (7.2)

$$A = \begin{bmatrix} -\frac{B}{J} & \frac{C\Phi_{s0}}{J} & \frac{CI_0}{J} \\[2mm] -\frac{C\Phi_{s0}}{L_r} & -\frac{R}{L_r} & -\frac{C\Omega_0}{L_r} \\[2mm] 0 & 0 & -R_s \end{bmatrix}, \quad B = \begin{bmatrix} 0 & 0 \\ \frac{1}{L_r} & 0 \\ 0 & 1 \end{bmatrix}, \quad C = \begin{bmatrix} 1 & 0 & 0 \\ 0 & 1 & 0 \end{bmatrix}.$$

The sample time interval is determined by $t_{k+1} - t_k = 0.01$ and $|\eta_\omega| \le 0.02$. The initial conditions for the given control system and for the dynamic controller are assumed to be zero. The control function is the bounded voltage (U_r, U_s). Evidently, the boundedness of the voltage implies the boundedness of Φ_s. For concrete simulation of the original nonlinear system (5.20) we have used here some concrete model parameters (see Table 5.1). The computational results are presented in Figure 5.1, which contains the state variables $x_1 = \Omega$, $x_2 = I_r$ and $x_3 = \Phi_s$.

As we can see, the trajectories remain inside the area delimited by the attractive ellipsoid after a certain time. This Lyapunov stable attractive ellipsoidal set has a concrete geometric characterization in the state space.

Finally, note that the computational experiments presented above were realized using the standard MATLAB and the SeDuMi/Yalmip toolboxes.

Table 5.1 The concrete parameters for the model of a DC motor

Parameter	Value	Unit
c	0.03	Wb/rad
J	0.001	kg/m^2
R_r	0.5	Ohms
R_s	85	Ohms
L_r	8.9	mH
L_s	50	H
B	0.009	Nm/rad

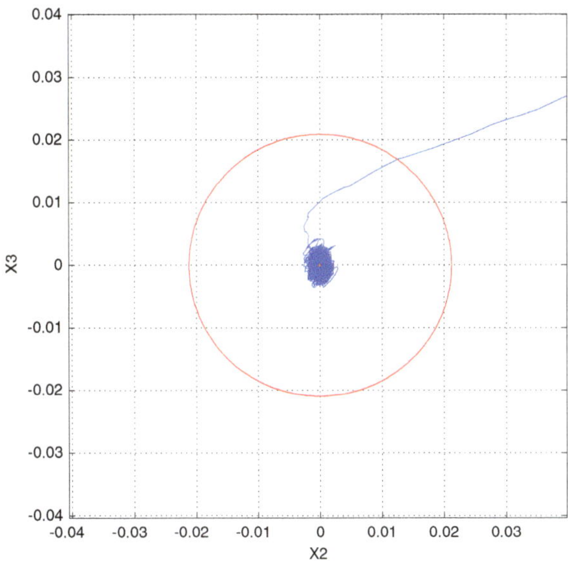

Fig. 5.1 The phase-portrait in the (x_2, x_3)-plane

5.4   Full-Order Robust Linear Dynamic Controller

Consider now the same dynamical system presented in the quasi-linear format (5.3).

5.4.1   The Structure of a Dynamic Controller

In this section, the set of admissible feedback controls $\mathcal{U}$ is determined as a class of full-order linear dynamic controllers with the following structure:

$$\left.\begin{aligned} u(t) &= C_r x_r(t) + D_r \bar{y}(t) \\[2mm] \dot{x}_r(t) &= A_r x_r(t) + B_r \bar{y}(t) \\[2mm] x_r(0) &= x_0^r \end{aligned}\right\} \tag{5.21}$$

The control design associated with (5.21) is completely determined by the selection of the matrix

$$\Theta := \begin{bmatrix} A_r & B_r \\ C_r & D_r \end{bmatrix} \in \mathbb{R}^{(n+m)\times(n+q)}.$$

We call Θ the *dynamic controller matrix*. The closed-loop realization of (5.3) with (5.21) can be compactly written as

$$\dot{z}(t) = \tilde{A}z(t) + F\omega(t) + \psi\Delta y(t), \ \forall t > 0,$$
$$z(0) = (x_0, x_0^r), \tag{5.22}$$

where

$$z(t) := [x(t)\ x_r(t)]^T, \quad \omega(t) := \left[\omega_y(t)\ \omega_x(t)\right]^T,$$

and

$$\Delta y(t) := \bar{y}(t) - CMz(t) + \omega_y(t), \ M := [I\ I].$$

We also use here the following additional notation:

$$\tilde{A} := \begin{bmatrix} A + BD_rC & BC_r \\ B_rC & A_r \end{bmatrix},$$
$$F := \begin{bmatrix} BD_r & I \\ B_r & 0 \end{bmatrix}, \ \psi := \begin{bmatrix} BD_r \\ B_r \end{bmatrix}.$$

Put $Q := (Q_f\ Q_\eta^1\ Q_\eta^2)$. From our basic assumptions, we obtain the simple observation

$$||\omega(t)||_Q^2 \le 1 + \delta + ||x(t)||_{Q_x}^2 \ \forall t \in R_+. \tag{5.23}$$

Stability Analysis of the Invariant Set

We consider again the Lyapunov–Krasovskii functional (5.9) and calculate its Lie derivative $dV(z(\cdot), \dot{z}(\cdot))(t)/dt$ on the trajectories of the dynamical system (5.22):

$$\left.\begin{array}{l} dV(z(\cdot), \dot{z}(\cdot))(t)/dt = 2z^T(t)P^{-1}\dot{z}(t) - a\int_{t-h}^t e^{a(s-t)}z^T(s)Sz(s)ds \\[2ex] + z^T(t)Sz(t) - e^{-ah}z(t-h)Sz(t-h) - ah\int_{-h}^0\int_{t+\theta}^t e^{a(s-t)}\dot{z}^T(s)R\dot{z}(s)dsd\theta \\[2ex] \qquad\qquad -h\int_{t-h}^t e^{a(s-t)}\dot{z}^T(s)R\dot{z}(s)ds + h^2\dot{z}^T(t)R\dot{z}(t) \end{array}\right\} \tag{5.24}$$

The Lie Derivative Estimation

We introduce the auxiliary symmetric matrix

$$W(P, S, R, \Theta, \Pi_a, \Pi_b, \Pi_c, \Pi_d, \Lambda) := \begin{bmatrix} w_{11} & w_{12} & w_{13} & w_{14} & w_{15} \\ \cdot & w_{22} & w_{23} & w_{24} & w_{25} \\ \cdot & \cdot & w_{33} & 0 & w_{35} \\ \cdot & \cdot & \cdot & w_{44} & w_{45} \\ \cdot & \cdot & \cdot & \cdot & w_{55} \end{bmatrix},$$

$$
\begin{aligned}
w_{11} &= aP^{-1} + \Pi_a \Lambda^{-1} \Pi_a + S + \tilde{A}^T \Pi_a + \Pi_a \tilde{A} \\
w_{12} &= P^{-1} - \Pi_a + \tilde{A}^T \Pi_b, \quad w_{13} = \tilde{A}^T \Pi_c, \quad w_{14} = \tilde{A}^T \Pi_d, \quad w_{15} = \Pi_a F, \\
w_{22} &= h^2 R - 2\Pi_b + \Pi_b \Lambda^{-1} \Pi_b, \quad w_{23} = -\Pi_c, \quad w_{24} = -\Pi_d, \quad w_{25} = \Pi_b F, \\
w_{33} &= -e^{-ah} S + \Pi_c \Lambda^{-1} \Pi_c, \quad w_{35} = \Pi_c F, \\
w_{44} &= -he^{-ah} R + \|CM\|_{Q_\eta}^2 + \Pi_d \Lambda^{-1} \Pi_d, \quad w_{45} = \Pi_d F, \quad w_{55} = -bQ_\eta,
\end{aligned}
$$

$$(5.25)$$

and $\Pi_a, \Pi_b, \Pi_c, \Pi_d$ are some symmetric matrices of suitable dimensions. We are now ready to formulate our first main result.

Theorem 5.3. *Let $a, b \in \mathbb{R}_+$ and let*

$$\eta(t) := \left(z(t), \ \dot{z}(t), \ z(t-h), \ \int_{t-h}^{t} \dot{z}(s)ds, \ \omega(t) \right)^T$$

be an extended state vector. We then have the following estimate for the Lie derivative $dV(z(\cdot), \dot{z}(\cdot))(t)/dt$:

$$
\left.
\begin{aligned}
\frac{dV}{dt}(z(\cdot), \dot{z}(\cdot))(t) &\le \eta^T W(P, S, R, \Theta, \Pi_a, \Pi_b, \Pi_c, \Pi_d, \Lambda)\eta \\
&\quad - aV(z(\cdot), \dot{z}(\cdot))(t) + (\tilde{b} + \psi^\mathsf{T} Q_\eta^{1/2} \Lambda Q_\eta^{1/2} \psi)
\end{aligned}
\right\}
$$

$$(5.26)$$

where $\tilde{b} := b(1 + \delta)$.

Proof. Considering the sixth term of the derivative $dV(z(\cdot), \dot{z}(\cdot))(t)/dt$, we obtain

$$-h \int_{t-h}^{t} e^{a(s-t)} \dot{z}^T(s) R\dot{z}(s)ds \le -he^{-ah} \int_{t-h}^{t} \dot{z}^T(s) R\dot{z}(s)ds.$$

Application of Jensen's inequality (see, e.g., Poznyak 2008) to the above integrals implies the estimate

$$he^{-ah} \int_{t-h}^{t} \dot{z}^T(s) R\dot{z}(s)ds \ge he^{-ah} \int_{t-h}^{t} \dot{z}^T(s)ds R \int_{t-h}^{t} \dot{z}^T(s)ds.$$

Hence

$$\frac{dV}{dt}(z(\cdot),\dot{z}(\cdot))(t) \leq 2z^T(t)P^{-1}\dot{z}(t) - a\int_{t-h}^{t} e^{a(s-t)}z^T(s)Sz(s)ds$$

$$+z^T(t)Sz(t) - e^{-ah}z^T(t-h)Sz(t-h) - ah\int_{-h}^{0}\int_{t+\theta}^{t} e^{a(s-t)}\dot{z}^T(s)R\dot{z}(s)dsd\theta$$

$$-he^{-ah}\int_{t-h}^{t}\dot{z}^T(s)dsR\int_{t-h}^{t}\dot{z}^T(s)ds + h^2\dot{z}^T(t)R\dot{z}(t).$$

One can add the term $\alpha V((z(\cdot),\dot{z}(\cdot))(t) - b||\omega(t)||_Q^2$, where $\alpha := a$, to both sides of the last inequality and deduce

$$\frac{dV}{dt}(z(\cdot),\dot{z}(\cdot))(t) + \alpha V(z(\cdot),\dot{z}(\cdot))(t) - b||\omega(t)||_Q^2 \leq 2z^T(t)P^{-1}\dot{z}(t)-$$

$$a\int_{t-h}^{t} e^{a(s-t)}z^T(s)Sz(s)ds + z^T(t)Sz(t) - e^{-ah}z^T(t-h)Sz(t-h)-$$

$$ah\int_{-h}^{0}\int_{t+\theta}^{t} e^{a(s-t)}\dot{z}^T(s)R\dot{z}(s)dsd\theta - he^{ah}\int_{t-h}^{t}\dot{z}^T(s)dsR\int_{t-h}^{t}\dot{z}^T(s)ds+$$

$$h^2\dot{z}^T(t)R\dot{z}(t) + \alpha z^T(t)P^{-1}z(t) + \alpha\int_{t-h}^{t} e^{a(s-t)}z^T(s)Sz(s)ds+$$

$$\alpha h\int_{-h}^{0}\int_{t+\theta}^{t} e^{a(s-t)}\dot{z}^T(s)R\dot{z}(s)dsd\theta - b||\omega(t)||_Q^2.$$

Since in view of (5.23), we have

$$b||\omega(t)||_Q^2 \leq b(\delta + 1) + b||Mz(t)||_Q^2,$$

we modify the right-hand side of the last inequality and obtain our next estimates:

$$2z^T(t)P^{-1}\dot{z}(t) + z^T(t)Sz(t) - e^{-ah}z^T(t-h)Sz(t-h)-$$

$$he^{ah}\int_{t-h}^{t}\dot{z}^T(s)dsR\int_{t-h}^{t}\dot{z}^T(s)ds + h^2\dot{z}^T(t)R\dot{z}(t)+$$

$$az^T(t)P^{-1}z(t) + z^T(s)Sz(s)ds - b\omega^T(t)Q\omega(t).$$

Recall that $\Delta y(t) = CM\Delta z(t) + \Delta\omega_y(t)$. This notation makes it possible to compute the quotient

$$\Delta z(t) = \int_{t-h}^{t} \dot{z}(s)ds$$

We introduce $\Delta\omega_y(t) := \omega_y(t) - \omega_{\bar{y}}(t)$ and calculate $\Delta z(t) = \int_{t-h}^{t}\dot{z}(s)ds$. Using the idea of the "descriptor method" (originally established for the stability of time-delay systems (Fridman 2006)), we now consider the following term (which is equal to zero in view of (5.22)):

$$2\left(z^T(t)\Pi_a + \dot{z}^T(t)\Pi_b + z^T(t-h)\Pi_c + \left(\int_{t-h}^t \dot{z}(s)ds\right)\Pi_d\right)$$

$$\times\left(\tilde{A}z(t) + F\omega(t) + \psi\left(CM\int_{t-h}^t \dot{z}(s)ds + \Delta\omega_y(t)\right) - \dot{z}(t)\right).$$

By the Λ-inequality, we also compute the upper bounds for the terms involving $\Delta\omega_y(t)$,

$$2z^T(t)\Pi_a(\psi\Delta\omega_y(t))^T \le z^T(t)\Pi_a\Lambda^{-1}\Pi_a z(t)+$$

$$(\psi\Delta\omega_y(t))^T\Lambda\psi\Delta\omega_y(t), 2\dot{z}^T(t)\Pi_b(\psi\Delta\omega_y(t))^T \le \dot{z}^T(t)\Pi_b\Lambda^{-1}\Pi_b\dot{z}(t)+$$

$$(\psi\Delta\omega_y(t))^T\Lambda\psi\Delta\omega_y(t), 2z^T(t-h)\Pi_c(\psi\Delta\omega_y(t))^T \le$$

$$z^T(t-h)\Pi_c\Lambda^{-1}\Pi_c z(t-h)+$$

$$(\psi\Delta\omega_y(t))^T\Lambda\psi\Delta\omega_y(t), 2\left(\int_{t-h}^t \dot{z}(s)ds\right)\Pi_d(\psi\Delta\omega_y(t))^T \le$$

$$\left(\int_{t-h}^t \dot{z}(s)ds\right)^T\Pi_d\Lambda^{-1}\Pi_d\left(\int_{t-h}^t \dot{z}(s)ds\right) + (\psi\Delta\omega_y(t))^T\Lambda\psi\Delta\omega_y(t),$$

and apply the estimate

$$\eta^T W(P, S, R, \Theta, \Pi_a, \Pi_b, \Pi_c, \Pi_d, \Lambda)\eta \ge$$

$$\frac{dV}{dt}(z(\cdot), \dot{z}(\cdot))(t) + \alpha V(z(\cdot), \dot{z}(\cdot))(t) - b\|\omega(t)\|_Q^2 \ge$$

$$\frac{dV}{dt}(z(\cdot), \dot{z}(\cdot))(t) + aV(z(\cdot), \dot{z}(\cdot))(t) - (\tilde{b} + \psi^\mathsf{T} Q_\eta^{1/2}\Lambda Q_\eta^{1/2}\psi),$$

which gives the desired result. This completes the proof of the theorem. ∎

This theorem and Lemma 2.5 allow us to make the following conclusion.

Conclusion 5.1. *If for $\alpha := a$ and $\beta := \tilde{b} + \psi^\mathsf{T}\Lambda\psi$, the matrix collection $P, S, R, \Theta, \Pi_a, \Pi_b, \Pi_c, \Pi_d$ provides the matrix inequality*

$$W(P, S, R, \Theta, \Pi_a, \Pi_b, \Pi_c, \Pi_d, \Lambda) \le 0,$$

then the composite function $\mathcal{V}(t) := V(z(\cdot), \dot{z}(\cdot))(t)$ (where $z(\cdot)$ is a trajectory of the closed-loop system (5.22)) satisfies all conditions of the storage function from Lemma 2.1.

5.4.2 The "Minimal-Size" Attractive Ellipsoid and LMI Constrained Optimization

In the context of Conclusion 5.1, we obtain the following practical interpretation of the following abstract optimization problem:

$$\text{minimize tr} \left\{ \frac{\beta}{\alpha} P \right\}$$
$$\text{subject to}$$
$$P, S, R > 0, \ P^T = P, \ S^T = S, \ R^T = R,$$
$$W(P, S, R, \Theta, \Pi_a, \Pi_b, \Pi_c, \Pi_d, \Lambda) \le 0. \tag{5.27}$$

Note that the optimization operation in (5.27) is considered with respect to the variables (P, S, R, Θ). The undetermined variables $(\Pi_a, \Pi_b, \Pi_c, \Pi_d, \Lambda)$ are some parameters in this optimization problem and must be selected from some other suitable algorithm (independently of (5.27)). Let us choose

$$\Lambda = \Pi_a = \Pi_b = \Pi_c = \Pi_d = P^{-1}, \tag{5.28}$$

In which case, the variable β from Conclusion 5.1 is a function of the free variable P. Due to the heavily nonlinear nature of this optimization problem (the nonlinear objective functional and the bilinear constraints), we try to reformulate the original problem (5.27) in a relaxed form. This relaxation will replace the bilinear matrix inequality constraints by some system of LMIs. Following this relaxation idea, we first introduce some additional notation:

$$E_1 := \begin{bmatrix} 0 & B \\ I & 0 \end{bmatrix}, \ E_2 := \begin{bmatrix} 0 & 0 \\ 0 & I \end{bmatrix}, \ G_1 := \begin{bmatrix} 0 & I \\ C & 0 \end{bmatrix}, \ G_2 := \begin{bmatrix} 0 \\ I \end{bmatrix},$$

and

$$A_0 := \begin{bmatrix} A & 0 \\ 0 & 0 \end{bmatrix}, \ F_0 := \begin{bmatrix} 0 & I \\ 0 & 0 \end{bmatrix}.$$

We now express the matrices $\tilde{A}$ and F in the following form:

$$\tilde{A} = A_0 + E_1 \Theta G_1, \ F = F_0 + E_1 \Theta E_2, \ \psi = E_1 \Theta G_2.$$

Define X as

$$X := P^{-1} E_1 \Theta.$$

We now introduce the relaxed matrix $\bar{W}(P, S, R, \Theta, X)$ associated with the matrix $W(P, S, R, \Theta, \Pi_a, \Pi_b, \Pi_c, \Pi_d, \Lambda)$ in (5.26) for the parameters (5.28):

$$\bar{W}(P, S, R, \Theta, X) := \begin{bmatrix} w_{11} & w_{12} & w_{13} & w_{14} & w_{15} \\ \cdot & w_{22} & w_{23} & w_{24} & w_{25} \\ \cdot & \cdot & w_{33} & 0 & w_{35} \\ \cdot & \cdot & \cdot & w_{44} & w_{45} \\ \cdot & \cdot & \cdot & \cdot & w_{55} \end{bmatrix},$$

$$w_{11} = A_0^T P^{-1} + G_1^T X^T + P^{-1} A_0 + X G_1 + (1 + a) P^{-1} + S,$$

$$w_{12} = w_{13} = w_{14} = A_0^T P^{-1} + G_1^T X^T,$$

$$w_{15} = w_{25} = w_{35} = w_{45} = P^{-1} F_0 + X E_2,$$

$$w_{22} = h^2 R - P^{-1}, \quad w_{23} = w_{24} = -P^{-1}, \quad w_{33} = -e^{-ah} S + P^{-1},$$

$$w_{44} = h e^{-ah} R + \|C M\|^2 + P^{-1}, \quad w_{55} = -b Q_\eta.$$

We are now able to formulate the relaxed minimization problem associated with (5.27):

$$\left.\begin{array}{c} \text{minimize tr} \left\{ \dfrac{\beta P^{-1}}{\alpha} \right\} \\ \text{subject to} \\[2mm] P, S, R > 0, \ P^T = P, \ S^T = S, \ R^T = R \\[2mm] \bar{W}(P, S, R, \Theta, X) \leq 0 \end{array}\right\} \qquad (5.29)$$

Assume that (5.29) has a solution $(\hat{P}, \hat{S}, \hat{R}, \hat{X})$. Denote by $\hat{\Theta}$ the corresponding element of an optimal solution to (5.27) for the above-mentioned concrete selection (5.28) of parameters $(\Pi_a, \Pi_b, \Pi_c, \Pi_d, \Lambda)$ in (5.27). The existence of an optimal solution to (5.27) has also been assumed. We obtain the following equivalence result.

Theorem 5.4. *The solution set of (5.29) defines the solution set $\hat{\Theta}$ of the original optimization problem (5.27) according to the relation*

$$\hat{\Theta} = (E_1^T E_1)^{-1} E_1^T \hat{P} \hat{X}.$$

Note that Theorem 5.4 is an immediate consequence of the following fact: the above definition of the auxiliary variable X is realized by a simple one-to-one relation. Note that $E_1^T E_1$ has the structure of a block-diagonal matrix $\text{diag}\{I, B^T B\}$ with $B^T B > 0$ (see the main assumptions in Sect. 7.2).

The resulting LMI-constrained optimization problem (5.29) can provide a basis for some constructive numerical solution procedures. An appropriate computational algorithm for the auxiliary problem (5.29) simultaneously provides a basis for a numerical approach to the robust control design for the initial system (5.3). This control design is determined by the practical evaluation of the optimal dynamic controller matrix $\hat{\Theta}$ from (5.29). Moreover, the optimal matrix $\hat{P}$ makes it possible to construct the attractive ellipsoid $\mathcal{E}^{\mathrm{opt}}$ that possesses the minimal-size property formalized in (5.29).

5.4.3   On Numerical Realization

In this section, we discuss the numerical extensions of the theoretical techniques developed in the previous sections. Our aim is to illustrate the implementability of the proposed robust control design and to test the corresponding computational aspects by means of simple examples. Observe that in general, the relaxed minimization problem (5.29) may have an optimal solution that does not satisfy Theorem 5.4. So for practical applications, the alternative numerical procedure presented below is preferable. It is based on the following restriction of the class of quadratic matrices P:

$$\tilde{P} = G_E^T \begin{pmatrix} P_1 & 0 \\ 0 & P_2 \end{pmatrix} G_E, \tag{5.30}$$

where $P_1 \in \mathbb{R}^{n-m \times n-m}$, $P_2 \in \mathbb{R}^{n+m \times n+m}$ are positive definite matrices, and G_E is an orthogonal matrix of the form

$$G_E \in \mathbb{R}^{2n \times 2n}, \ G_E G_E^T = I_{2n} :$$

$$G_E E_1 = \begin{bmatrix} 0 \\ \tilde{E}_1 \end{bmatrix}, \quad \tilde{E}_1 \in \mathbb{R}^{n+m \times n+m}, \quad \det(\tilde{E}_1) \neq 0. \tag{5.31}$$

Since the matrix $B^T B$ is positive, the orthogonal transformation G_E always exists and can be easily found in MATLAB using, for example, the function `null`. For a given matrix M, the function null(M) returns a matrix whose columns are an orthonormal basis of the null space of the matrix M. So the following representation of G_E holds:

$$G_E = \begin{pmatrix} E_1^\perp \\ E_1' \end{pmatrix},$$

$$E_1^\perp = (\mathrm{null}(E_1^T))^T, \ E_1' = (\mathrm{null}(E_1^\perp))^T.$$

In this case, the elements of the matrix $\bar{W}$ take the form

$$
\begin{aligned}
w_{11} &= A_0^T \tilde{P}^{-1} + G_1^T \begin{pmatrix} 0 \\ P_2^{-1}\tilde{E}_1\Theta \end{pmatrix}^T G_E + \tilde{P}^{-1} A_0 + \\
&\quad G_E^T \begin{pmatrix} 0 \\ P_2^{-1}\tilde{E}_1\Theta \end{pmatrix} G_1 + (1+a)\tilde{P}^{-1} + S,
\end{aligned}
$$

$$
w_{12} = w_{13} = w_{14} = A_0^T \tilde{P}^{-1} + G_1^T \begin{pmatrix} 0 \\ P_2^{-1}\tilde{E}_1\Theta \end{pmatrix}^T G_E, \tag{5.32}
$$

$$
w_{15} = w_{25} = w_{35} = w_{45} = \tilde{P}^{-1} F_0 + G_E^T \begin{pmatrix} 0 \\ P_2^{-1}\tilde{E}_1\Theta \end{pmatrix} E_2,
$$

$$
w_{22} = h^2 R - \tilde{P}^{-1}, \; w_{23} = w_{24} = -\tilde{P}^{-1}, \; w_{33} = -e^{-ah}S + \tilde{P}^{-1},
$$
$$
w_{44} = h e^{-ah} R + \|CM\|^2 + \tilde{P}^{-1}, \; w_{55} = -bQ_\eta.
$$

We set $Y = P_2^{-1}\tilde{E}_1\Theta$ and obtain the following minimization problem:

$$
\begin{aligned}
&\text{minimize tr} \left\{ \frac{\beta}{\alpha}\tilde{P} \right\} \\
&\text{subject to} \\
&\tilde{P}, S, R > 0, \; \tilde{P}^T = \tilde{P}, \; S^T = S, \; R^T = R, \\
&\tilde{W}(\tilde{P}, S, R, Y) \leq 0,
\end{aligned} \tag{5.33}
$$

where the elements of the matrix $\tilde{W}$ are defined by (5.32). If the tuple

$$
(\tilde{P}^{\text{opt}}, S^{\text{opt}}, R^{\text{opt}}, Y^{\text{opt}})
$$

is the optimal solution of (5.33), then the control gain matrix can be found in the form

$$
\Theta^{\text{opt}} = \tilde{E}_1^{-1} P_2 Y_{\text{opt}},
$$

where P_2 is the corresponding block of $\tilde{P}$ (5.30), and E_1 is defined by (5.31). In contrast to (5.29), the obtained control gain matrix Θ^{opt} will always satisfy the initial constraints (5.27). However, the restriction on the form of the matrix P (5.30) makes the obtained solution only suboptimal.

5.5 Conclusion

- Here we demonstrated a new analytic and computational method for robust feedback control design associated with a class of affine uncertain systems. We incorporated into our consideration sample-data outputs and the full-order dynamic controller. The main control design procedure proposed in this chapter constitutes an extension of the conventional invariant ellipsoid method (the attractive ellipsoid approach). From a theoretical point of view, the approach presented here generates an admissible linear feedback control law that guarantees the existence and a concrete characterization of a minimal-size invariant ellipsoid for the corresponding closed-loop system. Moreover, this ellipsoidal invariant set is asymptotically stable in the sense of Lyapunov.

- The generic computational part of the method discussed here leads to an auxiliary nonlinear minimization problem with matrix constraints. We propose an effective relaxation of this initial optimization problem in the form of an LMI-constrained program. Finally, we obtained an attractive ellipsoid with some minimal properties (a minimal "size") that can be interpreted as a maximal robustness of the closed-loop system. The effectiveness of the proposed computational schemes and the associated control design was demonstrated by some illustrative examples.

- Finally, note that our main idea can be easily generalized for some alternative classes of nonlinear control systems with bounded uncertainties and complex discrete–continuous dynamic behavior. It seems also to be possible to apply the control design techniques presented here in combination with some nonlinear feedback-type control strategies.

Chapter 6
Sample Data and Quantifying Output Control

Abstract In this chapter, we consider the analysis and design of an output feedback controller for a perturbed nonlinear system in which the output is sampled and quantized. Using the attractive ellipsoid method, which is based on Lyapunov analysis techniques, together with the relaxation of a nonlinear optimization problem, sufficient conditions for the design of a robust control law are obtained. Since the original conditions result in nonlinear matrix inequalities, a numerical algorithm to obtain the solution is presented. The obtained control ensures that the trajectories of the closed-loop system will converge to a minimal (in a sense to be made specific) ellipsoidal region. Finally, numerical examples are presented to illustrate the applicability of the proposed design method.

Keywords Quantization • Sampled-data systems • Output-based control

In this chapter, we consider the analysis and design of an output feedback controller for a perturbed nonlinear system in which the output is sampled and quantized. Using the attractive ellipsoid method, which is based on Lyapunov analysis techniques, together with the relaxation of a nonlinear optimization problem, sufficient conditions for the design of a robust control law are obtained. Since the original conditions result in nonlinear matrix inequalities, a numerical algorithm to obtain the solution is presented. The obtained control ensures that the trajectories of the closed-loop system will converge to a minimal (in a sense to be made specific) ellipsoidal region. Finally, numerical examples are presented to illustrate the applicability of the proposed design method.

6.1 Introduction

Motivated by emerging applications in networked control systems (see, e.g., Peng & Tian 2007; Peng et al. 2011; Zhang & Yu 2007), the control community has witnessed a renewed interest in phenomena that are inherent to the digital implementation of continuous-time control systems, such as sampling and quantization. A major line of research in this area incorporates the information-theoretical aspects

© Springer International Publishing Switzerland 2014

A. Poznyak et al., *Attractive Ellipsoids in Robust Control*, Systems & Control:
Foundations & Applications, DOI 10.1007/978-3-319-09210-2_6

(such as channel capacity) of the networked control problem and aims at a theory that parallels the celebrated mathematical theory of communication (Shannon 1948).

Interesting results have been obtained by following this direction. It is now possible, for example, to relate the absolute value of the unstable eigenvalues of a system and the minimum channel capacity that is required in order to stabilize it (see Nair & Evans 2003; Tatikonda & Mitter 2004; Matveev & Savkin 2007). While certainly of great theoretical interest, most of these results so far have been limited to linear systems. The problem statement is cast in a stochastic framework, and emphasis is given to the coding and decoding aspects of the communication channel (see Phat, Jiang, Savkin, & Petersen 2004 for a coding scheme).

From a different point of view, quantization can be regarded either as a deterministic noise or as a deterministic perturbation, depending on whether quantization affects the control or the output signals. A robust-control approach, such as $\mathcal{H}_\infty$ (Gao & Chen 2008) or the sector bound (Fu & Xie 2005), can then be applied to cope with the quantization problem. Again, most of the results using this approach are limited to linear systems. In this chapter, we deal with the quantization problem by applying the invariant ellipsoid method (see Glover & Schweppe 1971; Kurzhanski & Varaiya 2006; Polyak et al. 2004; Polyak and Topunov 2008; Davila & Poznyak 2011). This allows us to design dynamic feedback control laws for a class of *nonlinear* systems satisfying a quasi-Lipschitz condition (Azhmyakov, Poznyak, & Juárez 2013; Azhmyakov, Poznyak, & Gonzalez 2013). The class of systems is fairly large because it includes systems with hard or even discontinuous nonlinearities.

We consider static and time-invariant quantizers. Because of its time-invariant nature, the required quantizer has an infinite number of quantization levels, and practical stability is obtained instead of asymptotic stability (see Brockett & Liberzon 2000) for a finite dynamic quantizer achieving asymptotic stability. The invariant ellipsoid method delivers an estimated region of convergence in the form of an ellipsoid. Using numerical methods, a controller is chosen with a clear performance criterion: to minimize the size of the ellipsoid.

To deal with the sampling problem, it is typically assumed that the system is already in discrete-time form. We do not make such an assumption. In the spirit of Tian, Yue, and Chen (2008) and Fridman and Dambrine (2009), we consider continuous-time systems and approach the sampling problem from a time-delay systems perspective. To compute the aforementioned ellipsoid, we construct a Lyapunov–Krasovskii functional instead of the usual Lyapunov function. In this regard, the present work can be seen as an extension of the work presented in Mera, Poznyak, Azhmyakov and Fridman (2009), to the case in which quantization phenomena are present.

6.2 Problem Formulation

Consider the nonlinear system

$$\dot{x}(t) = f(t, x(t)) + Bu(t) + \upsilon_x(t), \tag{6.1}$$

where $x(t) \in \mathbb{R}^n$, $u(t) \in \mathbb{R}^m$, and $\upsilon_x(t) \in \mathbb{R}^n$ are, respectively, the state vector, control input, and perturbation at time $t \in \mathbb{R}_+$. We use the following model to describe a noisy sampled and quantized output:

$$\bar{\bar{y}}(t) = Cx(t) + \omega_y(t), \tag{6.2}$$

$$\bar{y}(t) = \sum_{t_k} \bar{\bar{y}}(t_k)\chi_{[t_k,t_{k+1})}(t), \tag{6.3}$$

$$y(t) = \pi(\bar{y}(t)). \tag{6.4a}$$

The vector $\omega_y(t) \in R^q$ in (6.2) is the deterministic noise. The symbol $\chi_{[t_k,t_{k+1})}$ in (6.3) denotes the characteristic function of the time interval $[t_k, t_{k+1})$, that is,

$$\chi_{[t_k,t_{k+1})}(t) := \begin{cases} 1 & \text{if } t \in [t_k, t_{k+1}), \\ 0 & \text{otherwise,} \end{cases} \qquad k = 0, 1, 2, \ldots .$$

Thus, $\bar{y} : \mathbb{R}_+ \to \mathbb{R}^q$ is the piecewise constant function obtained by sampling and holding $\bar{\bar{y}}$ at the discrete instants t_k. The actual system output at time t is $y(t) \in \mathbb{R}^q$, and it is obtained by quantizing the sampled signal $\bar{y}$. Formally, let $Y \subset \mathbb{R}^q$ be a countable set of possible output values. Then $\pi : \mathbb{R}^q \to Y$ in (6.4a) is defined as a projection operator, that is as an operator that satisfies

$$\pi \circ \pi(\bar{y}) \equiv \pi(\bar{y}).$$

The image of π is a discrete subset of $\mathbb{R}^q$. The components of the measurable output $y(t)$ have the form depicted in Fig. 6.1.

Let us now formulate our basic assumptions.

Assumptions

(1) The perturbation and noise are unknown but bounded. More precisely, there are known positive definite matrices $Q_x \in \mathbb{R}^n$ and $Q_y \in \mathbb{R}^q$ such that

$$\|\upsilon_x(t)\|_{Q_x}^2 + \|\omega_y(t)\|_{Q_y}^2 \leq 1 \quad \text{for all } t \in \mathbb{R}_+. \tag{6.5}$$

Here, $\| \cdot \|_{Q_x}$ and $\| \cdot \|_{Q_y}$ are weighted norms given by Q_x and Q_y.

Fig. 6.1 The components of the measurable output

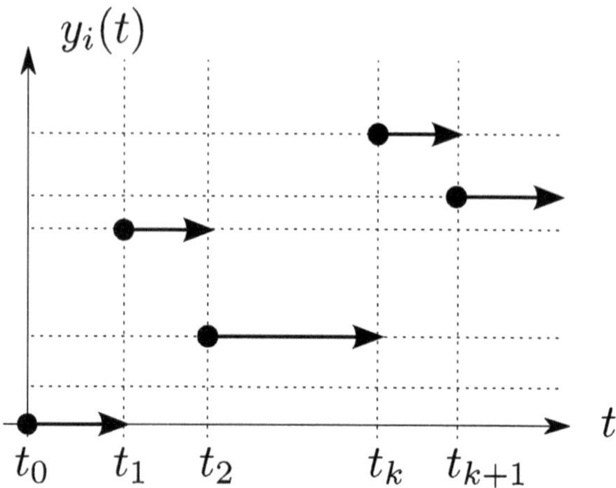

(2) The function f is also unknown but satisfies the *quasi-Lipschitz bound*

$$\|f(t,x) - Ax(t)\|_{Q_x}^2 \le \delta + \|x(t)\|_Q^2 \quad \text{for all } (t,x) \in \mathbb{R}^+ \times \mathbb{R}^n, \tag{6.6}$$

where $\delta > 0$ is a scalar and $Q > 0$ and A are known $n \times n$ matrices.

(3) The pair (A, B) is controllable, and (A, C) is observable.

(4) The sampling intervals need not be regular, but there exists a maximum sampling interval

$$h := \max_k |t_{k+1} - t_k|.$$

(5) The quantization error is bounded, i.e., the positive scalar

$$c := \max_{\bar{y} \in \mathbb{R}^q} \|\pi(\bar{y}) - \bar{y}\|_{Q_y}^2 \tag{6.7}$$

is finite.

Notice that 6.6 is not restrictive, and it comprises a large class of unknown nonlinear functions. By defining the auxiliary function

$$\omega_x(t) := \upsilon_x(t) + f(t, x(t)) - Ax(t),$$

we can rewrite (6.1) in the quasilinear format

$$\dot{x}(t) = Ax(t) + Bu(t) + \omega_x(t). \tag{6.8}$$

We approach the partial-information problem using a conventional Luenberger observer

$$\dot{\hat{x}}(t) = A\hat{x}(t) + Bu(t) + L(y(t) - C\hat{x}(t)), \tag{6.9}$$

where $L \in \mathbb{R}^{n \times q}$ is the observer gain. The control law is taken as

$$u(t) = K\hat{x}(t), \tag{6.10}$$

with $K \in \mathbb{R}^{m \times n}$ the control gain.

Let us now introduce the estimation error

$$e(t) := x(t) - \hat{x}(t)$$

and the auxiliary variable

$$\Delta y(t) := y(t) - \bar{\bar{y}}(t).$$

It can be seen readily that $e(t)$ satisfies the dynamic equation

$$\dot{e}(t) = Ax(t) + Bu(t) + \omega_x(t) - \\ \left[A\hat{x}(t) + Bu(t) + L(\bar{\bar{y}} + \Delta y - C\hat{x}(t)) \right],$$

or equivalently,

$$\dot{e}(t) = (A - LC)e(t) - L(\Delta y(t) + \omega_y(t)) + \omega_x(t). \tag{6.11}$$

It is possible to write the closed-loop equations (6.9) and (6.11) more compactly as

$$\dot{z}(t) = \tilde{A}z(t) + F\omega(t) + \psi(t), \tag{6.12}$$

where we have defined the vectors

$$z(t) := \begin{pmatrix} \hat{x}(t) \\ e(t) \end{pmatrix}, \quad \omega(t) := \begin{pmatrix} \omega_x(t) \\ \omega_y(t) \end{pmatrix}, \quad \text{and } \psi(t) := \begin{pmatrix} L \\ -L \end{pmatrix}, \Delta y(t)$$

and the matrices

$$\tilde{A} := \begin{pmatrix} A + BK & LC \\ 0 & A - LC \end{pmatrix} \quad \text{and } F := \begin{pmatrix} 0 & L \\ I_{n \times n} & -L \end{pmatrix}.$$

Because of the presence of ω and ψ, it is unreasonable to expect $z(t)$ to converge to the origin as $t \to \infty$. On the other hand, if K and L are properly chosen, it is reasonable to expect $z(t)$ to converge to a "small" set containing the origin. Our problem is first to find an estimate of such a set and then to find L and K that minimize (in a sense to be defined later) its "size."

6.3   A Lyapunov–Krasovskii Functional

Since sampling entails delays, instead of a regular function we suggest using the *Lyapunov–Krasovskii functional*.

More precisely, let $\mathcal{C}^0(\mathbb{R}, \mathbb{R}^{2n})$ be the space of all continuous functions of $\mathbb{R}$ into $\mathbb{R}^{2n}$, differentiable almost everywhere; let $R > 0$ and $P > 0$ be $2n \times 2n$ matrices; and let $\alpha > 0$ be a scalar. We propose the functional

$$V : \mathbb{R} \times \mathcal{C}^0(\mathbb{R}, \mathbb{R}^{2n}) \to \mathbb{R}_+,$$

defined as[1]

$$V(t, z(\cdot)) := z^{\top}(t) P^{-1} z(t) + h \int_{\theta=-h}^{0} \int_{s=t+\theta}^{t} e^{\alpha(s-t)} \dot{z}^{\top}(s) R \dot{z}(s) \, ds \, d\theta. \qquad (6.13)$$

Our primary goal is to derive sufficient conditions for $V(t, z(\cdot))$ to satisfy Lemma 2.5 with $\alpha > 0$ and $\beta \geq 0$ when z is a solution of (6.12). Let us begin with the case in which z is arbitrary.

Theorem 6.1. *For given*

$$z(\cdot) \in \mathbb{C}, \quad h, \alpha, b \in \mathbb{R}, \quad P, R \in \mathbb{R}^{2n \times 2n}$$

such that $h > 0$, $\alpha > 0$, $P > 0$, and $R > 0$, the time derivative of $V(t, z(\cdot))$ in (6.13) satisfies the bound

$$\dot{V}(t, z(\cdot)) \leq -\alpha V(t, z(\cdot)) + b\bar{\delta} + \eta(t, z(\cdot))^{\top} W \eta(t, z(\cdot)), \qquad (6.14)$$

where

$$\eta(t, z(\cdot)) := \begin{pmatrix} z(t) \\ \dot{z}(t) \\ z(t) - z(t_k) \\ \omega(t) \end{pmatrix}, \quad W := \begin{pmatrix} \alpha P^{-1} + b Q_z & P^{-1} & 0 & 0 \\ P^{-1} & h^2 R & 0 & 0 \\ 0 & 0 & -h e^{-\alpha h} R & 0 \\ 0 & 0 & 0 & -b \bar{Q} \end{pmatrix},$$

$$\bar{Q} := \begin{pmatrix} Q_x & 0 \\ 0 & Q_y \end{pmatrix}, \quad Q_z := \begin{pmatrix} I \\ I \end{pmatrix} Q \begin{pmatrix} I & I \end{pmatrix}, \quad \text{and } \bar{\delta} := \delta + 1. \qquad (6.15)$$

Before giving the proof of the theorem, let us state a pair of simple lemmas.

---

[1]This functional does not contain a single-integral term as in Fridman 2001 and Mera et al. 2009.

Lemma 6.1. *Under the given assumptions, the perturbation ω satisfies the bound*

$$\|\omega(t)\|_{\bar{Q}}^2 \le \bar{\delta} + \|x(t)\|_Q^2. \tag{6.16}$$

Proof. Direct computation of the norm gives

$$\begin{aligned}
\|\omega(t)\|_{\bar{Q}}^2 &= \|\omega_x(t)\|_{Q_x}^2 + \|\omega_y(t)\|_{Q_y}^2 = \\
&\|v_x(t) + f(t, x(t)) - Ax(t)\|_{Q_x}^2 + \|\omega_y(t)\|_{Q_y}^2 \le \\
&\|v_x(t)\|_{Q_x}^2 + \|f(t, x(t)) - Ax(t)\|_{Q_x}^2 + \|\omega_y(t)\|_{Q_y}^2.
\end{aligned} \tag{6.17}$$

Substitution of (6.5) and (6.6) into (6.17) leads to

$$\|\omega(t)\|_{\bar{Q}}^2 \le 1 + \delta + \|x\|_Q^2. \qquad \blacksquare$$

Lemma 6.2. *For given $z(\cdot) \in \mathbb{C}$, $h > 0$, $\alpha > 0$, $R > 0$, we have*

$$\begin{aligned}
&-h \int_{t-h}^t e^{\alpha(s-t)} \dot{z}^\top(s) R \dot{z}(s) \mathrm{d}s \le \\
&-h e^{-\alpha h} \int_{t_k}^t \dot{z}^\top(s) \mathrm{d}s \, R \int_{t_k}^t \dot{z}(s) \mathrm{d}s.
\end{aligned} \tag{6.18}$$

Proof. Since $e^{-\alpha h} \le e^{\alpha(s-t)}$ for all $s \in [t-h, t]$, and R is positive definite, we have

$$-h \int_{t-h}^t e^{\alpha(s-t)} \dot{z}^\top(s) R \dot{z}(s) \mathrm{d}s \le -h e^{-\alpha h} \int_{t-h}^t \dot{z}^\top(s) R \dot{z}(s) \mathrm{d}s. \tag{6.19}$$

By splitting the interval of integration at the time $t_k \in [t - h, t)$, we obtain

$$-h e^{-\alpha h} \int_{t-h}^t \dot{z}^\top(s) R \dot{z}(s) \mathrm{d}s = -h e^{-\alpha h} \int_{t-h}^{t_k} \dot{z}^\top(s) R \dot{z}(s) \mathrm{d}s$$

$$-h e^{-\alpha h} \int_{t_k}^t \dot{z}^\top(s) R \dot{z}(s) \mathrm{d}s \le -h \int_{t_k}^t \dot{z}^\top(s) R \dot{z}(s) \mathrm{d}s \tag{6.20}$$

$$\le -h e^{-\alpha h} \int_{t_k}^t \dot{z}^\top(s) \mathrm{d}s \, R \int_{t_k}^t \dot{z}(s) \mathrm{d}s,$$

where the first inequality follows from the fact that h is positive, and the second from Jensen's inequality (Poznyak 2008). Combining (6.19) and (6.20) yields (6.18). $\blacksquare$

Now we are ready to give a proof of Theorem 6.1

Proof of Theorem 6.1. We begin by directly computing $\dot{V}$:

$$\dot{V}(t, z(\cdot)) = 2 z^\top(t) P^{-1} \dot{z}(t)$$

$$-\alpha h \int_{-h}^0 \int_{t+\theta}^t e^{\alpha(s-t)} \dot{z}^\top(s) R \dot{z}(s) \mathrm{d}s \mathrm{d}\theta \tag{6.21}$$

$$-h \int_{t-h}^t e^{\alpha(s-t)} \dot{z}^T(s) R \dot{z}(s) \mathrm{d}s + h^2 \dot{z}^T(t) R \dot{z}(t).$$

By adding and subtracting $\alpha V(t, z(\cdot))$ to the right-hand side of (6.21), we obtain

$$\dot{V}(t, z(\cdot)) = 2z^\top(t) P^{-1} \dot{z}(t) + \alpha z^\top(t) P^{-1} z(t)$$

$$-h \int_{t-h}^{t} e^{\alpha(s-t)} \dot{z}^\top(s) R \dot{z}(s) \mathrm{d}s + h^2 \dot{z}^\top(t) R \dot{z}(t) - \alpha V(t, z(\cdot)). \tag{6.22}$$

The following upper bound for $\dot{V}$ can be easily obtained from (6.22) and (6.18):

$$\dot{V}(t, z(\cdot)) \leq -\alpha V(t, z(\cdot)) + \eta_1(t, z(\cdot))^\top W_1 \eta_1(t, z(\cdot)), \tag{6.23}$$

where

$$\eta_1(t, z(\cdot)) := \begin{pmatrix} z(t) \\ \dot{z}(t) \\ \int_{t_k}^{t} \dot{z}(s)\mathrm{d}s \end{pmatrix} = \begin{pmatrix} z(t) \\ \dot{z}(t) \\ z(t) - z(t_k) \end{pmatrix}$$

and W_1 is a symmetric matrix defined by

$$W_1 := \begin{pmatrix} \alpha P^{-1} & P^{-1} & 0 \\ P^{-1} & h^2 R & 0 \\ 0 & 0 & -h e^{-\alpha h} R \end{pmatrix}.$$

By adding and subtracting $b\|\omega(t)\|_{\bar{Q}}^2$ to and from the right-hand side of (6.23), we can rewrite the upper bound as

$$\dot{V}(t, z(\cdot)) \leq -\alpha V(t, z(\cdot)) + b\|\omega(t)\|_{\bar{Q}}^2$$

$$+ \eta(t, z(\cdot))^\top W_2 \eta(t, z(\cdot)), \tag{6.24}$$

where

$$W_2 := \begin{pmatrix} \alpha P^{-1} & P^{-1} & 0 & 0 \\ P^{-1} & h^2 R & 0 & 0 \\ 0 & 0 & -h e^{-\alpha h} R & 0 \\ 0 & 0 & 0 & -b\bar{Q} \end{pmatrix}.$$

From (6.16), we have

$$\dot{V}(t, z(\cdot)) \leq -\alpha V(t, z(\cdot)) + b(\bar{\delta} + \|x(t)\|_{Q}^2)$$

$$+ \eta(t, z(\cdot))^\top W_2 \eta(t, z(\cdot)). \tag{6.25}$$

Since

$$\|x(t)\|_Q^2 = \|\hat{x}(t) + e(t)\|_Q^2 = \left\|\begin{pmatrix} I & I \end{pmatrix} z(t)\right\|_Q^2 = z(t)^\top Q_z z(t),$$

we can finally rewrite (6.25) as (6.14). ∎

Now we will refine the bound given in Theorem 6.1 by restricting $z(\cdot)$ to the set of solutions of (6.12). In order to do so, we follow the idea presented in Fridman 2006 and in Fridman and Niculescu (2008), which, originally devised for systems in descriptor form, consists in adding a term (the *descriptor term*) to the expression for $\dot{V}$. The descriptor term has to be zero for every solution z of the system. In our case, we will add the term

$$\mathcal{D}(t, z(\cdot)) := 2\left(z(t)^\top \Pi_a + \dot{z}(t)^\top \Pi_b\right) \cdot \left(\tilde{A}z(t) + F\omega(t) + \psi(t) - \dot{z}(t)\right),$$

where Π_a and Π_b are in $\mathbb{R}^{2n}$. Obviously, $\mathcal{D}$ is zero along the solutions of (6.12).

Theorem 6.2. *Let ρ be a positive scalar satisfying*

$$L^\top L \le \rho_1 I. \tag{6.26}$$

Then, for every

$$z(\cdot) \in \mathbb{C}, \ h, \alpha, b, \varepsilon \in \mathbb{R}, \ P, R, \Pi_a, \Pi_b \in \mathbb{R}^{2n \times 2n}$$

such that z is a solution of (6.12), $h > 0$, $\alpha > 0$, $P > 0$, and $R > 0$, the time derivative of $V(t, z(\cdot))$ in (6.13) satisfies

$$\dot{V}(t, z(\cdot)) \le -\alpha V(t, z(\cdot)) + \beta$$
$$+ \xi(t, z(\cdot))^\top \Omega \xi(t, z(\cdot)), \tag{6.27}$$

where

$$\Omega := \begin{pmatrix} \alpha P^{-1} + bQ_z \\ + 2\Pi_a \tilde{A} & P^{-1} - \Pi_a \\ & + \Pi_b \tilde{A} & 0 & \Pi_a F & \Pi_a \\ P^{-1} - \Pi_a + \Pi_b \tilde{A} & h^2 R - 2\Pi_b & 0 & \Pi_b F & \Pi_b \\ 0 & 0 & \begin{array}{c} -he^{-\alpha h}R \\ + \varepsilon\rho Q_c \end{array} & 0 & 0 \\ \Pi_a F & 0 & 0 & -b\bar{Q} & 0 \\ \Pi_a & \Pi_b & 0 & 0 & -\varepsilon I \end{pmatrix} \tag{6.28}$$

and

$$\xi(t, z(\cdot)) := \begin{pmatrix} z(t) \\ \dot{z}(t) \\ z(t) - z(t_k) \\ \omega(t) \\ \psi(t) \end{pmatrix},$$

$$Q_c := \begin{pmatrix} I \\ I \end{pmatrix} C^\top Q_y C \begin{pmatrix} I & I \end{pmatrix}, \ \beta := b\bar{\delta} + \varepsilon\rho(2 + c),$$

$$\rho := 2\rho_1 / \lambda_{\min}(Q_y). \tag{6.29}$$

The following lemma will be needed before the proof of the theorem.

Lemma 6.3. *The uncertainty $\psi(t)$ resulting from noise, sampling, and quantization is bounded by*

$$\|\psi(t)\|^2 \leq \rho\left((z(t) - z(t_k))^\top Q_c(z(t) - z(t_k)) + 2 + c\right). \tag{6.30}$$

Proof. We will begin by computing an upper bound for Δy. We have

$$\|\Delta y(t)\|^2_{Q_y} = \|y(t) - \bar{\bar{y}}(t)\|^2_{Q_y} \leq$$

$$\|y(t) - \bar{y}(t)\|^2_{Q_y} + \|\bar{y}(t) - \bar{\bar{y}}(t)\|^2_{Q_y}. \tag{6.31}$$

Notice that

$$\bar{y}(t) - \bar{\bar{y}}(t) = C(x(t) - x(t_k)) + \omega_y(t) - \omega_y(t_k)$$

$$= C \begin{pmatrix} I & I \end{pmatrix} (z(t) - z(t_k)) + \omega_y(t) - \omega_y(t_k).$$

So

$$\|\bar{y}(t) - \bar{\bar{y}}(t)\|^2_{Q_y} \leq (z(t) - z(t_k))^\top Q_c(z(t) - z(t_k)) + 2, \tag{6.32}$$

where we have used (6.16) to establish

$$\|\omega_y(t)\|^2_{Q_y} + \|\omega_y(t_k)\|^2_{Q_y} \leq 2.$$

Substituting (6.32) and (6.7) into (6.31) gives

$$\|\Delta y(t)\|^2_{Q_y} \leq (z(t) - z(t_k))^\top Q_c(z(t) - z(t_k)) + 2 + c. \tag{6.33}$$

The norm of ψ then satisfies

$$\|\psi(t)\|^2 = \left\|\begin{pmatrix} I \\ I \end{pmatrix} L\Delta y(t)\right\|^2 = 2\Delta y(t)^\top L^\top L\Delta y(t)$$

$$\leq 2\rho_1 \|\Delta y(t)\|^2 \leq \frac{2\rho_1}{\lambda_{\min(Q_y)}}\|\Delta y(t)\|^2_{Q_y}. \tag{6.34}$$

From (6.34) and (6.33), we obtain (6.30). ∎

Now we can begin our proof of Theorem 6.2.

Proof of Theorem 6.2.   Adding the null term

$$\mathcal{D}(t, z(\cdot)) + \varepsilon\|\psi(t)\|^2 - \varepsilon\|\psi(t)\|^2$$

to (6.14) gives

$$\begin{aligned}
\dot{V}(t, z(\cdot)) &\leq -\alpha V(t, z(\cdot)) + b\bar{\delta} + \varepsilon\|\psi(t)\|^2 + \eta(t, z(\cdot))^\top W\eta(t, z(\cdot)) \\
&\quad + 2\left(z(t)^\top \Pi_a + \dot{z}(t)^\top \Pi_b\right)\left(\tilde{A}z(t) + F\omega(t) + \psi(t) - \dot{z}(t)\right) \\
&\quad - \varepsilon\|\psi(t)\|^2.
\end{aligned} \tag{6.35}$$

Substituting (6.30) into (6.35) establishes

$$\dot{V}(t, z(\cdot)) \leq -\alpha V(t, z(\cdot)) + \beta + \varepsilon\rho(z(t) - z(t_k))^\top Q_c(z(t) - z(t_k))$$

$$+ \eta(t, z(\cdot))^\top W\eta(t, z(\cdot)) + 2\left(z(t)^\top \Pi_a + \dot{z}(t)^\top \Pi_b\right) \times \tag{6.36}$$

$$\left(\tilde{A}z(t) + F\omega(t) + \psi(t) - \dot{z}(t)\right) + \varepsilon\|\psi(t)\|^2.$$

Equation (6.27) is nothing but (6.36) written in compact form. ∎

6.3.1   Main Result

The following corollary follows from Theorem 6.2 and Lemma 2.5.

Corollary 6.1. *Let*

$$\alpha > 0, \ b > 0, \ \varepsilon > 0, \ \rho_1 > 0,$$

$$P^{-1} > 0, \ R > 0, \ \Pi_a, \ \Pi_b, \ L, \ K \tag{6.37}$$

be a set of control parameters such that

$$\Omega \le 0 \ \textit{and} \ L^{\top} L \le \rho_1 I \tag{6.38}$$

with Ω defined by (6.28), Q_z, Q_c, $\bar{Q}$, and let ρ be given by (6.29). Then the ellipsoid

$$\mathcal{E} := \left\{ z \in \mathbb{R}^{2n} : z^{\top} P^{-1} z \le \frac{\beta}{\alpha} \right\}$$

with β given by (6.29) and $\alpha = a$ is an attractive and invariant set.

6.4 Numerical Aspects

Given Corollary (6.1), it is natural to look for a set of parameters (6.37) such that the attractive ellipsoid is minimal in some sense. An obvious objective function to minimize is $\mathrm{tr} P$. Unfortunately, such a problem is strongly nonlinear and difficult to solve, even numerically, so we will have to settle for a suboptimal solution. Our goal here is to find a numerically tractable expression that ensures

$$\Omega \le 0.$$

More precisely, we seek an expression that is linear in the matrix parameters, so that the well-known convex tools for matrix inequalities can be applied.

The first step involves applying the Schur complement to several blocks of the original matrix Ω. In order to achieve this, we first simplify our parameter space by setting

$$\Pi_a = \Pi_b = P^{-1}$$

and restricting P^{-1} and R to the class of block diagonal matrices of the form

$$P^{-1} = \mathrm{diag}\left(P_1^{-1}, P_2^{-1}\right) \ \text{and} \ R = \mathrm{diag}\left(R_1, R_2\right).$$

Let us define $\Omega_A := T\Omega T^{\top}$ with

$$T = \mathrm{diag}\left(P_2^{-1} P_1, I, P_2^{-1} P_1, I, \cdots, I\right) \in \mathbb{R}^{10n \times 10n}.$$

Since T is nonsingular, $\Omega \le 0$ is equivalent to

$$\Omega_A \le 0.$$

To ease notation, we will write Ω_A in terms of the block matrices $A_{ij} \in \mathbb{R}^{n \times n}$, with $i, j = 1, \ldots, 10$. Notice that the following elements of Ω_A are nonlinear with respect to P_2 and P_1:

$$A_{11} = P_2^{-1}\left(bP_1QP_1 + aP_1 + P_1A^\top + AP_1 + BKP_1 + P_1K^\top B^\top\right)P_2^{-1},$$
$$A_{12} = A_{21}^\top = P_2^{-1}LC + bP_2^{-1}QP_1,$$
$$A_{13} = A_{31}^\top = P_2^{-1}\left(AP_1 + BKP_1\right)P_2^{-1},$$
$$A_{33} = P_2^{-1}\left(P_1h^2R_1P_1 - 2P_1\right)P_2^{-1}.$$

By defining the matrix

$$J := \left(P_2^{-1}P_1 \ I \ 0 \cdots 0\right),$$

it is possible to express Ω_A as

$$\Omega_A = \Omega_B + J^\top(bQ)J. \tag{6.39}$$

The matrix subblocks of Ω_B are the same as those of Ω_A: $B_{ij} = A_{ij}$ with the exception of

$$B_{11} = P_2^{-1}\left(aP_1 + P_1A^\top + AP_1 + BKP_1 + P_1K^\top B^\top\right)P_2^{-1},$$
$$B_{12} = B_{21}^\top = P_2^{-1}LC,$$

which are now simpler than A_{11} and A_{12}. Using a Schur complement argument and defining $\mathcal{O} = \left(0 \ I \ 0 \cdots 0\right)$, it can be seen that Ω_A is negative semidefinite if and only if

$$\begin{pmatrix} \Omega_B & J^\top \\ J & -\frac{1}{b}Q^{-1} \end{pmatrix} = \begin{pmatrix} \Omega_B & \mathcal{O}^\top \\ \mathcal{O} & -\frac{1}{b}Q^{-1} \end{pmatrix} + $$
$$\begin{pmatrix} 0 \\ \vdots \\ 0 \\ P_1 \end{pmatrix} \left(P_2^{-1} \ 0 \cdots 0\right) + \begin{pmatrix} P_2^{-1} \\ 0 \\ \vdots \\ 0 \end{pmatrix} \left(0 \cdots 0 \ P_1\right) \le 0. \tag{6.40}$$

According to the Λ-inequality (Poznyak 2008; see also Chapter 2 of this book),

$$\mathcal{X}\mathcal{Y}^\top + \mathcal{Y}\mathcal{X}^\top \le \mathcal{X}^\top \Lambda \mathcal{X} + \mathcal{Y}^\top \Lambda^{-1}\mathcal{Y},$$

valid for every $\mathcal{X}$, $\mathcal{Y}$, and nonsingular Λ with compatible dimensions, and setting

$$\mathcal{X} = \left(0 \cdots 0 \ P_1\right)^\top, \quad \mathcal{Y} = \left(P_2^{-1} \ 0 \cdots 0\right)^\top$$

with $\Lambda = \Lambda_1$ in (6.40), we obtain

$$\begin{pmatrix} \Omega_B & J^\top \\ J & -\frac{1}{b}Q^{-1} \end{pmatrix} \le \begin{pmatrix} \Omega_C & \mathcal{O}^\top \\ \mathcal{O} & -\frac{1}{b}Q^{-1} + P_1\Lambda_1^{-1}P_1 \end{pmatrix}, \tag{6.41}$$

where the subblocks of Ω_C are the same as those of Ω_B, that is, $C_{ij} = B_{ij}$, with the exception of C_{11}, which is

$$C_{11} = B_{11} + P_2^{-1}\Lambda_1 P_2^{-1} =$$
$$P_2^{-1}\left(aP_1 + P_1 A^\top + AP_1 + BKP_1 + P_1 K^\top B^\top + \Lambda_1\right)P_2^{-1}.$$

Let us now introduce a new variable, G_f, which will serve as an upper bound for the nonlinear term $\left(-\frac{1}{b}Q^{-1} + P_1\Lambda_1^{-1}P_1\right)$ in (6.41), namely

$$-\frac{1}{b}Q^{-1} + P_1\Lambda_1^{-1}P_1 \le G_f. \tag{6.42}$$

This implies

$$\begin{pmatrix} \Omega_C & \mathcal{O}^\top \\ \mathcal{O} & -\frac{1}{b}Q^{-1} + P_1\Lambda_1^{-1}P_1 \end{pmatrix} \le \begin{pmatrix} \Omega_C & \mathcal{O}^\top \\ \mathcal{O} & G_f \end{pmatrix}. \tag{6.43}$$

Using Schur complements again, it can be readily shown that (6.42) is equivalent to

$$\begin{pmatrix} -G_f - \frac{1}{b}Q^{-1} & P_1 \\ P_1 & -\Lambda_1 \end{pmatrix} \le 0. \tag{6.44}$$

We want now to obtain a linear upper bound for Ω_C, i.e., a matrix Ω_D such that

$$\Omega_C \le \Omega_D. \tag{6.45}$$

Notice that all the subblocks of Ω_C are linear in P_1 and P_2^{-1}, except for C_{11}, C_{13}, and C_{33}, which are of the form $\mathcal{P}\mathcal{M}\mathcal{P}$, where $\mathcal{P}$ can take the value of P_1 or P_2^{-1}, and $\mathcal{M}$ depends on P_1. These terms can be majorized using the Λ-inequality again. Set

$$\mathcal{X} = \mathcal{P}, \ \mathcal{Y} = I, \ \Lambda = \mathcal{M},$$

and introduce a new term, $\mathcal{G}$, that will serve as an upper bound

$$\mathcal{P}\mathcal{M}\mathcal{P} \le \mathcal{G}.$$

Equivalently, by the Schur complement, we can rewrite it as

$$\begin{pmatrix} -\mathcal{G} - 2\mathcal{P} & I \\ I & \mathcal{M} \end{pmatrix} \le 0. \tag{6.46}$$

It is natural to propose an Ω_D with subblocks equal to those of Ω_C with the exception of

$$D_{11} = G_1,$$
$$D_{13} = D_{13}^{\top} = G_2,$$
$$D_{33} = G_3,$$

where G_1, G_2, and G_3 are matrix block variables that satisfy some additional restrictions. To obtain these restrictions, let us write the upper left-hand $3n \times 3n$ subblock matrix of the difference $\Omega_C - \Omega_D$:

$$
\begin{pmatrix} C_{11}-G_1 & 0 & C_{13}-G_2 \\ 0 & 0 & 0 \\ C_{31}-G_2 & 0 & C_{33}-G_3 \end{pmatrix} = \begin{pmatrix} I \\ 0 \\ 0 \end{pmatrix} (C_{11}-G_1-C_{13}+G_2) \begin{pmatrix} I & 0 & 0 \end{pmatrix} +
$$
$$
\begin{pmatrix} 0 \\ 0 \\ I \end{pmatrix} (C_{33}-G_3-C_{13}+G_2) \begin{pmatrix} 0 & 0 & I \end{pmatrix} + \begin{pmatrix} I \\ 0 \\ I \end{pmatrix} (C_{13}-G_2) \begin{pmatrix} I & 0 & I \end{pmatrix}.
$$

It is clear that

$$\Omega_C - \Omega_D \leq 0$$

if

$$C_{11} - C_{13} \leq G_1 - G_2,\ C_{33} - C_{13} \leq G_3 - G_2 \text{ and } C_{13} \leq G_2.$$

Applying again the Schur complement to these inequalities, we get

$$C_{11} - C_{13} = P_2^{-1} \left(aP_1 + P_1 A^{\top} + P_1 K^{\top} B^{\top} + \Lambda_1 \right) P_2^{-1} \leq G_1 - G_2,$$

which is equivalent to

$$
\begin{pmatrix} -2P_2^{-1} - G_1 + G_2 & P_2^{-1} \\ P_2^{-1} & -(aP_1 + P_1 A^{\top} + P_1 K^{\top} B^{\top} + \Lambda_1)^{-1} \end{pmatrix} \leq 0. \qquad (6.47)
$$

Now proposing a lower bound for the term

$$(aP_1 + P_1 A^{\top} + P_1 K^{\top} B^{\top} + \Lambda_1)^{-1}$$

as

$$(aP_1 + P_1 A^{\top} + P_1 K^{\top} B^{\top} + \Lambda_1)^{-1} \geq H_1$$

and by the Schur complement, we obtain the first additional linear restriction:

$$
\begin{pmatrix} H_1 & I \\ I & aP_1 + P_1 A^{\top} + P_1 K^{\top} B^{\top} + \Lambda_1 \end{pmatrix} \geq 0.
$$

Clearly, the left-hand side of the inequality

$$\begin{pmatrix} -2P_2^{-1} - G_1 + G_2 & P_2^{-1} \\ P_2^{-1} & -H_1 \end{pmatrix} \le 0$$

is linear (with respect to P_1 and P_2^{-1}), implying (6.47). Likewise,

$$C_{33} - C_{13} = P_2^{-1} \left(P_1 h^2 R_1 P_1 - 2P_1 - AP_1 - BKP_1 \right) P_2^{-1}$$
$$\le P_2^{-1} \left(G_0 - AP_1 - BKP_1 \right) P_2^{-1} \le G_3 - G_2$$

if

$$\begin{pmatrix} H_2 & I \\ I & h^2 R_1 \end{pmatrix} \ge 0,$$

$$\begin{pmatrix} -2P_1 - G_0 & P_1 \\ P_1 & -H_2 \end{pmatrix} \le 0,$$

and using (6.46),

$$\begin{pmatrix} -2P_2^{-1} - G_3 + G_2 & I \\ I & G_0 - P_1 A^\top - AP_1 - BKP_1 - P_1 K^\top B^\top \end{pmatrix} \le 0.$$

Finally,

$$C_{13} = P_2^{-1} \left(P_1 A + BKP_1 \right) P_2^{-1} \le G_2$$

if

$$\begin{pmatrix} -2P_1 - G_2 & I \\ I & AP_1 + BKP_1 \end{pmatrix} \le 0. \tag{6.48}$$

With these restrictions, inequality (6.45) holds, so

$$\begin{pmatrix} \Omega_C & \mathcal{O}^\top \\ \mathcal{O} & G_f \end{pmatrix} \le \begin{pmatrix} \Omega_D & \mathcal{O}^\top \\ \mathcal{O} & G_f \end{pmatrix}.$$

Our constraint set is thus given by (6.44), (6.48), and

$$\begin{pmatrix} \Omega_D & \mathcal{O}^\top \\ \mathcal{O} & G_f \end{pmatrix} \le 0. \tag{6.49}$$

It is noteworthy that Ω_D is linear in P_1 and P_2^{-1}. However, there exist some bilinear matrix terms in all of P_1, P_2^{-1}, K, and L. To deal with these and to obtain an LMI (in the matrix variables), we proceed to define

$$X_1 := P_1, \; Y_1 := KP_1, \; X_2 := P_2^{-1} \text{ and } Y_2 := P_2^{-1}L.$$

So now all the inequalities are linear in the matrix arguments.

A natural objective for the controller is to minimize the volume of the ellipsoid, i.e., to minimize the trace of P, which amounts to minimizing the objective function $tr(X_1) + tr(X_2^{-1})$. This is still a nonlinear problem. By including the last linear constraint,

$$\begin{pmatrix} H & I \\ I & X_2 \end{pmatrix} \geq 0, \tag{6.50}$$

we can now state the numerically tractable suboptimal problem:

$$\text{minimize } [\text{tr}(X_1) + \text{tr}(H)]$$

subject to the constraints

$$X_1, X_2, H, R_1, R_2 > 0$$

and (6.44), (6.48), (6.49), and (6.50) with respect to the matrix variables

$$X_1, X_2, H, R_1, R_2, G_1, G_2, G_3, G_f, G_0 \in \mathbb{R}^{n \times n}, \; Y_1 \in \mathbb{R}^{n \times m}, \; Y_2 \in \mathbb{R}^{q \times n}$$

and the scalar variables $a, b, \rho, \rho_q, \varepsilon$. The suboptimal ellipsoid is defined by

$$P^{-1} = \begin{pmatrix} X_1 & 0 \\ 0 & X_2^{-1} \end{pmatrix}.$$

The controller and observer gains can be obtained uniquely as

$$K = Y_1 X_1^{-1} \text{ and } L = X_2 Y_2.$$

Observe that this problem is still altogether bilinear in the matrix and scalar variables, so we propose the following algorithm:

- fix ρ, ρ_q and ϵ;
- set a^* to a very small value a_0;
- set b^* to a very small value b_0;
- set T^* to a very large value T_0;
- for $j = 1$ to m

- for $i = 1$ to n

  - try to solve LMI minimization problem
  - increase a^* by STEP1 until a solution is feasible
  - set T_i to trace(P^{-1}) evaluated in the solution
  - if $\{T_i < T^*\}$ set $T^* = T_i$
  - divide STEP1 by 2

- end for
- for $i = 1$ to n

  - try to solve LMI minimization problem
  - increase b by STEP2 until a solution is feasible
  - set T_i to trace(P^{-1}) evaluated in the solution
  - if $\{T_i < T^*\}$ set $b^* = b_i$
  - divide STEP2 by 2

- end for

- end for
- return a^* and b^*

This algorithm can be implemented easily using off-the-shelf software.

6.5   Numerical Examples

The previous algorithm was implemented in MATLAB to exemplify the applicability of our method. The objective was to design a robust controller based on the previously described method for two different systems. The first is a two-dimensional nonlinear system, whose dynamics include a sign function. It also has bounded state and output perturbations. The second is a four-dimensional linear system with bounded uncertainties and perturbations. Its dynamics are modeled as a pair of double integrators. As in similar algorithms, the time consumed to obtain a solution by this particular method depends heavily on the initial values of the variables. The closer the initial values to the solution, the shorter the time used by the algorithm. In the following section, we present the details of the systems used as examples and the results of the implementation of our method.

6.5.1   Example 1

Consider the following discontinuous system:

$$\dot{x}_1 = \text{sign}(x_2) + \upsilon_1,$$

$$\dot{x}_2 = x_1 + 2u(t) + \upsilon_2,$$

$$\bar{\bar{y}} = x_1 + 2x_2 + \omega_y.$$

Let us assume that $|\upsilon_1|, |\upsilon_2| \leq 0.1$ and that $|\omega_y| \leq 0.2$. These bounds satisfy the accepted Assumption 1 above with $Q_x = Q_y = I_{2\times2}$. Using the equivalent transformations discussed in the previous section, we can write the equivalent system (6.12) as

$$\dot{x} = \begin{pmatrix} 0 & 1 \\ 1 & 0 \end{pmatrix} x + \begin{pmatrix} 1 \\ 2 \end{pmatrix} u + \begin{pmatrix} \omega_1 \\ \omega_2(t) \end{pmatrix},$$

$$\bar{\bar{y}} = \begin{pmatrix} 1 & 2 \end{pmatrix} x + \omega_y.$$

By defining

$$Q_x := \begin{pmatrix} q_{11} & q_{12} \\ q_{21} & q_{22} \end{pmatrix},$$

it can be seen that

$$\|f(t, x) - Ax\|_{Q_x}^2 = q_{11} \left(\text{sign}(x_2) - x_2\right)^2 \leq q_{11} \left(x_2^2 + 1\right).$$

Choosing $Q_x = Q = I_{2\times2}$ and selecting $\delta = 1$, Assumption 2 is satisfied. Also, the resulting system is controllable and observable as needed in Assumption 3. The numerical treatment of the minimization problem was stated using the following parameters:

- the sample time interval is fixed at 0.01 seconds, so we can choose directly $h = 0.01$;
- the initial conditions for the dynamic system are $x_1(0) = x_2(0) = 10$;
- the quantization constant selected was $c = 1$;
- the constants c and h satisfy Assumptions 4 and 5;
- with respect to the observer, the initial conditions were chosen as the origin.

Using the algorithm, the observer and the controller gains were obtained as

$$K = \begin{pmatrix} -2.1463 & -0.6364 \end{pmatrix} \text{ and } L = \begin{pmatrix} 0.7821 & 0.7819 \end{pmatrix}^{\mathsf{T}}.$$

The simulated trajectories are shown in Figs. 6.2 and 6.3. The estimated ellipsoidal region is also shown in Fig. 6.2. It can be appreciated that the ellipsoidal region is positively invariant. Notice that the estimate is accurate enough, since the ellipsoid encloses the trajectories tightly. Figure 6.3 shows how the estimated states converge to the actual ones. Finally, Fig. 6.4 shows a comparison between the control input u

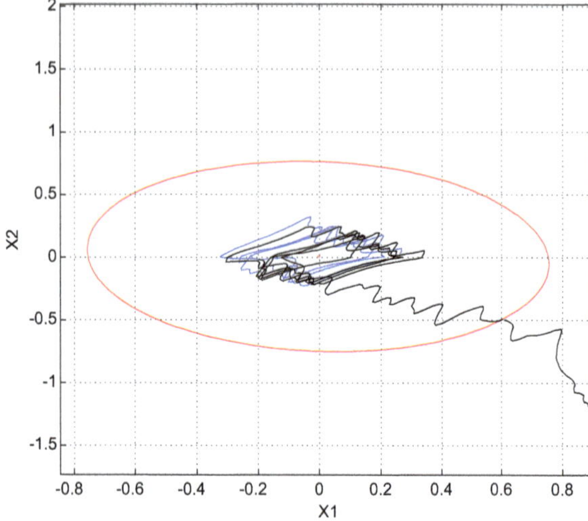

Fig. 6.2 Estimated ellipsoid and system trajectories for Example 1

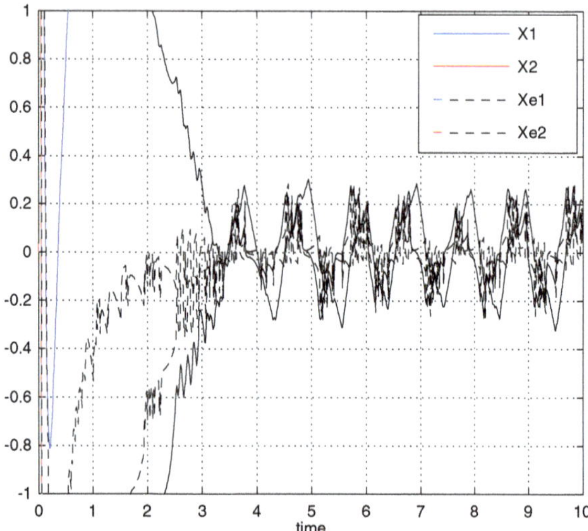

Fig. 6.3 Actual and estimated states for Example 1

and the measurable output y. It is clear that this output is quantized. The effect of the quantization and the measurement noise can be appreciated.

6.5.2   Example 2

For the second example, an optical disk drive system, is used. The dual-actuator disk drive system was modeled as a pair of double integrators. The four states in this case were the relative position error and its derivative and the tracking error

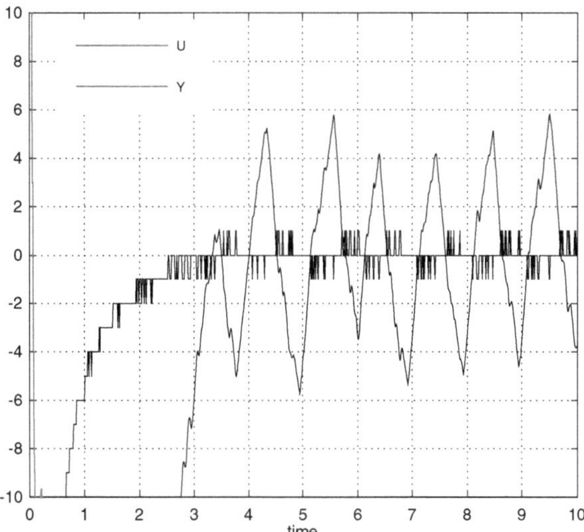

Fig. 6.4 Input and output signals for Example 1

and its derivative. The system is perturbed by $\|\omega_x\| \leq 0.1$ and $\|\omega_y\| \leq 0.2$. So the model is as follows:

$$\dot{x} = \begin{pmatrix} 0 & 1 & 0 & 0 \\ 0 & 0 & 0 & 0 \\ 0 & 0 & 0 & 1 \\ 0 & 0 & 0 & 0 \end{pmatrix} x + \begin{pmatrix} 0 & 0 \\ 10 & 0 \\ 0 & 0 \\ -10 & -20 \end{pmatrix} u + \omega_x,$$

$$\bar{\bar{y}} = \begin{pmatrix} 1 & 0 & 0 & 0 \\ 0 & 0 & 1 & 0 \end{pmatrix} x + \omega_y.$$

The simulation was run using the same sample time $h = 0.01$, and $x_i(0) = 5$ with $i = 1, 2, 3, 4$ for initial conditions. Two different quantization constants were considered. The first was $c = 1$, and the second was $c = 2$.

Figure 6.5 shows the ellipsoid projection on the x_1-x_2 plane with the respective trajectories when $c = 1$. The first three states and their estimates can be seen in Fig. 6.6. Both inputs and outputs are shown in Fig. 6.7. It is worth pointing out that the outputs in steady state use only the first level of quantization ($\pm c$).

The remaining figures were obtained using a quantization constant $c = 2$. Certain differences between the two choices of quantization constants are worth commenting on. First, the estimated ellipsoidal region and the actual convergence region are obviously larger due to a greater quantization constant effect, as can be seen in Fig. 6.8. Also, the estimation error is larger (Fig. 6.9). Finally, in Fig. 6.10, although the steady-state output still uses only the first level of quantization $\pm c$, it is twice as large as in the first case.

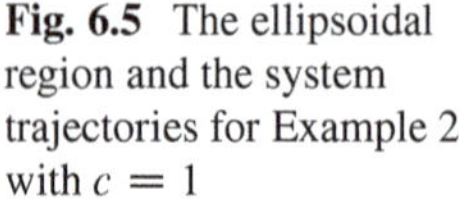

Fig. 6.5 The ellipsoidal region and the system trajectories for Example 2 with $c = 1$

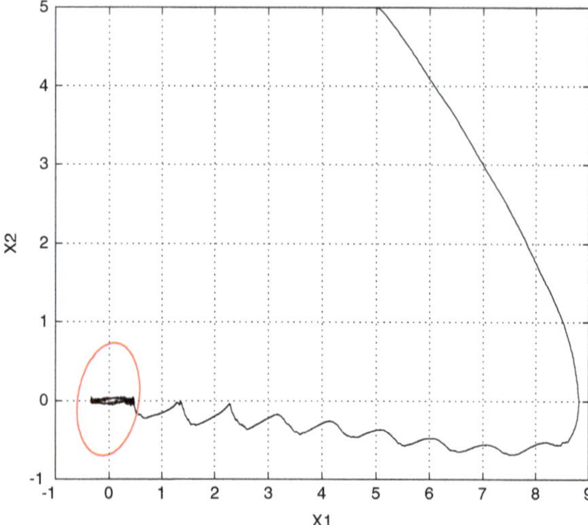

Fig. 6.6 First two actual and estimated states for Example 2 with $c = 1$

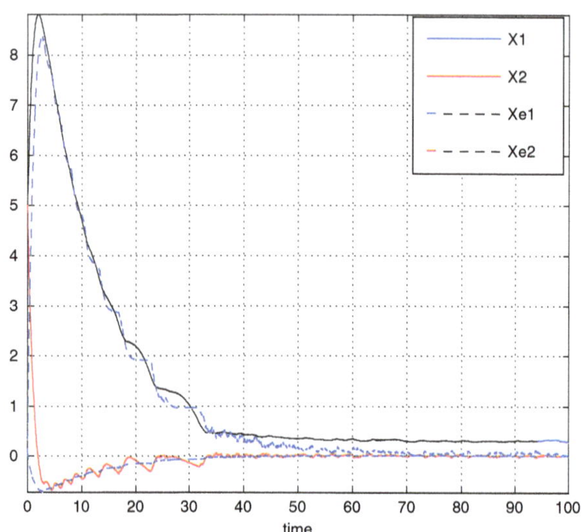

6.6 Conclusions

In this chapter, a new analytic and numerical methodology for robust control design associated with nonlinear perturbed systems was developed. We also considered sampled data and quantization at the output of these systems. In the model we used, the quantization error was bounded. The control design strategy proposed in this paper is an extension of the invariant ellipsoid method. This approach produces a control law such that the existence and an actual characterization of a minimal-size invariant ellipsoid for the closed-loop system can be guaranteed. The computational

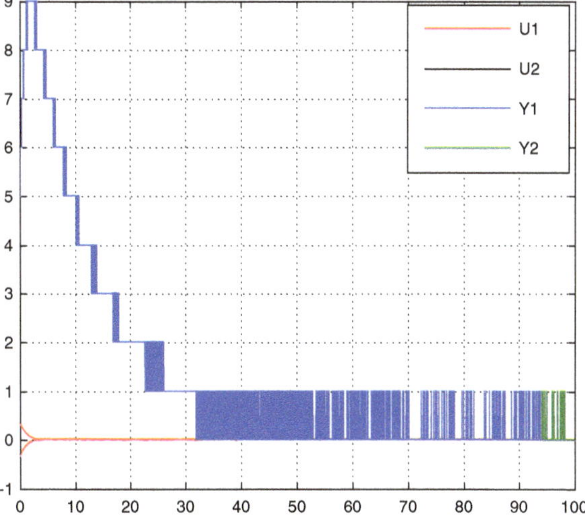

Fig. 6.7 Input and output signals for Example 2 with $c = 1$

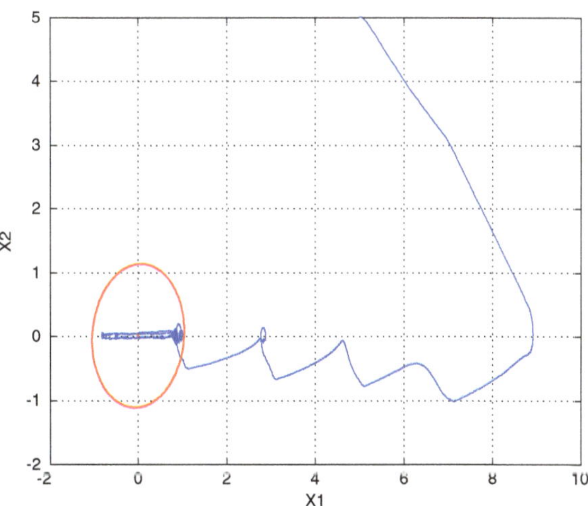

Fig. 6.8 Ellipsoid and system trajectories for Example 2 with $c = 2$

implementation of the aforementioned method led us to a complex nonlinear minimization problem with nonlinear matrix constraints. In this contribution, we proposed an effective relaxation from this initial optimization problem to an LMI (linear in the matrix variables). The final product was an attractive ellipsoidal region with minimal "size."

Finally, this approach can be generalized to systems with delays and networked control systems with relative ease. It also seems possible to apply the presented control design techniques in combination with some nonlinear feedback control strategies.

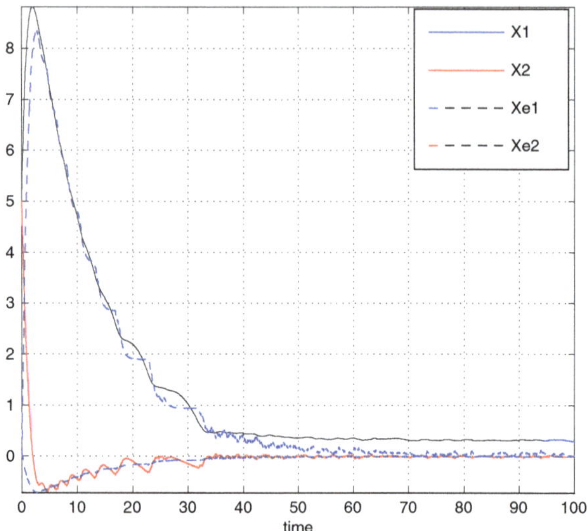

Fig. 6.9 Actual states and estimated states for Example 2 with $c = 2$

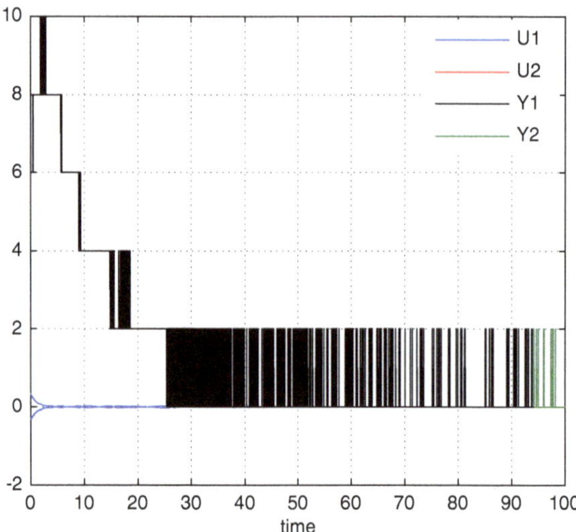

Fig. 6.10 Input and output signals for Example 2 with $c = 2$

Chapter 7
Robust Control of Implicit Systems

Abstract This chapter deals with a new approach to robust control design for a class of nonlinearly affine control systems. The dynamic models under consideration are described by implicit differential equations in the presence of additive bounded uncertainties. The proposed robust feedback design procedure is based on an extended version of the classical invariant ellipsoid technique. In this book, this extension is called the attractive ellipsoid method. The stability/robustness analysis of the resulting closed-loop system involves a modified descriptor approach associated with the usual Lyapunov-type methodology. The theoretical schemes elaborated in our contribution are finally illustrated by a simple computational example.

Keywords Implicit systems • Sampled-data systems • Attractive ellipsoids

This chapter deals with a new approach to robust control design for a class of nonlinearly affine control systems. The dynamic models under consideration are described by implicit differential equations in the presence of additive bounded uncertainties. The proposed robust feedback design procedure is based on an extended version of the classical invariant ellipsoid technique. In this book, this extension is called the attractive ellipsoid method (AEM). The stability/robustness analysis of the resulting closed-loop system involves a modified descriptor approach associated with the usual Lyapunov-type methodology. The theoretical schemes elaborated in our contribution are finally illustrated by a simple computational example.

7.1 Introduction

Recently, interest in new and powerful robust control design approaches to unconventional control systems has significantly increased (see, e.g., Basin & Calderon-Alvarez 2010; Basin et al. 2007; Dahleh et al. 1988; Fridman 2006; Haddad & Chellaboina 2008; Khalil 2002; Kurzhanski & Veliov 1994; Karafyllis & Jiang 2011; Mera et al. 2011; Michel et al. 2007; Polyak and Topunov 2008; Poznyak

© Springer International Publishing Switzerland 2014
A. Poznyak et al., *Attractive Ellipsoids in Robust Control*, Systems & Control:
Foundations & Applications, DOI 10.1007/978-3-319-09210-2_7

et al. 2011). In consequence, the robust stabilization of dynamic models described by implicit differential equations has emerged as a challenging problem of modern systems theory (Dai 1989; Takaba 2002; Kunkel & Mehrmann 2006). During the last decade, a vast body of research on implicit control systems has been produced, drawing its motivation from the fact that many engineering application domains involve that class of processes (see Kunkel & Mehrmann 2006; Reis & Stykel 2011; Takaba 2002 and the references therein). Implicit control systems generalize the classical modeling framework for ordinary differential equation (ODE) control systems and have proved a useful tool for capturing more general real-world phenomena (Kunkel & Mehrmann 2006).

Here we consider a particular family of control processes governed by implicit systems in the presence of bounded uncertainties and restrict our consideration to mathematical models of nonlinear systems with affine structures. Our aim is to elaborate a constructive control design scheme that guarantees the practical stability (stabilization with respect to a prespecified bounded "small" region) of the closed-loop system realizations. The concrete robust control synthesis approach proposed in this chapter is a consequence of the application of the AEM discussed in this book.

Given the view of an implicit control system with a class of restricted additive uncertainties as a nonlinearly affine model, it is quite intuitive to exploit Lyapunov-type theoretical tools in the design of stabilizing controls. However, a direct application of the conventional Lyapunov stability theory in that case implies some natural computational difficulties. Here we investigate a particular family of dynamic processes characterized by implicit differential equations with so-called quasi-Lipschitz right-hand sides that can be reduced to differential algebraic equations (DAEs) of index 1 (see, e.g., Kunkel & Mehrmann 2006, for details). We apply the above-mentioned AEM to a class of control systems governed by implicit differential equations, generate a practically stabilizing control, and describe the corresponding invariant set of the closed-loop system. Note that an abstract existence question of an invariant set for a general dynamical system in fact constitutes a sophisticated mathematical question. The same is also true with respect to possible constructive characterizations of the invariant set. We choose this set, as everywhere throughout of this book, in the form of an ellipsoid (called an attractive ellipsoid). We construct the resulting ellipsoidal invariant set in such a way that it possesses some optimal (minimal) properties and is effectively used in the main feedback-type control design procedure. It is unsurprising that a robust synthesis problem for the given nonlinearly affine system can finally be reduced to an auxiliary LMI-constrained optimization problem (see, e.g., Boyd et al. 1994). Roughly speaking, we choose the necessary gain matrices that minimize the size of the obtained attractive ellipsoid for the resulting implicit closed-loop control system.

As mentioned above, the main mathematical tool used in this book is an appropriate extension of the classical Lyapunov-based techniques (consult Haddad & Chellaboina 2008; Khalil 2002). However, we apply these techniques in combination with the descriptor method. Recall that the descriptor method was initially created for stability analysis of time-delayed systems (Fridman 2006).

The simple structure of the Lyapunov-like ("energetic") functions we apply and the well-established numerical algorithms for LMI-constrained problems make it possible to obtain simple computational implementations of the theoretical schemes elaborated in this chapter. The effectiveness of the method we propose is illustrated by means of an example of a strictly nonlinear implicit control system.

7.2 Some Preliminaries

7.2.1 Model Description

Let us consider the following initial value problem for a nonlinear implicit system with affine structure

$$E\dot{x}(t) = f(t, x(t)) + Bu(x(t)),$$

$$x(0) = x_0, \ t \in \mathbb{R}_+,$$

(7.1)

where

- $E \in \mathbb{R}^{n \times n}$ is a given singular matrix and $B \in \mathbb{R}^{n \times m}$,
- $x(t) \in \mathbb{R}^n$ is the n-dimensional state vector at time $t \in \mathbb{R}_+$,
- $u(x) \in \mathbb{R}^m$ is an m-dimensional feedback-type control vector.

Problem (7.1) constitutes an adequate modeling framework for a wide class of control systems characterized by an implicit state equation.

Following (Masubuchi, Kamitane, Ohara, & Suda 1997), we assume for simplicity

$$E = \begin{pmatrix} I_{n_1} & 0 \\ 0 & 0 \end{pmatrix} \quad \text{and} \quad E^{\perp} = \begin{pmatrix} 0 & 0 \\ 0 & I_{n_2} \end{pmatrix},$$

where $n_1 + n_2 = n$.

Let us now make the basic technical assumptions associated with (7.1):

- $f : \mathbb{R}^n \to \mathbb{R}^n$ is a smooth function with locally bounded derivatives and moreover, it belongs to the class of the quasi-Lipschitz functions:

$$||f(t, x) - Ax||^2_{Q_f} \leq \delta + ||x||^2_{Q_x} \forall \ x \in \mathbb{R}^n, t > \mathbb{R}_+,$$

where $A \in \mathbb{R}^{n \times n}$ is a matrix, Q_f and Q_x are symmetric positive definite matrices, and by $||\cdot||_H$ (where H is an $n \times n$ symmetric positive definite matrix) we denote the weighted Euclidean norm

$$||x||_H := \sqrt{x^T H x}.$$

– $u(x)$ is chosen in the form of a linear feedback, namely,

$$u(x) = Kx,$$

where $K \in \mathbb{R}^{m \times n}$.

The class of right-hand sides of (7.1) characterized by the above technical assumptions is quite general. Throughout this chapter, functions $f(\cdot)$ that satisfy the second basic hypothesis are called *quasi-Lipschitz* functions (the second hypothesis from the above main conditions).

7.2.2   Useful Concepts and Facts

Following (Chistyakov 2007), we can establish the existence of solutions to (7.1) with corresponding right-hand sides that satisfy our basic assumptions. For example, the existence of a solution is guaranteed in the case of differentiable bounded functions $\eta(\cdot)$. On the other hand, the existence of solutions of more general implicit systems was investigated in Rheinbolt 1981 and Rheinbolt 1988 from the differential-geometric point of view.

We now use the basic assumptions described above and rewrite (7.1) in the following equivalent form:

$$E\dot{x}(t) = Ax(t) + BKx(t) + [f(t, x(t)) - Ax(t)],$$

$$x(0) = x_0, t \in \mathbb{R}_+.$$

$$(7.2)$$

Regular Matrix Pairs and Their Properties

Let us introduce some useful concepts and facts associated with a matrix pair (Γ_1, Γ_2), where $\Gamma_1, \Gamma_2 \in \mathbb{R}^{n \times n}$. We refer to Kunkel and Mehrmann (2006) for additional technical details.

Definition 7.1. A matrix pair (Γ_1, Γ_2) is said to be **regular** if the characteristic polynomial

$$p(\lambda) = \det(\lambda \Gamma_1 - \Gamma_2), \ \lambda \in \mathbb{R},$$

is not the zero polynomial. A matrix pair that is not regular is called **singular**.

Definition 7.2. Two pairs of matrices (Γ_1, Γ_2) and (Σ_1, Σ_2), $(\Sigma_1, \Sigma_2) \in \mathbb{R}^{n \times n}$ are called **strongly equivalent** if there exist nonsingular matrices $\Theta, \Delta \in \mathbb{R}^{n \times n}$ such that

$$\Sigma_1 = \Theta \Gamma_1 \Delta, \quad \Sigma_2 = \Theta \Gamma_2 \Delta.$$

If this is the case, we write

$$(\Gamma_1, \Gamma_2) \sim (\Sigma_1, \Sigma_2).$$

Recall that a matrix N is said to be *nilpotent* of index $\nu \in \mathbb{N}$ if

$$N^\nu = 0$$

and

$$N^{\nu-k} \neq 0$$

for all

$$1 \leq k \leq \nu - 1, \ k \in \mathbb{N}.$$

Theorem 7.1 ((Kunkel & Mehrmann 2006)). *Let* (Γ_1, Γ_2) *be regular. Then there exist matrices J and N such that*

$$(\Gamma_1, \Gamma_2) \sim \left(\begin{bmatrix} I & 0 \\ 0 & N \end{bmatrix}, \begin{bmatrix} J & 0 \\ 0 & I \end{bmatrix} \right),$$

where I is the unit matrix of suitable dimension, J is a matrix in Jordan canonical form, and N is a nilpotent matrix.

Theorem 7.1 gives rise to a further transformation of the original closed-loop system (7.2). Let us additionally assume that the given matrix pair (E, A) is regular, namely

$$(E, A) \sim \left(\begin{bmatrix} I & 0 \\ 0 & N \end{bmatrix}, \begin{bmatrix} J & 0 \\ 0 & I \end{bmatrix} \right).$$

7.2.3 Transformation to Differential-Algebraic Form

We now introduce a coordinate transformation Ψ determined by

$$x = \Psi z$$

and multiply both sides of the equation in (7.2) by a matrix Π (from the left). Here Ψ and Π are matrices of suitable dimension. We can rewrite (7.2) in the following form:

$$\dot{z}_1(t) = J z_1(t) + \Pi_1(f(t, \Psi z(t)) - A\Psi z(t) + BK\Psi z(t)),$$

$$N\dot{z}_2(t) = z_2(t) + \Pi_2(f(t, \Psi z(t)) - A\Psi z(t) + BK\Psi z(t)), \qquad (7.3)$$

$$z(0) = \Psi^{-1}x_0,$$

where $z := (z_1, z_2)^T$ such that

$$\dim\{z_1\} = k := \dim\{J\},$$

$$\dim\{z_2\} = n - \dim\{z_1\},$$

and

$$\Pi = \begin{pmatrix} \Pi_1 \\ \Pi_2 \end{pmatrix}, \quad \Pi_1 \in \mathbb{R}^{k \times n}, \quad \Pi_2 \in \mathbb{R}^{(n-k) \times n}.$$

Assume that the nilpotent matrix N in (7.3) has nilpotency index ν. In that case, the second equation in (7.3) involves the following algebraic equation:

$$0 = N^{\nu-1}z_2(t) + N^{\nu-1}\Pi_2(f(t, \Psi z(t)) - A\Psi z(t) + BK\Psi z(t)).$$

Finally, we obtain a system of DAEs

$$\dot{z}_1(t) = J z_1(t) + \Pi_1(f(t, \Psi z(t)) - A\Psi z(t) + BK\Psi z(t)),$$

$$0 = N^{\nu-1}z_2(t) + N^{\nu-1}\Pi_2(f(t, \Psi z(t)) - A\Psi z(t) + BK\Psi z(t)), \qquad (7.4)$$

$$z(0) = \Psi^{-1}x_0,$$

that is equivalent to the original initial value problem (7.2).

In this chapter, we also restrict our consideration to the implicit systems (7.2) of finite differentiation index. Recall this definition.

Definition 7.3. The *differentiation index* of the DAE (7.4) is the minimal number $L \in \mathbb{N}$ such that the equation

$$\frac{d^{(l)}}{dt^{(l)}}\left[N^{\nu-1}z_2(t) + N^{\nu-1}\Pi_2(f(t, \Psi z(t)) - A\Psi z(t) + BK\Psi z(t))\right] = 0$$

holds with $l = 1, \ldots, L$, implies an explicit differential equation with respect to z_2.

In that case, we can use all the conventional systematic concepts associated with the dynamic behavior of the implicit systems of type (7.2). In particular, the above assumptions on the finite differentiation index of the DAE (7.4) or of the initial implicit system (7.2) make it possible to use all the classical concepts of Lyapunov stability and set stability. One can consider these definitions and the corresponding techniques for the resulting classical system of ODEs that is obtained after an adequate number of differentiations of the algebraic part of (7.4).

7.2.4 Problem Formulation

From the point of view of an implemented control application, we are interested in constructing an attractive ellipsoid

$$\mathcal{E}(P) = \left\{ x \in \mathbb{R}^n : x^\mathsf{T} E^T P^{-1} E x \leq 1 \right\},$$
$$P = P^T \geq 0, P \in \mathbb{R}^{n \times n}, \tag{7.5}$$

of minimal size (in a suitable sense). This requirement can be formalized as a specific minimization problem subject to the characteristic parameters of $\mathcal{E}(P)$. Here we will define P in such an "optimal" way. The associated constrained minimization problem will incorporate some natural additional restrictions for P and for the gain matrix K.

Write the matrix P^{-1} in the block form

$$P^{-1} = \begin{pmatrix} H_{11} & H_{12} \\ H_{12}^\mathsf{T} & H_{22} \end{pmatrix},$$

where $H_{11} \in \mathbb{R}^{n_1 \times n_1}$, $H_{12} \in \mathbb{R}^{n_1 \times n_2}$, and $H_{22} \in \mathbb{R}^{n_2 \times n_2}$.

Since

$$E^\mathsf{T} P^{-1} E = \begin{pmatrix} H_{11} & 0 \\ 0 & 0 \end{pmatrix},$$

this "minimality property" can be formalized by the following optimization problem.

Problem 7.1. *Our problem of the interest is*

$$minimize \ \mathrm{tr}\{H_{11}^{-1}\}$$

$$\tag{7.6}$$

subject to $P \in \mathbb{R}^{n \times n}, P = P^T > 0, K \in \Upsilon$,

where $\Upsilon \subset \mathbb{R}^{m \times n}$ is the set of matrices under which we may ensure the invariance of the attractive ellipsoid $\mathcal{E}(P)$.

The basic optimization problem (7.6) determines a suitable control function of linear feedback type that guarantees attractivity and the invariance of the designated attractive ellipsoid.

Next, we evidently need to give a constructive characterization of the set Υ and propose a method for the computational treatment of the optimization problem (7.6). The above-mentioned concrete specification of the set Υ will be carried out (in the following sections) by means of a Lyapunov-like analysis using the technique of linear matrix inequalities (LMIs).

7.3　Attractive Ellipsoid for Implicit Systems

As mentioned above, the basic results from Lyapunov set stability theory cannot be applied directly to the implicit nonlinear system (7.2). Moreover, the analogous theory for general implicit systems (see, e.g., Dai 1989; Kunkel & Mehrmann 2006; Rheinbolt 1981, 1988; Reis & Stykel 2011) is quite sophisticated in the context of practical implementations.

Our aim is to show that under some additional assumptions the combined function $V(x(\cdot))$, where $x(\cdot)$ is a solution to (7.2) for a suitable gain matrix K, possesses the properties of a Lyapunov type function.

$$V : \mathbb{R}^n \to \mathbb{R}_+$$

associated with the original implicit system (7.2):

$$V(x) := x^\mathsf{T} E^T P^{-1} E x, \quad P = P^\mathsf{T}, \quad P > 0.$$

Note that the symmetric positive definite matrix P here is the same matrix as in (7.6).

7.3.1　Descriptive Method Application

Consider the above-defined energetic function $V(x(t))$ evaluated around the trajectories of the initial system (7.2). We use the additional notation

$$\sigma(t, x) := f(t, x) - Ax.$$

We are now ready to formulate the main theoretical result of this chapter.

Theorem 7.2. *Let*

$$\tau_1 > 0, \tau_2 > 0.$$

Then the time derivative of $V(x(\cdot))$ satisfies the following inequality:

$$\dot{V}(x(t)) + \tau_1 V(x(t)) - \delta\tau_2 \leq z^{\mathsf{T}}(t)\tilde{W}(P, K, \alpha)z(t),$$

where

$$z(t) = (x^{\mathsf{T}}(t), \sigma(t, x(t)))^{\mathsf{T}}$$

is the extended state vector, $P \in \mathbb{R}^{n \times n} : P = P^{\mathsf{T}} > 0, i = 1, 2,$ and $\tilde{W}(P, K, \tau_1, \tau_2)$ is the matrix function

$$\tilde{W} := \begin{pmatrix} \tilde{W}_{11} & \tilde{W}_{12} \\ \tilde{W}_{12}^{\mathsf{T}} & \tilde{W}_{22} \end{pmatrix}$$

with

$$\tilde{W}_{11} := \Psi(A + BK) + (A + BK)^{\mathsf{T}}\Psi^{\mathsf{T}} + \tau_1 E^{\mathsf{T}}P^{-1}E + \tau_2 Q_x,$$

$$\tilde{W}_{12} := \Psi, \qquad \tilde{W}_{22} := -\tau_2 Q_f,$$

where $\Psi = E^{\mathsf{T}}P^{-1} + P^{-1}E^{\perp}$.

Proof. Using the idea of the "descriptor method" (see Fridman 2006, 2010 for details), we can represent $\dot{V}(x(t))$ as follows:

$$\dot{V}(x(t)) = 2x^{\mathsf{T}}(t)E^{\mathsf{T}}P^{-1}(A + BK)x(t) + 2x^{\mathsf{T}}(t)E^{\mathsf{T}}P\sigma(t, x(t))$$
$$+ 2x^{T}(t)P^{-1}E^{\perp}(-E\dot{x}(t) + (A + BK)x(t) + \sigma(t, x(t))),$$

where the last term is an effective zero.

Using the boundedness conditions from Sect. 7.2,

$$\tau_2 \left(\sigma^{\mathsf{T}}Q_f\sigma - \delta - \|x\|_{Q_x}^2\right) \leq 0, \quad \forall\tau_2 > 0,$$

and taking into account $E^{\mathsf{T}}P^{-1} + P^{-1}E^{\perp} = \Psi$, we get

$$\dot{V}(x(t)) + \tau_1 V(x(t)) - \delta\tau_2 \leq z^{\mathsf{T}}(t)\tilde{W}(P, K, \tau_1, \tau_2)z(t). \qquad \blacksquare$$

Using Theorem 7.2, we are now able to characterize the abstract constraint set Γ in the main optimization problem (7.6) in a constructive way. Recall that

$$\Gamma \subset \mathbb{R}^{n \times n} \times \mathbb{R}^{m \times n}$$

is the set of the admissible gain matrices that guarantee the attractivity of the ellipsoid $\mathcal{E}(P)$. Therefore, the main problem (7.6) has now the following form:

$$\text{minimize } \operatorname{tr}\{H_{11}^{-1}\}$$

$$\text{subject to } P = P^{\mathsf{T}} > 0,$$

$$\tilde{W}(P, K, \alpha) \le 0, \tag{7.7}$$

$$\tau_1 > 0, \quad \tau_2 > 0, \quad \tau_1 \ge \delta \tau_2,$$

with a symmetric block matrix $\tilde{W}(P, K, \alpha)$. Let us note that due to the nonlinear structure of $\tilde{W}(P, K, \alpha)$, the negative definiteness of this matrix function constitutes a numerically sophisticated question.

7.3.2   Reduction of Nonlinear Matrix Inequalities to LMIs

The nonlinear nature of the obtained multidimensional inequality in (7.7) evidently involves a possible relaxation approach. We next replace the above bilinear matrix inequality with an auxiliary system of linear matrix inequalities (LMIs). This expected relaxed system will provide a basis for the further numerical treatment of the main optimization problem (7.7).

In this case,

$$\Psi = E^{\mathsf{T}} P^{-1} + P^{-1} E^{\perp} = \begin{pmatrix} H_{11} & 2H_{12} \\ 0 & H_{22} \end{pmatrix}.$$

If $P > 0$, then $P^{-1} > 0$, $H_{11} > 0$, and $H_{22} > 0$. The inverse matrix to Ψ can be represented as follows:

$$\Psi^{-1} = \begin{pmatrix} H_{11}^{-1} & -2H_{11}^{-1} H_{12} H_{22}^{-1} \\ 0 & H_{22}^{-1} \end{pmatrix}.$$

We introduce the symmetric matrix function

$$W(X, Y, \tau_1, \tau_2) := M \tilde{W}(P, K, \tau_1, \tau_2) M^{\mathsf{T}},$$

where

$$X = \begin{pmatrix} X_{11} & X_{12} \\ 0 & X_{22} \end{pmatrix} := \Psi^{-1} \text{ and } Y := K X^{\mathsf{T}} \in \mathbb{R}^{m \times n},$$

$$X_{11} \in \mathbb{R}^{n_1 \times n_1}, \quad X_{11} = X_{11}^{\mathsf{T}} > 0,$$
$$X_{22} \in \mathbb{R}^{n_2 \times n_2}, \quad X_{22} = X_{22}^{\mathsf{T}} > 0,$$
$$X_{12} \in \mathbb{R}^{n_1 \times n_2},$$

are artificial variables, and

$$M := \mathrm{diag}(\Psi^{-1}, I_n) \in \mathbb{R}^{2n \times 2n}$$

is an auxiliary matrix. Taking into account

$$\Psi^{-1} E^\mathsf{T} P^{-1} E \Psi^{-\mathsf{T}} = \Psi^{-1} \begin{pmatrix} H_{11} & 0 \\ 0 & 0 \end{pmatrix} \Psi^{-\mathsf{T}} = \begin{pmatrix} X_{11} & 0 \\ 0 & 0 \end{pmatrix},$$

we see that the componentwise definition of $W(P, Y, \tau_1, \tau_2)$ can be given by the corresponding expressions

$$W_{11} = A X^\mathsf{T} + X A^\mathsf{T} + B Y + Y^\mathsf{T} B^\mathsf{T} + \tau_1 \begin{pmatrix} X_{11} & 0 \\ 0 & 0 \end{pmatrix} + \tau_2 X Q_x X^\mathsf{T},$$

$$W_{12} = I.$$

Applying the Schur complement to the inequality $W \le 0$, we derive $\mathcal{W} \le 0$ with

$$\mathcal{W} := \begin{pmatrix} A X^\mathsf{T} + X A^\mathsf{T} + B Y + Y^\mathsf{T} B^\mathsf{T} + \alpha_1 \begin{pmatrix} X_{11} & 0 \\ 0 & 0 \end{pmatrix} + \alpha_2 Q_f^{-1} & X \\ X^\mathsf{T} & -\alpha_2 Q_x^{-1} \end{pmatrix},$$

where

$$\alpha_1 = \tau_1 \quad \text{and} \quad \alpha_2 = \frac{1}{\tau_2}.$$

Using the Schur complement, it can easily be shown that the inequality $\mathcal{W} \le 0$ implics $W \le 0$ and consequently, $\tilde{W} \le 0$.

The main optimization problem (7.6) now obtains the equivalent form

$$\text{minimize } \mathrm{tr}\{X_{11}\}$$

$$\text{subject to } \mathcal{W}(X, Y, \alpha_1, \alpha_2) \le 0,$$
$$X = \begin{pmatrix} X_{11} & X_{12} \\ 0 & X_{22} \end{pmatrix} \text{ and } Y \in \mathbb{R}^{m \times n},$$
$$X_{11} \in \mathbb{R}^{n_1 \times n_1}, \quad X_{11} = X_{11}^\mathsf{T} > 0,$$
$$X_{22} \in \mathbb{R}^{n_2 \times n_2}, \quad X_{22} = X_{22}^\mathsf{T} > 0,$$
$$X_{12} \in \mathbb{R}^{n_1 \times n_2},$$
$$\alpha_1 > 0 \quad \alpha_2 > 0 \quad \alpha_1 \alpha_2 \ge \delta.$$
\tag{7.8}

Theorem 7.3. *Assume that the relaxed optimization problem (7.8) has an optimal solution* $(X^{\mathrm{opt}}, Y^{\mathrm{opt}}, \alpha_1^{\mathrm{opt}}, \alpha_2^{\mathrm{opt}})$. *Then the 5-tuple* $(P^{\mathrm{opt}}, K^{\mathrm{opt}}, \tau_1^{\mathrm{opt}}, \tau_2^{\mathrm{opt}})$ *is also an optimal solution to the initial nonlinearly constrained problem (7.7) with*

$$K^{\mathrm{opt}} := Y^{\mathrm{opt}} \left(X^{\mathrm{opt}} \right)^{-1},$$

$$P^{\mathrm{opt}} = \begin{pmatrix} (X_{11}^{\mathrm{opt}})^{-1} & -\tfrac{1}{2}(X_{11}^{\mathrm{opt}})^{-1} X_{12}^{\mathrm{opt}} (X_{22}^{\mathrm{opt}})^{-1} \\ -\tfrac{1}{2}(X_{22}^{\mathrm{opt}})^{-1} (X_{12}^{\mathrm{opt}})^{\mathsf{T}} (X_{11}^{\mathrm{opt}})^{-1} & (X_{22}^{\mathrm{opt}})^{-1} \end{pmatrix}^{-1},$$

and $\tau_1 = \alpha_1$ *and* $\tau_2 = \frac{1}{\alpha_2}$.

The last result is a direct consequence of the one-to-one transformations (introduced above).

Therefore, the formal relaxation of the initial problem does not really extend the solution set of the original problem (7.7). The obtained equivalent optimization problem (7.8) with linear-type restrictions (LMIs) provides an analytic basis for the implementable numerical algorithms associated with the given problem (7.6) initially formulated in abstract form. Recall that (7.6) determines the zone-stabilizing linear feedback control strategy for the implicit dynamical system (7.2). This control design is based on the minimal-size attractive ellipsoid $\mathcal{E}(P)$ and is in fact defined by the selection of the optimal gain matrix K^{opt}.

Example 7.1. Consider a system of the type (7.1) with the following concrete parameters:

$$E = \begin{pmatrix} 1 & 0 & 0 \\ 0 & 1 & 0 \\ 0 & 0 & 0 \end{pmatrix}, \quad f(t,x) = \begin{pmatrix} x_2 + \frac{\sin^2(tx_1)x_3 + 0.1\sin(at)}{\sqrt{20}} \\ \frac{\cos(tx_1)x_2}{\sqrt{10}} + x_3 \\ ax_1 - 2x_2 - x_3 \end{pmatrix}, \quad B = \begin{pmatrix} 0 \\ 0 \\ 1 \end{pmatrix},$$

where $a \in \mathbb{R}$ is an unknown value such that $|a| \leq 1/\sqrt{10}$. Selecting

$$A = \begin{pmatrix} 0 & 1 & 0 \\ 0 & 0 & 1 \\ 0 & -2 & -1 \end{pmatrix},$$

we evidently have the following simple estimates that correspond to our main assumptions (see Section 7.2):

$$1000 \| f(t,x) - Ax \|^2 \leq 1 + 100(x_1^2 + x_2^2 + x_3^2).$$

Solving the optimization problem (7.8) under additional the constraint $KK^{\mathsf{T}} \leq 10$ (see Proposition 3.1), we obtain

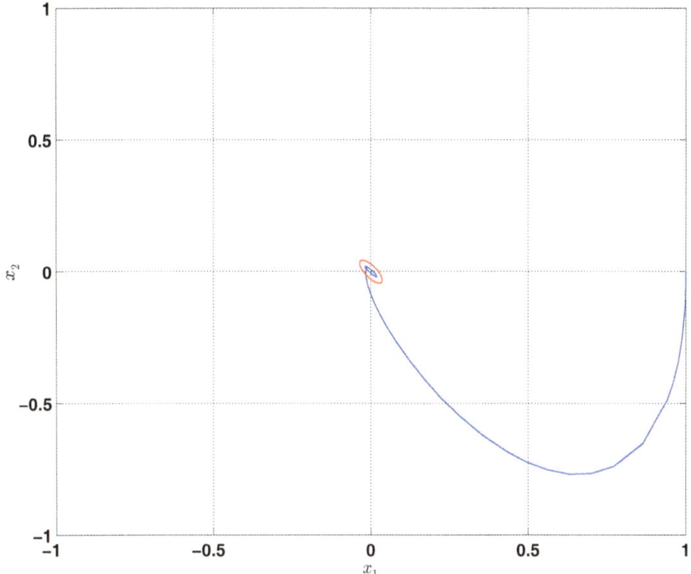

Fig. 7.1 Phase portrait of the considered nonlinear system

$$P^{\mathrm{opt}} = \begin{pmatrix} 0.0017 & -0.0017 & -0.0001 \\ -0.0017 & 0.0051 & -0.0117 \\ -0.0001 & -0.0117 & 0.0618 \end{pmatrix},$$

$$K^{\mathrm{opt}} = (-7.0943, -4.7830, -1.1341).$$

Using the obtained value K^{opt} and applying the standard MATLAB toolboxes Ode23t, Ode15s, and Ode15i, we are able to calculate the trajectory $x(\cdot)$ of the closed-loop system (7.2). The unknown parameter a is selected as 0.2.

The results of numerical simulations are displayed in Figs. 7.1 and 7.2. Figure 7.1 represents the state coordinates of the closed-loop system (7.2). The calculated optimal matrix P^{opt} determines the corresponding attractive ellipsoid of minimal size that contains the origin. Figure 7.2 shows the time evolution of the state components.

Finally, note that the applied theoretical method and corresponding computational results constitute robust behavior of the resulting system in the sense of the effective rejection of disturbances.

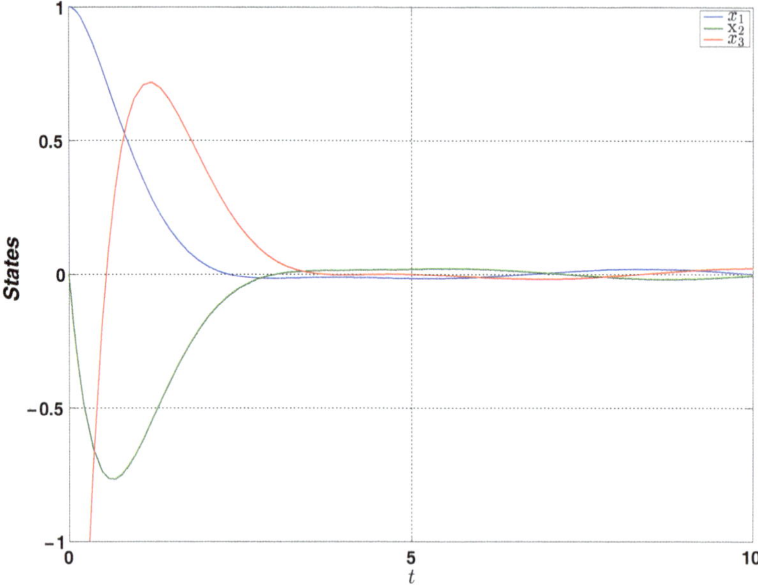

Fig. 7.2 The behavior of trajectory components x_1 and x_2

7.4   Concluding Remarks

- In this chapter, we discussed a new analytical and computational method for the robust feedback control design of nonlinear implicit dynamical systems closed by a linear control law.
- The proposed design procedure in our contribution constitutes an extension of the conventional attractive ellipsoid technique.
- From the analytical point of view, the elaborated methodology generates an admissible linear feedback control strategy that guarantees the existence and a concrete characterization of a minimal-size invariant ellipsoid associated with the system realization.
- The obtained ellipsoidal invariant set is asymptotically stable in the sense of Lyapunov.
- The generic computational part of the discussed method leads to an auxiliary nonlinear minimization problem with some nonlinear matrix constraints.
- The main numerical procedure leads to an effective relaxation scheme proposed for this initial optimization problem. This relaxation implies a consistent LMI-constrained program.
- Finally, we not only constructed an adequate robust feedback-type control law but also generated the corresponding attractive ellipsoid (the invariant set). This ellipsoidal asymptotically stable invariant set of the given closed-loop system possesses some strong minimal properties that are expressed in the

main optimization problem framework. In this way, we have incorporated the robustness or practical stability property into the resulting closed-loop dynamics.

- The effectiveness of the proposed computational schemes and the associated concrete controller design were illustrated with a computational example.

Finally, note that the main analytical idea and the associated numerical treatment can be easily applied to some alternative linear and nonlinear control processes that incorporate some natural bounded uncertainties. It seems also be possible to use the control design techniques presented here in combination with generally nonlinear feedback-type control strategies.

Chapter 8
Attractive Ellipsoids in Sliding Mode Control

Abstract In this chapter, a new sliding mode control design algorithm for a linear and a class of nonlinear quasi-Lipschitz disturbed systems is presented. It is based on the appropriate selection of a sliding surface via the invariant ellipsoid method. The designed control guarantees minimization of unmatched disturbance effects to system motions in a sliding mode. The theoretical results are verified by numerical simulations. Additionally, a methodology for the design of sliding mode controllers for linear systems subjected to matched and unmatched perturbations is proposed. It is considered that the control signal is applied through a first-order low-pass filter. The technique is based on the existence of an attracting (invariant) ellipsoid such that the convergence to a quasiminimal region of the origin using the suboptimal control signal is guaranteed. The design procedure is given in terms of the solution of a set of matrix inequalities. Benchmark examples illustrating the design are given.

Keywords Sliding mode control • Unmatched perturbations • Disturbance reduction

In this chapter, a new sliding mode control design algorithm for a linear and a class of nonlinear quasi-Lipschitz disturbed systems is presented. It is based on the appropriate selection of a sliding surface via the invariant ellipsoid method. The designed control guarantees minimization of unmatched disturbance effects to system motions in a sliding mode. The theoretical results are verified by numerical simulations. Additionally, a methodology for the design of sliding mode controllers for linear systems subjected to matched and unmatched perturbations is proposed. It is considered that the control signal is applied through a first-order low-pass filter. The technique is based on the existence of an attracting (invariant) ellipsoid such that the convergence to a quasiminimal region of the origin using the suboptimal control signal is guaranteed. The design procedure is given in terms of the solution of a set of matrix inequalities. Benchmark examples illustrating the design are given. The last section of this chapter extends the attractive ellipsoid technique to time-delay control systems with predictor-based feedback. It presents a method of minimizing the effects produced by disturbances and time delays in control input.

© Springer International Publishing Switzerland 2014

A. Poznyak et al., *Attractive Ellipsoids in Robust Control*, Systems & Control: Foundations & Applications, DOI 10.1007/978-3-319-09210-2_8

8.1 Minimization of Unmatched Uncertainties Effect in Sliding Mode Control

This section extends some fundamental principles of the attractive ellipsoid method to a class of systems with sliding modes. It addresses the problem of minimization of unmatched uncertainties and disturbances in sliding mode control systems and presents a control design scheme in accordance with the concept of attractive ellipsoids.

8.1.1 Problem Statement

Consider the uncertain nonlinear controlled system

$$\dot{x}(t) = Ax(t) + Bu(t) + Df(t, x), \tag{8.1}$$

where

- $x \in \mathbb{R}^n$ is the system state vector,
- $A \in \mathbb{R}^{n \times n}$ is the system matrix,
- $u \in \mathbb{R}^m$ is the vector of the control inputs,
- $B \in \mathbb{R}^{n \times m}$ is the matrix of control gains, $\mathrm{rank}(B) = m$,
- the matrix $D \in \mathbb{R}^{n \times k}$ describes the gains of the disturbances, and
- the vector function $f : \mathbb{R}_+ \times \mathbb{R}^n \to \mathbb{R}^k$ characterizing the external system disturbances is assumed to be bounded as

$$f^T Q_f f \leq f_0 + x^T Q_x x, \tag{8.2}$$

where positive definite matrices Q_f, Q_x and the positive constant f_0 are given. In fact, the nonlinearity f may be considered a quasi-Lipschitz function from the class $C(0, c_0, c_1)$ (see Definition 2.1 in Chap. 2).

Suppose that

- the state space vector x is completely measurable;
- the pair $\{A, B\}$ is controllable;
- we deal with the stabilization problem of the system (8.1), where the *sliding mode controller* (Edwards & Spurgeon 1998)

$$u(t) = -\left(\tilde{C}B\right)^{-1}\tilde{C}Ax(t) - M(x(t))\mathrm{Sign}\left(\tilde{C}x(t)\right), \quad M(x) > 0, \tag{8.3}$$

is applied. Here

$$\mathrm{Sign}(\sigma) := \left(\mathrm{sign}(\sigma_1), \mathrm{sign}(\sigma_2), \ldots, \mathrm{sign}(\sigma_m)\right)^T,$$

$$\text{sign}(\sigma_i) := \begin{cases} 1, & \sigma_i > 0, \\ -1, & \sigma_i < 0, \\ \in [-1, 1], & \sigma_i = 0, \end{cases}$$

and the matrix $\tilde{C} \in \mathbb{R}^{m \times n}$ defines a sliding surface such that

$$\det(\tilde{C} B) \neq 0$$

and the positive control gain function $M(x)$ has the form

$$M(x) = \sqrt{\alpha + x^T R x}, \quad \alpha \in \mathbb{R}, \ R \in \mathbb{R}^{n \times n},$$

where the number α and the positive semidefinite matrix R are bounded control parameters, namely

$$0 < \alpha_{\min} \leq \alpha \leq \alpha_{\max} \text{ and } 0 \leq \|R\| \leq \beta. \tag{8.4}$$

Observe that the sliding mode control gain form and the restrictions on system disturbances are similar to conventional ones (Edwards & Spurgeon 1998; Utkin, Guldner, & Shi 1999):

$$M(x) = \mu_0 + \mu_1 \|x\|$$

and

$$\|f\| \leq f_0 + f_1 \|x\|.$$

Here we just use weighted Euclidean norms.

We assume that the matching condition does not hold.[1] In this case, the system (8.1) may have so-called *unmatched disturbances*, which cannot be suppressed by any sliding mode control acting only within the region range(B).

So the main problem with which we are dealing here is to design a sliding mode control of the form (8.3) that minimizes the effects of unmatched uncertainties. The minimization concept will be based on the attractive ellipsoid method.

In previous chapters, the method of linear matrix inequalities (LMIs) has been used for finding an attractive ellipsoid of systems with linear controls. Since the sliding mode control design consists largely in the selection of an appropriate *linear* sliding surface

$$\tilde{C} x = 0,$$

we may expect that similar LMIs will appear in this case.

---

[1] The matching condition for the system (8.1) has the form $Df \in \text{range}(B)$.

8.1.2 LMI-Based Sliding Mode Control Design

Since $\text{rank}(B) = m$, the matrix B can be partitioned (perhaps after reordering the state vector components) as

$$B = \begin{pmatrix} B_1 \\ B_2 \end{pmatrix},$$

where

$$B_1 \in \mathbb{R}^{(n-m)\times m}, \ B_2 \in \mathbb{R}^{m\times m}$$

with

$$\det(B_2) \neq 0.$$

In this case, the nonsingular coordinate transformation

$$\begin{pmatrix} x_1 \\ x_2 \end{pmatrix} = Gx, \ \text{where } G = \begin{pmatrix} I_{n-m} & -B_1 B_2^{-1} \\ 0 & B_2^{-1} \end{pmatrix},$$

reduces the system (8.1) to the regular form (Utkin et al. 1999)

$$\begin{cases} \dot{x}_1 = A_{11}x_1 + A_{12}x_2 + D_1 f, \\[2mm] \dot{x}_2 = A_{21}x_1 + A_{22}x_2 + u(t) + D_2 f, \end{cases} \tag{8.5}$$

where

$$x_1 \in \mathbb{R}^{n-m}, \ x_2 \in \mathbb{R}^m$$

are blocks of the system state vector, and

$$A_{11} \in \mathbb{R}^{(n-m)\times(n-m)}, \ A_{12} \in \mathbb{R}^{(n-m)\times m}, \ A_{21} \in \mathbb{R}^{m\times(n-m)}, \ A_{22} \in \mathbb{R}^{m\times m}$$

are blocks of the system matrix

$$\begin{pmatrix} A_{11} & A_{12} \\ A_{21} & A_{22} \end{pmatrix} = GAG^{-1} \ \text{and} \ \begin{pmatrix} D_1 \\ D_2 \end{pmatrix} = GD,$$

where

$$D_1 \in \mathbb{R}^{(n-m)\times k}, \ D_2 \in \mathbb{R}^{m\times k}.$$

Observe that if the system (8.1) is controllable, then the pair $\{A_{11}, A_{12}\}$ is controllable as well (Utkin et al. 1999). We have the following result.

Lemma 8.1. *If the number $\alpha \in \mathbb{R}$, the matrix $R \in \mathbb{R}^{n \times n}$, and the matrix $C \in \mathbb{R}^{m \times (n-m)}$ satisfy the system of LMIs*

$$\begin{pmatrix} \dfrac{\alpha}{f_0} Q_f & D^T G^T (C \;\vdots\; I_m)^T \\ (C \quad I) G D & I_m \end{pmatrix} > 0, \tag{8.6}$$

$$\begin{pmatrix} \beta^2 I_n & R \\ R & I_n \end{pmatrix} \geq 0, \quad R \geq \frac{\alpha}{f_0} Q_x, \tag{8.7}$$

$$\alpha_{\min} \leq \alpha \leq \alpha_{\max}, \tag{8.8}$$

then the hyperplane

$$\tilde{C} x = 0$$

with

$$\tilde{C} = (C \;\vdots\; I_m) G$$

*is the **sliding manifold** for the system (8.1) with a controller of the form (8.3)–(8.4).*

Proof. Observe that for $\tilde{C} = (C \;\vdots\; I_m) G$, we have $\tilde{C} B = I_m$. Consider the following Lyapunov function candidate:

$$V(s) = \frac{1}{2} s^T s,$$

where $s = \tilde{C} x$. Its total time derivative along the trajectories of the system (8.1) is given by

$$\begin{aligned} \dot{V}(s) &= s^T \tilde{C} \dot{x} = s^T \tilde{C} A x(t) + s^T \tilde{C} B u \\ &+ s^T \tilde{C} D f = s^T \tilde{C} D f - M(x) |s|, \end{aligned}$$

where

$$|s| = \sum_{i=1}^{m} |s_i|.$$

Obviously, the inequality

$$\left\| \tilde{C} D f \right\| < M(x) \quad \text{for all } x \in \mathbb{R}, \tag{8.9}$$

in view of $\|s\| \leq |s|$, guarantees

$$\dot{V} \leq -k|s| \leq -\sqrt{2} k \sqrt{V} \quad \text{for some } k > 0,$$

implying the existence of the sliding surface $s = 0$.

The inequality (8.9) can be rewritten as

$$f^T D^T \tilde{C}^T \tilde{C} D f < \alpha + x^T R x.$$

Hence (8.3) implies (8.9) if

$$\frac{1}{\alpha} R > \frac{1}{f_0} Q_x \quad \text{and} \quad \frac{1}{f_0} Q_f > \frac{1}{\alpha} D^T \tilde{C}^T \tilde{C} D.$$

The Schur complement applied to the last matrix inequality gives (8.6). Finally, the same transformation helps us to rewrite the inequality $\|R\| < \beta$ in the form (8.7). ∎

Note that for sufficiently large β and $\alpha_{\max}$, the system of matrix inequalities (8.6)–(8.8) is feasible.

8.1.3 Optimal Sliding Surface

When the sliding mode appears in the hyperplane

$$\tilde{C} x = 0,$$

the sliding motion equation takes the form

$$\dot{x}_1 = (A_{11} - A_{12} C) x_1 + D_1 f(t, x), \tag{8.10}$$

where

$$x \in \mathbb{R}^n : \tilde{C} x = C x_1 + x_2 = 0.$$

The following lemma explains how to find the invariant ellipsoid for sliding motion governed by Eq. (8.10).

Lemma 8.2. *If the tuple* $(\tau_1, \bar{\tau}_2, C, P)$, *where*

$$\tau_1, \bar{\tau}_2 \in \mathbb{R}, \ P \in \mathbb{R}^{(n-m)\times(n-m)}, \ C \in \mathbb{R}^{m\times(n-m)},$$

is a solution of the system of matrix inequalities

$$\begin{pmatrix} \begin{array}{c} P(A_{11} - A_{12}C)^T + (A_{11} - A_{12}C)P \\ + \tau_1 P + \bar{\tau}_2 D_1 Q_f^{-1} D_1 \\ \begin{pmatrix} P \\ -CP \end{pmatrix} \end{array} & \begin{array}{c} \begin{pmatrix} P \vdots - PC^T \end{pmatrix} \\ \\ -\bar{\tau}_2 G Q_x^{-1} G^T \end{array} \end{pmatrix} \le 0, \qquad (8.11)$$

$$\tau_1 \bar{\tau}_2 \ge f_0, \ P > 0, \qquad (8.12)$$

then $\varepsilon(P)$ *is the invariant and attractive ellipsoid of the system (8.10).*

Proof. Consider the quadratic function

$$V(x_1) = x_1^T P^{-1} x_1, \ P > 0.$$

The total derivative of this function calculated along the trajectories of the system (8.10) has the form

$$\dot{V}(x_1) = x_1^T \left([A_{11} - A_{12} C]^T P^{-1} + P^{-1}[A_{11} - A_{12} C]\right)x_1 + 2f^T D_1^T P^{-1} x_1 =$$

$$\begin{pmatrix} x_1 \\ f \end{pmatrix}^T \begin{pmatrix} (A_{11} - A_{12} C)^T P^{-1} + P^{-1}(A_{11} - A_{12} C) & P^{-1} D_1 \\ D_1^T P^{-1} & 0 \end{pmatrix} \begin{pmatrix} x_1 \\ f \end{pmatrix}.$$

Obviously, the ellipsoid $\varepsilon(P)$ is state-invariant if and only if the condition

$$\dot{V}(x_1) \le 0$$

holds for every

$$x_1 \in \mathbb{R}^{n-m} : x_1^T P^{-1} x_1 \ge 1$$

and for every uncertain vector function f satisfying (8.2).

Let us rewrite the restriction (8.2) under the assumption that the sliding mode in the surface

$$\tilde{C}x = (C \vdots I)Gx = 0$$

appears. In this case,

$$x_2 = -Cx_1$$

and

$$\begin{pmatrix} x \\ f \end{pmatrix}^T \begin{pmatrix} -Q_x & 0 \\ 0 & Q_f \end{pmatrix} \begin{pmatrix} x \\ f \end{pmatrix} =$$

$$\begin{pmatrix} x_1 \\ x_2 \\ f \end{pmatrix}^T \begin{pmatrix} -(G^{-1})^T Q_x G^{-1} & 0 \\ 0 & Q_f \end{pmatrix} \begin{pmatrix} x_1 \\ x_2 \\ f \end{pmatrix} =$$

$$\begin{pmatrix} x_1 \\ f \end{pmatrix}^T \begin{pmatrix} -\begin{pmatrix} I_{n-m} \\ -C \end{pmatrix}^T G^{-T} Q_x G^{-1} \begin{pmatrix} I_{n-m} \\ -C \end{pmatrix} & 0 \\ 0 & Q_f \end{pmatrix} \begin{pmatrix} x_1 \\ f \end{pmatrix} \leq f_0.$$

We define

$$A_0 := \begin{pmatrix} (A_{11} - A_{12}C)^T P^{-1} + P^{-1}(A_{11} - A_{12}C) & P^{-1}D_1 \\ D_1^T P^{-1} & 0 \end{pmatrix},$$

$$A_1 := \begin{pmatrix} -P^{-1} & 0 \\ 0 & 0 \end{pmatrix},$$

$$A_2 := \begin{pmatrix} -\begin{pmatrix} I_{n-m} \\ -C \end{pmatrix}^T G^{-T} Q_x G^{-1} \begin{pmatrix} I_{n-m} \\ -C \end{pmatrix} & 0 \\ 0 & Q_f \end{pmatrix},$$

and

$$g_i(y) := y^T A_i y \quad \text{for } y \in R^{n-m+k}.$$

According to the S-procedure, the inequalities

$$g_1(y) \leq \alpha_1 = -1$$

and

$$g_2(y) \leq \alpha_2 = f_0$$

imply

$$g_0(y) \leq \alpha_0 = 0$$

if and only if there exist $\tau_1, \tau_2 \geq 0$ such that

$$A_0 \leq \tau_1 A_1 + \tau_2 A_2 \quad \text{and} \quad \alpha_0 \geq \alpha_1 \tau_1 + \alpha_2 \tau_2.$$

Hence, using the Schur's complement, we derive

$$
\begin{pmatrix}
\begin{array}{c}(A_{11} - A_{12}\,C)^T\,P^{-1} + \\ P^{-1}\,(A_{11} - A_{12}\,C) + \tau_1\,P^{-1}\end{array} & P^{-1}\,D_1 & \left(I_{n\text{-}m} \;\vdots\; -\,C^T\right) \\[2mm]
D_1^T\,P^{-1} & -\,\tau_2\,Q_f & 0 \\[2mm]
\begin{pmatrix}I_{n\text{-}m}\\ -C\end{pmatrix} & 0 & -\frac{1}{\tau_2}\,G\,Q_x^{-1}\,G^T
\end{pmatrix} < 0.
$$

Multiplying the matrix from the last inequality on both sides by

$$
\begin{pmatrix} P & 0 \\ 0 & I_{n+k} \end{pmatrix}
$$

and applying Schur's complement again, we obtain (8.11)–(8.12) with

$$
\overline{\tau}_2 := 1/\tau_2. \qquad\blacksquare
$$

Finally, we can formulate the main result for the considered system with sliding mode control.

Theorem 8.1. *If the tuple* $(\alpha, \tau_1, \overline{\tau}_2, R, C, P, Z)$ *is a solution of the optimization problem*

$$
\mathrm{tr}(Z) \to \min
$$

subject to (8.6)–(8.8), (8.11), (8.12), and

$$
\begin{pmatrix}
P & \left(P \;\vdots\; -\,PC^T \right) \\[2mm]
\begin{pmatrix} P \\ -CP \end{pmatrix} & GZG^T
\end{pmatrix} \geq 0,
\tag{8.13}
$$

$$
Z \in \mathbb{R}^{n\times n},\ Z > 0,
$$

then the corresponding controller of the form (8.3)–(8.4) guarantees that every trajectory of the system (8.1) converges to a "minimal" attractive ellipsoid $\varepsilon(Z)$.

Proof. To avoid consideration of the sliding motion equation with respect to the variable x_2, let us use the sliding mode property

$$
x_2 = -C x_1.
$$

Then each ellipsoid $\varepsilon(Z)$ (where $Z \in \mathbb{R}^{n\times n}$ is some positive definite matrix) for the variable x can be represented in the form

$$x^T Z^{-1} x = x_1^T (I \;\vdots\; - C^T)(G^{-1})^T Z^{-1} G^{-1} \begin{pmatrix} I \\ -C \end{pmatrix} x_1 < 1.$$

From Lemma 8.2, it follows that the attractive ellipsoid for the variable x_1 is $\varepsilon(P)$. Hence if Z satisfies

$$(I \;\vdots\; - C^T)(G^{-1})^T Z^{-1} G^{-1} \begin{pmatrix} I \\ -C \end{pmatrix} \leq P^{-1},$$

then $\varepsilon(Z)$ is the *attractive* ellipsoid of the system (8.1). Finally, Schur's complement together with the equivalent transformation

$$T = \begin{pmatrix} P & 0 \\ 0 & I_n \end{pmatrix}$$

applied to the resulting matrix inequality yields the inequality (8.13).   ∎

The previous theorem gives a theoretical answer to the question how to design an "optimal" (in the ellipsoidal sense) sliding mode controller. Unfortunately, the result presented allows us to find only a suboptimal invariant ellipsoid, since Lemma 8.1 gives only a sufficient sliding mode existence condition based on the quadratic Lyapunov function $V(s) = \frac{1}{2} s^T s$ (see Utkin et al. 1999 for details). To find an optimal invariant ellipsoid, we need to use a condition that is necessary and sufficient for sliding mode existence in the system (8.1). But this is an extremely hard task under the given assumptions.

8.1.4   Numerical Aspects of Sliding Surface Design

Our main theoretical result, namely Theorem 8.1, is related to the optimization problem with a linear functional and bilinear constraints. A formal resolution of such a problem is a nontrivial task (Toker & Ozbay 1995). To simplify the situation, we rewrite all the constraints in the form of a two-parametric family of LMIs. Let us set

$$Y := CP.$$

In this case, the inequalities (8.11) and (8.13) can be rewritten in the form

$$\begin{pmatrix} \begin{matrix} PA_{11}^T + A_{11}P - Y^T A_{12}^T - A_{12}Y \\ + \tau P + \overline{\tau}_2 D_1 Q_f^{-1} D_1^T \end{matrix} & (P \;\vdots\; - Y^T) \\ \begin{pmatrix} P \\ -Y \end{pmatrix} & -\overline{\tau}_2 Q_x^{-1} \end{pmatrix} \leq 0, \qquad (8.14)$$

$$\begin{pmatrix} P & \begin{pmatrix} P \vdots -Y^T \end{pmatrix} \\ \begin{pmatrix} P \\ -Y \end{pmatrix} & GZG^T \end{pmatrix} \geq 0. \tag{8.15}$$

For every C and $P > 0$, one can find a number $\delta > 0$ and a matrix $L \in \mathbb{R}^{m \times m}$: $L > 0$ such that the upper estimate

$$\begin{pmatrix} \frac{1}{\delta}P^{-1} & 0 \\ 0 & L^{-1} \end{pmatrix} \geq \begin{pmatrix} C^T \\ I_m \end{pmatrix} (C \vdots I_m) \tag{8.16}$$

holds. Consider δ and L as the new variables. In this case, the inequalities

$$\begin{pmatrix} \frac{\alpha}{f_0} Q_f & D^T G^T \\ GD & \begin{pmatrix} \delta P & 0 \\ 0 & L \end{pmatrix} \end{pmatrix} > 0 \tag{8.17}$$

and (8.16) imply (8.6).

Using the equivalent transformation $T_2 = \mathrm{diag}(P, I)$ and Schur's complement, we represent (8.16) in the form

$$\begin{pmatrix} \frac{1}{\delta}P & Y^T \\ Y & I_m - L \end{pmatrix} \geq 0. \tag{8.18}$$

We have thus proved the following theorem.

Theorem 8.2. *The solution* $(\alpha, \tau_1, \overline{\tau}_2, \delta, L, Y, P, R, Z)$ *of the minimization problem*

$$\mathrm{tr}(Z) \to \min$$

subject to the inequalities (8.7), (8.8), (8.14), (8.15), (8.17), and (8.18) with

$$\tau \geq 0, \ P > 0, \ L > 0, \ \delta > 0$$

provides the "quasiminimal" invariant ellipsoid $\varepsilon(Z^{-1})$ *for the system (8.1) with the control (8.3)–(8.4), where*

$$\tilde{C} = (YP^{-1} \vdots I_m)G.$$

For fixed parameters τ and δ, the system of the matrix inequalities presented becomes a linear form that can be easily solved using MATLAB toolboxes SeDuMi and YALMIP. To find the optimal solution numerically, we denote the minimal value of the functional $\mathrm{tr}(Z)$ subject to the same constraints with fixed τ and δ by $g(\tau, \delta)$ and search for the minimum of the function $g(\tau, \delta)$ using a derivative-free method.

8.1.5 Numerical Example

Consider the following benchmark example:

$$A = \begin{pmatrix} 1 & 2 & 0.4 \\ -1.5 & 0.7 & 2 \\ 0.5 & -0.6 & 1 \end{pmatrix}, \ B = \begin{pmatrix} 0.3 \\ 0 \\ 1.2 \end{pmatrix}, \ D = \begin{pmatrix} 1 & 0 \\ 1 & -1 \\ 0 & 1 \end{pmatrix}.$$

The disturbances f are as follows:

$$f = \begin{pmatrix} 0.0028 \cos(0.4t) - 0.0879 \sin(0.4t) \\ 0.0499 \cos(0.4t) + 0.0049 \sin(0.4t) \end{pmatrix},$$

which satisfy the matrix inequality

$$f^T \begin{pmatrix} 130 & 15 \\ 15 & 400 \end{pmatrix} f < 1,$$

and $u(t)$ is the scalar sliding mode controller having the form

$$u(t) = -(\tilde{C} B)^{-1} \tilde{C} A x - 0.5 \mathrm{Sign}\left(\tilde{C} x\,(t)\right), \quad \tilde{C} \in \mathbb{R}^{1 \times 3}.$$

The numerical procedure described above gives

$$\tau = 1.5265, \ \delta = 28.08,$$

and

$$Z = \begin{pmatrix} 0.0053 & -0.0073 & 0.0089 \\ -0.0073 & 0.0103 & -0.0148 \\ 0.0089 & -0.0148 & 0.0488 \end{pmatrix},$$

$$\tilde{C} = (2.6974, 2.1442, 0.1590).$$

Figure 8.1 shows the corresponding "quasiminimal" invariant ellipsoid $\varepsilon(Z)$ and presents the convergence process of the system trajectory into this ellipsoid. Evolution of the control law is presented in Fig. 8.2.

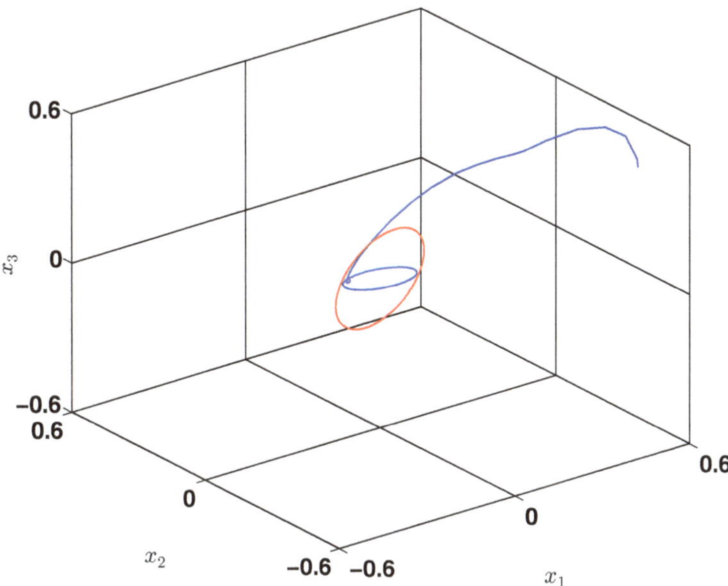

Fig. 8.1 Convergence of the system trajectory into the invariant ellipsoid

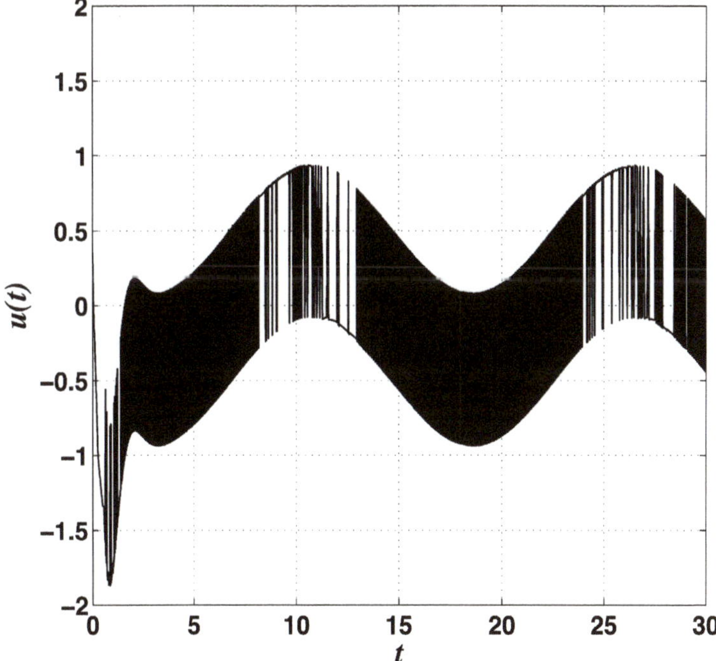

Fig. 8.2 The control law

Similar results have been obtained for "quasi-Lipschitz" disturbances. All simulation results were implemented using MATLAB with the `ode23` solver and relative tolerance of 10^{-2}.

8.2   Gain Matrix Tuning in Dynamic Actuators

8.2.1   *Problem Statement*

Following (Davila & Poznyak 2011), consider a perturbed linear system given by the equation

$$\dot{x}(t) = Ax(t) + Bu(t) + Df(t, x, u), \tag{8.19}$$

where

- $x(t) \in \mathbb{R}^n$ is the state variable,
- $u(t) \in \mathbb{R}^m$ is the control inputs vector,
- $f(t, x, u) \in \mathbb{R}^k$ is the function describing system uncertainties and disturbances,
- $A \in \mathbb{R}^{n \times n}$, $B \in \mathbb{R}^{n \times m}$, $D \in \mathbb{R}^{n \times k}$, are system matrices, assumed to be known.

The notation x, u, f will be introduced for simplicity. The differential equations are understood in the Filippov sense (Filippov 1988) in order to provide the possibility of discontinuous signals in controls. Filippov's solutions coincide with the usual solutions when the right-hand sides are continuous. It is assumed also that all considered inputs allow the existence of solutions and their extension to the whole semiaxis $t \geq 0$.

Let the sliding variable $\sigma \in \mathbb{R}^m$ be given by

$$\sigma = \left[C \vdots I_m \right] \begin{bmatrix} x \\ u \end{bmatrix} = Cx + u, \tag{8.20}$$

where $C \in \mathbb{R}^{m \times n}$. Suppose that the sliding mode controller is applied through an *actuator* whose dynamics are given by

$$H\dot{u} = -u - M(x, u)\mathrm{Sign}\,(\sigma\,(t)) + \xi(t, x, u), \tag{8.21}$$

where

- $H \in \mathbb{R}^{m \times m}$ is a diagonal matrix of the actuators' time constants,

$$H = \mathrm{diag}(h_1, h_2, \ldots, h_m),\ h_i > 0,\ i = 1, 2, \ldots, m,$$

- $M : \mathbb{R}^n \times \mathbb{R}^m \to \mathbb{R}^{m \times m}$ is a matrix-valued function of the relay control gains such that

$$M(x, u) = \mathrm{diag}(M_1(x, u), M_2(x, u), \ldots, M_m(x, u)), \ M_i : \mathbb{R}^n \times \mathbb{R}^m \to \mathbb{R},$$

- the function $\xi(t, x, u)$ describes uncertainties and disturbances of the actuator.

We assume that the uncertain functions f and ξ belong to the class of quasi-Lipschitz functions, i.e

$$f^T Q_f f \leq f_0 + x^T Q_x x + u^T Q_u u, \tag{8.22}$$

where $f_0 \in \mathbb{R}$ is a nonnegative number, and the matrices

$$Q_f \in \mathbb{R}^{k \times k}, Q_x \in \mathbb{R}^{n \times n}, Q_u \in \mathbb{R}^{m \times m}$$

are assumed to be positive definite, i.e.,

$$Q_f > 0, Q_x > 0$$

and

$$\xi^T Q_\xi \xi \leq \xi_0 + x^T Q'_x x + u^T Q'_u u,$$

where $\xi_0 \in \mathbb{R}$ is a nonnegative number and the matrices

$$Q_\xi \in \mathbb{R}^{m \times m}, Q'_x \in \mathbb{R}^{n \times n}, Q'_u \in \mathbb{R}^{m \times m}$$

are positive definite, too, i.e.,

$$Q_\xi > 0, Q'_x > 0$$

and

$$Q_u > 0.$$

The variable control gains $M_i(x, u)$ have the form

$$M_i(x, u) = \sqrt{\alpha_i + x^T R_i x + u^T K_i u}, \tag{8.23}$$

where

$$\alpha_i \in \mathbb{R}, \alpha_i > 0, R_i \in \mathbb{R}^{n \times n}, R_i > 0$$

and

$$K_i \in \mathbb{R}^{m \times m}, K_i > 0$$

are control parameters to be designated satisfying

$$0 < \alpha_i^{\min} \le \alpha_i \le \alpha_i^{\max},$$

$$0 \le ||R_i|| \le \beta_i, \quad 0 \le ||K_i|| \le \gamma_i,$$

(8.24)

where $i = 1, 2, \ldots, m$, and $||R|| := \sqrt{\lambda_{\max}(R^T R)}$ is the Euclidian matrix norm.

The aim of this section is to propose a design methodology for the control parameters α_i, R_i, K_i, C, to ensure global convergence of the state trajectories to a minimum attractive ellipsoid.

8.2.2   Controller Design

The controller will be designed in two steps. First, the gains of the controller are chosen such that the sliding variable converges to zero. In the second step, we select the sliding surface in such way that the dynamics of the closed loop system over the sliding surface tend to a smallest attractive ellipsoid. We have the following result.

Lemma 8.3. *If the matrices*

$$R_i \in \mathbb{R}^{n \times n}, K_i \in \mathbb{R}^{m \times m}, c_i \in \mathbb{R}^{1 \times n}$$

and the numbers $\alpha_i > 0$ satisfy the system of LMIs

$$\begin{bmatrix} R_i - \mu_1 Q_x \\ -\mu_2 Q'_x & 0 & 0 & 0 & \left(h_i A^\top + I_n\right) c_i^\top \\ 0 & \begin{matrix} K_i - \mu_1 Q_u \\ -\mu_2 Q'_u \end{matrix} & 0 & 0 & h_i B^\top c_i^\top \\ 0 & 0 & \mu_1 Q_f & 0 & h_i D^\top c_i^\top \\ 0 & 0 & 0 & \mu_2 Q_\xi & e_i \\ c_i(h_i A + I_n) & c_i h_i B & c_i h_i D & e_i^\top & 1 \end{bmatrix} > 0,$$

(8.25)

$$\alpha_i \ge \mu_1 f_0 + \mu_2 \xi_0, \quad \alpha_i^{\min} \le \alpha_i \le \alpha_i^{\max}, \quad \mu_1 \ge 0, \quad \mu_2 \ge 0,$$

(8.26)

and

$$\begin{bmatrix} \beta_i I_n & R_i \\ R_i & I_n \end{bmatrix} > 0, \quad \begin{bmatrix} \gamma_i I_n & K_i \\ K_i & I_n \end{bmatrix} > 0,$$

(8.27)

where

$$e_i = (0, \ldots, 0, 1, 0, \ldots, 0)^\top \in \mathbb{R}^m$$

is the identity vector, $i = 1, \ldots, m$, then the hyperplane

$$Cx + u = 0, \quad \text{where } C := [c_1^\mathsf{T}, c_2^\mathsf{T}, \ldots, c_m^\mathsf{T}]^\mathsf{T},$$

is the sliding manifold of the system (8.19)–(8.24).

Proof. Consider the quadratic function

$$V_\sigma = \frac{1}{2}\sigma^T \sigma. \tag{8.28}$$

Its time derivative is given by

$$\dot{V}_\sigma = \sigma^T (C\dot{x} + \dot{u}) = \sigma^T (CAx + CBu - H^{-1}u + CDf$$

$$+ H^{-1}\xi - H^{-1}M \operatorname{Sign}(\sigma(t))) = -\sigma^T H^{-1}\sigma +$$

$$\sigma^T (CAx + CBu + H^{-1}Cx + CDf + H^{-1}\xi - H^{-1}M \operatorname{Sign}(\sigma(t))) \le$$

$$\sum_{i=1}^{m} \sigma_i h_i^{-1}(c_i(h_i A + I_n)x + h_i c_i Bu + h_i c_i Df$$

$$+ e_i^T \xi - M_i(x, u)\operatorname{Sign}(\sigma(t))),$$

where

$$e_i = (0, \ldots, 0, 1, 0, \ldots, 0)^T \in \mathbb{R}^m$$

is the identity vector and c_i is the ith row of the matrix C. Hence the inequality $\dot{V}_\sigma < 0$ is satisfied if

$$\|c_i(h_i A + I_n)x + c_i h_i Bu + c_i h_i Df + e_i^T \xi\| < M_i(x, u),$$

or equivalently,

$$\begin{pmatrix} x \\ u \\ f \\ \xi \end{pmatrix}^T A_0 \begin{pmatrix} x \\ u \\ f \\ \xi \end{pmatrix} < \alpha_i b \tag{8.29}$$

where

$$A_0 = \begin{bmatrix} (h_i A + I_n)^T c_i^T \\ h_i B^T c_i^T \\ h_i D^T c_i^T \\ e_i \end{bmatrix} \begin{bmatrix} c_i(h_i A + I_n) \vdots c_i h_i B \vdots c_i h_i D \; e_i^T \end{bmatrix}$$

$$+ \begin{bmatrix} -R_i & 0 & 0 & 0 \\ 0 & -K_i & 0 & 0 \\ 0 & 0 & 0 & 0 \\ 0 & 0 & 0 & 0 \end{bmatrix}.$$

Taking into account the quasi-Lipschitz property of the functions f and ξ, we conclude that the inequality (8.29) must hold under the following conditions:

$$\begin{pmatrix} x \\ u \\ f \\ \xi \end{pmatrix}^T A_1 \begin{pmatrix} x \\ u \\ f \\ \xi \end{pmatrix} < f_0, \quad A_1 = \begin{bmatrix} -Q_x & 0 & 0 & 0 \\ 0 & -Q_u & 0 & 0 \\ 0 & 0 & Q_f & 0 \\ 0 & 0 & 0 & 0 \end{bmatrix},$$

and

$$\begin{pmatrix} x \\ u \\ f \\ \xi \end{pmatrix}^T A_2 \begin{pmatrix} x \\ u \\ f \\ \xi \end{pmatrix} < \xi_0, \quad A_2 = \begin{bmatrix} -Q_x' & 0 & 0 & 0 \\ 0 & -Q_u' & 0 & 0 \\ 0 & 0 & 0 & 0 \\ 0 & 0 & 0 & Q_\xi \end{bmatrix}.$$

By the S-procedure, we obtain that the required condition (8.29) is equivalent to

$$A_0 \leq \mu_1 A_1 + \mu_2 A_2 \text{ and } \alpha_i \geq \mu_1 f_0 + \mu_2 \xi_0,$$

where $\mu_1, \mu_2 \in \mathbb{R}$ are some nonnegative numbers that will be considered new variables hereinafter. The inequality referring to R_i and K_i in (8.24) can be written as

$$\beta_i^2 I_n - R_i^T R_i \geq 0, \quad \gamma_i^2 I_n - K_i^T K_i \geq 0.$$

By application of Schur's complements, the last inequalities give the LMIs of (8.27). ∎

Once the systems begins to slide on the surface, the equality $\sigma = 0$ is satisfied, which implies that the equivalent control becomes

$$u = -Cx,$$

and the initial sliding mode control problem (8.19)–(8.24) transforms into a linear one, namely (3.1)–(3.3). According to Lemmas 3.1, 3.2, and 8.3 the quasiminimal attractive ellipsoid of the system (8.19)–(8.24) can be found by resolving the optimization procedure

$$\min_{P,C,R_i,K_i,\alpha_i,\tau_1,\tau_2,\mu_1,\mu_2} \operatorname{tr}(P)$$

$$\text{subject to} \quad (8.25)-(8.27) \text{ and}$$

$$\begin{bmatrix} AP + BCP + PA^T + PC^T B^T + \tau_1 P & D \\ + \tau_2 PQ_x P + \tau_2 PC^T Q_u CP & \\ D^T & -\tau_2 Q_f \end{bmatrix} \leq 0,$$

$$\tau_1, \tau_2 \in \mathbb{R}, \tau_1 \geq 0, \tau_2 \geq 0, \tau_1 \geq f_0 \tau_2.$$

The restrictions of the optimization problem obtained have the form of bilinear matrix inequalities. Repeating the considerations of Sect. 8.1.4, we restrict them to LMIs.

First of all, using Schur's complement twice, we rewrite the matrix inequality (8.25) in the form

$$\begin{bmatrix} R_i - \mu_1 Q_x - \mu_2 Q_x' & 0 & 0 \\ 0 & K_i - \mu_1 Q_u - \mu_2 Q_u' & 0 \\ 0 & 0 & \mu_1 Q_f \end{bmatrix} >$$

$$\begin{bmatrix} h_i A^T + I_n \\ h_i B^T \\ h_i D^T \end{bmatrix} c_i^T \frac{\mu_2}{\mu_2 - e_i^T Q_\xi^{-1} e_i} c_i \begin{bmatrix} h_i A + I_n \vdots h_i B \vdots h_i D \end{bmatrix}.$$

We introduce the new scalar variable $\delta > 0$ such that the inequality

$$\delta P^{-1} \geq c_i^T \frac{\mu_2}{\mu_2 - c_i^T Q_\xi^{-1} e_i} c_i$$

holds. Obviously, it can be rewritten in the form

$$\begin{bmatrix} \delta P & P c_i^T & 0 \\ c_i P & 1 & e_i^T \\ 0 & e_i & \mu_2 Q_\xi \end{bmatrix} > 0. \tag{8.30}$$

In this case, the last inequality together with

$$
\begin{bmatrix}
R_i - \mu_1 Q_x - \mu_2 Q'_x & 0 & 0 \\
0 & K_i - \mu_1 Q_u - \mu_2 Q'_u & 0 \\
0 & 0 & \mu_1 Q_f
\end{bmatrix} >
$$

$$
\begin{bmatrix}
h_i A^\top + I_n \\
h_i B^\top \\
h_i D^\top
\end{bmatrix}
\delta P^{-1}
\begin{bmatrix}
h_i A + I_n \vdots h_i B \vdots h_i D
\end{bmatrix}
$$

implies (8.25).

Setting

$$
Y = (y_1^T, y_2^T, \ldots, y_m^T)^T := CP,
$$

we rewrite the optimization problem in the restricted form

$$
\min_{P,Y,R_i,K_i,\alpha_i,\tau_1,\tau_2,\mu_1,\mu_2,\delta} \operatorname{tr}(P)
$$

subject to   (8.26), (8.27) and

$$
\begin{bmatrix}
AP + BY + PA^T + Y^T B^T + \tau_1 P + \bar{\tau}_2 DQ_f^{-1} D^T & P & Y^T \\
P & -\bar{\tau}_2 Q_x^{-1} & 0 \\
Y & 0 & -\bar{\tau}_2 Q_u^{-1}
\end{bmatrix} \leq 0,
$$

$$
\begin{bmatrix}
R_i - \mu_1 Q_x - \mu_2 Q'_x & 0 & 0 & h_i A^\top + I_n \\
0 & K_i - \mu_1 Q_u - \mu_2 Q'_u & 0 & h_i B^\top \\
0 & 0 & \mu_1 Q_f & h_i D^\top \\
h_i A + I_n & h_i B & h_i D & \delta^{-1} P
\end{bmatrix} > 0,
$$

$$
\begin{bmatrix}
\delta P & y_i^T & 0 \\
y_i & 1 & e_i^T \\
0 & e_i & \mu_2 Q_\xi
\end{bmatrix} > 0,
$$

$$
\tau_1 \geq 0, \ \bar{\tau}_2 \geq 0, \ \tau_1 \bar{\tau}_2 \geq f_0, \ \delta > 0, \ P > 0.
$$

8.2.3 Example

Consider the linear system

$$
\dot{x} = Ax + Bu + Df
$$

with the matrixes A, B, C be defined as

$$A = \begin{bmatrix} -10 & 2 & -40 \\ -1.5 & -200 & 20 \\ 50 & -6 & 100 \end{bmatrix}, \quad B = \begin{bmatrix} 3/10 \\ 0 \\ 6/5 \end{bmatrix}, \quad D = \begin{bmatrix} 1 & 0 \\ 1 & -1 \\ 0 & 1 \end{bmatrix}.$$

The perturbation f is given by

$$f = \begin{pmatrix} 0.0028\cos(0.4t) - 0.0879x_1\sin(0.4t) \\ 0.0499\cos(0.4t) + 0.0049x_2\sin(0.4t) \end{pmatrix},$$

which satisfies the inequality

$$f^T \begin{bmatrix} 130 & 15 \\ 15 & 400 \end{bmatrix} f \le 1 + x^T \begin{bmatrix} 0.02 & 0 & 0 \\ 0 & 0.0001 & 0 \\ 0 & 0 & 0.0001 \end{bmatrix} x.$$

Let the sliding mode controller be applied through the dynamics given by

$$\dot{u} = -H(u + M\,sign\,\sigma) + \xi.$$

The dynamic perturbation in the actuator is

$$\xi = 0.05\sin(\cos(0.9t) + \sin(0.4t)) - 0.0479\sin(0.4t),$$

with constant $K_\xi = 100$. The controller restrictions are given by $\alpha_{min} = 0.1$, $\alpha_{max} = 400$, $\beta_x = 40$, $\beta_u = 40$. The controller gains designed to resolve the above optimization procedure, with the selection of parameters $\tau_1 = 0.0001$, $\tau_2 = 0.0001$, $\tau_3 = 10$, $\tau_4 = 0.0001$, $\tau_5 = 2$, $\tau_6 = 0.7$, $\gamma = 0.7$, are given by

$$\sigma - 6.9251x_1 + 48.9162x_2 + 164.9354x_3 + u, \quad H = 0.6,$$

$$M = \sqrt{25.4 + \alpha_x||x||^2 + \alpha_u||u||^2},$$

where

$$\alpha_x = 348710, \quad \alpha_u = 50402.$$

The problem is solved with the quasioptimal solution $tr\ P_z = 614.33$.

Let the initial conditions for the system be $x(0) = [3 \quad -2 \quad 1]^T$. The state trajectory of the controlled system is shown in Fig. 8.3, and a zoom of the trajectory is presented in Fig. 8.4. The convergence to the invariant ellipsoid can be seen in Fig. 8.5. Notice that even in the presence of perturbations, the trajectory converges to a small bounded region of the origin. A zoom on a vicinity of the origin is shown

Fig. 8.3 State trajectory

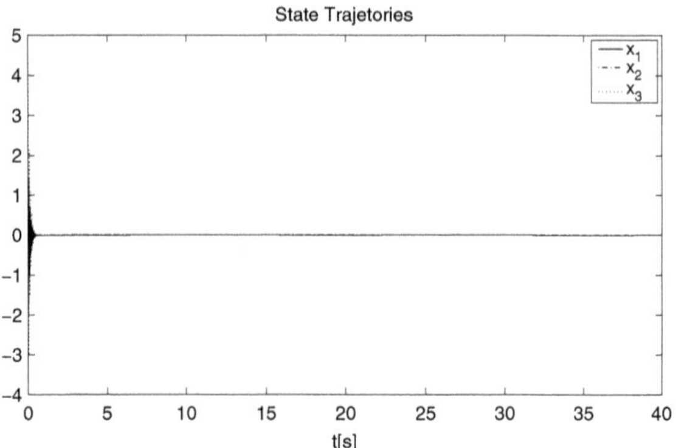

Fig. 8.4 Zoom of the state
trajectory

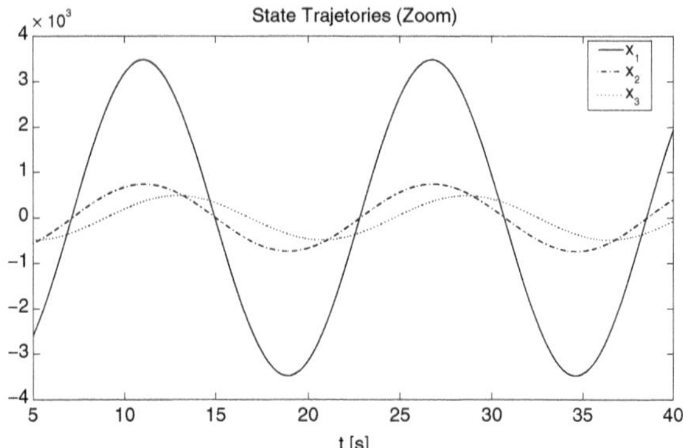

Fig. 8.5 Convergence to the
invariant ellipsoid

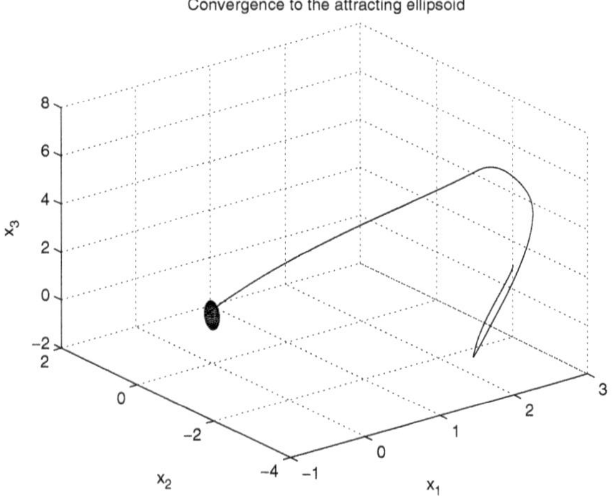

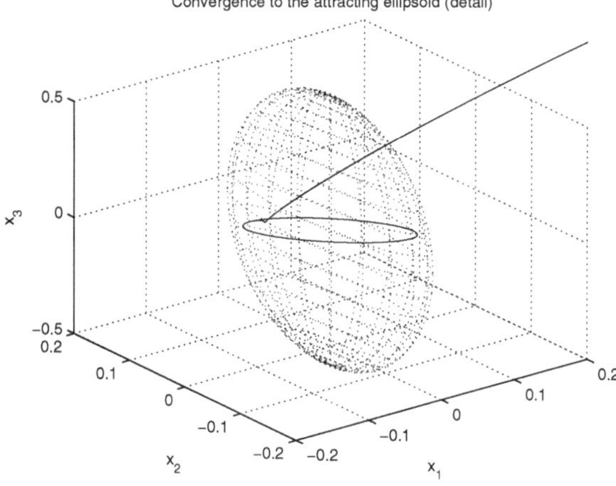

Fig. 8.6 Detail of the convergence of the state trajectories (*solid line*) to the attracting invariant ellipsoid (*dashed line*)

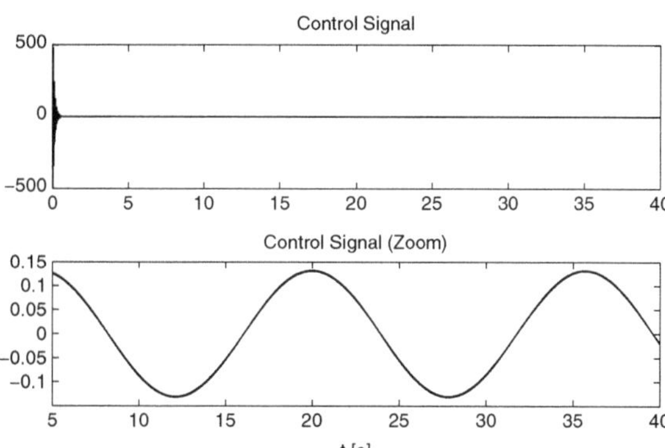

Fig. 8.7 Control signal (*above*) and detail of the control signal (*below*)

in Fig. 8.6. In that figure, one can see the convergence to the quasioptimal invariant ellipsoid $\varepsilon(\frac{\gamma}{\beta} P_x)$. The control signal is presented in Fig. 8.7.

8.3 Conclusion

In this chapter, the sliding mode control approach was introduced using the attractive ellipsoid method. It provides minimization of the effects of system disturbances and helps in tuning the sliding mode control gains for systems with actuators.

Chapter 9
Robust Stabilization of Time-Delay Systems

Abstract In this chapter, we consider the class of uncertain time-delay affine-controlled systems in which a delay is admitted in state variables, and we show that the attractive ellipsoid method allows us to create a feedback that provides the convergence of any state trajectory of the controlled system from a given class to an ellipsoid whose "size" depends on the parameters of the applied feedback. Finally, we present a method for numerical calculation of these parameters that provides the "smallest" zone convergence for controlled trajectories.

Keywords Input time-delay • Robust control • Unknown delay

In this chapter, we consider the class of uncertain time-delay affine-controlled systems in which a delay is admitted in state variables, and we show that the attractive ellipsoid method (AEM) allows us to create a feedback that provides the convergence of any state trajectory of the controlled system from a given class to an ellipsoid whose "size" depends on the parameters of the applied feedback. Finally, we present a method for numerical calculation of these parameters that provides the "smallest" zone convergence for controlled trajectories.

9.1 Time-Delay Systems with Known Input Delay

9.1.1 Brief Historical Remark

The presence of time-delay effects in the feedback of a control system is a frequent obstacle to controlling realization (Edwards & Spurgeon 1998; Fridman, Acosta, & Polyakov 2001; Loukianov, Espinosa-Guerra, Castillo-Toledo, & Utkin 2006; Utkin 1992; Utkin, Guldner, & Shi 1999). Such effects may produce oscillations of a closed-loop system and provoke its instability (Malek-Zaverei & Jamshidi 1987).

Time-delay compensation is the common control design for systems with time delays (Fridman et al. 2001; Li & Yurkovitch 1999; Loukianov et al. 2006; Manitius & Olbrot 1979; Roh & Oh 1999; Polyakov 2012). It is based on a possible prediction of system state for time-delay interval. Unfortunately, the corresponding predictor-

© Springer International Publishing Switzerland 2014

A. Poznyak et al., *Attractive Ellipsoids in Robust Control*, Systems & Control: Foundations & Applications, DOI 10.1007/978-3-319-09210-2_9

based control algorithms may lose properties typical of the delay-free case. For example, predictor-based sliding mode control algorithms are not robust with respect to system disturbances and variations in plant parameters, even when the so-called *matching condition* holds (Fridman et al. 2001). The matching condition requires that disturbances act in the same subspace as an admissible control. For prediction systems, the matching condition almost always fails, and disturbances can be only partially rejected by enforcing a sliding mode.

In this section, we extend the attractive ellipsoid technique to time-delay control systems with predictor-based feedback. It gives an exact method of minimization of the effects produced by disturbances and time delays in control input.

9.1.2 System Description and Problem Statement

Let us consider a time-delay control system of the form

$$\dot{x}(t) = Ax(t) + Bu(t - h) + Df(t), \tag{9.1}$$

where

- $x \in \mathbb{R}^n$ is the vector of the system state,
- $A \in \mathbb{R}^{n \times n}$ is the system matrix,
- $u \in \mathbb{R}^m$ is the vector of control inputs,
- $B \in \mathbb{R}^{n \times m}$ is the matrix of control gains,
- $h > 0$ is the constant input time delay, assumed to be known, and
- $D \in \mathbb{R}^{n \times k}$ is the gain matrix of the system disturbances,

Moreover, the vector-valued function $f(t)$ describing the perturbations is assumed to be Lebesgue measurable and bounded as

$$f^T Q_f f < 1, \tag{9.2}$$

where the positive definite symmetric matrix $Q_f \in \mathbb{R}^{k \times k}$ is given.

The system (9.1) is considered with the following initial conditions:

$$x(0) = x_0 \in \mathbb{R}^n \quad \text{and} \quad u(t) = v_0(t) \text{ for } t \in [-h, 0],$$

where $v_0(t)$ is some given function, which is usually assumed to equal zero.

Let us suppose additionally that the pair $\{A, B\}$ is controllable and the matrix B has full rank, that is,

$$\text{rank}(B) = m \leq n.$$

To stabilize the time-delay control system (9.1), the prediction approach (see Li & Yurkovitch 1999; Loukianov et al. 2006; Manitius & Olbrot 1979; Roh & Oh

1999) is used. The typical prediction equation (Li & Yurkovitch 1999; Loukianov et al. 2006) for the system (9.1) has the form

$$y(t) = e^{Ah}x(t) + \int_{-h}^{0} e^{-\theta A} B u(t + \theta) d\theta. \tag{9.3}$$

Obviously, the calculation of the *prediction variable* $y(t)$ requires knowing the control function $u(t)$ on the time interval $[t-h, t)$. It is assumed that this information is admissible and can be used for control design.

It can be easily checked that the predictor variable $y(t)$ satisfies the following delay-free equation:

$$\dot{y}(t) = Ay(t) + Bu(t) + e^{Ah} Df(t). \tag{9.4}$$

According to the predictor method, stabilization of the original system (9.1) can be ensured by designing the stabilizing controller for the prediction system (9.4). Unfortunately, in the disturbed case, the property $y(t) \to 0$ does not imply $x(t) \to 0$, even when the matching condition

$$\text{range}(D) \subseteq \text{range}(B)$$

holds. The integral term in the formula (9.3) is nonzero in general (see, e.g., Fridman et al. 2001). Particularly for the sliding mode control case, we have

$$u(t) = u_{eq}(t) \neq 0,$$

where $u_{eq}(t)$ is an *equivalent control* (Utkin 1992). Moreover, the property $y(t) \to 0$ can be guaranteed by implementing the sliding mode control algorithm for the predictor system (9.4) in the very restrictive case

$$\text{range}(e^{Ah} D) \subseteq \text{range}(B).$$

Therefore, the time-delay control system (9.1) may be *practically stabilized* only in some zone (attractive set).

So our aim is to develop a predictor-based control design scheme that minimizes (in some sense) the attractive set of the system (9.1). For this purpose, we will use the AEM as described in previous chapters. The combination of the AEM and the ellipsoidal calculus (Kurzhanski & Valyi 1997) will help us to introduce an optimality criterion and to give a geometric estimate for the attractive set of the original system (9.1).

Assume that the standard sliding mode controller of the form (Edwards & Spurgeon 1998; Utkin 1992; Utkin et al. 1999)

$$u(t) = -(\tilde{C} B)^{-1}\tilde{C} Ay(t) - K \operatorname{sign}\left(\tilde{C} y(t)\right) \tag{9.5}$$

is applied. The matrix $\tilde{C}$ defines the sliding surface

$$\tilde{C} y = 0$$

in the space of the predictor variables, so that $\tilde{C} B$ is not degenerate, that is,

$$\det(\tilde{C} B) \neq 0.$$

The matrix of control parameters K has the diagonal form

$$K = \begin{pmatrix} k_1 & 0 & 0 & \ldots & 0 \\ 0 & k_2 & 0 & \ldots & 0 \\ \multicolumn{5}{c}{\ldots\ldots\ldots\ldots\ldots} \\ 0 & 0 & \ldots & \ldots & k_m \end{pmatrix}$$

and

$$\mathrm{Sign}\,[v] = (\mathrm{sign}\,[v_1], \ldots, \mathrm{sign}\,[v_m])^T \quad \forall v = (v_1, \ldots, v_m)^T \in \mathbb{R}^m.$$

The linear part of the control function (9.5) is related to the corresponding component of the equivalent control (see Edwards & Spurgeon 1998; Utkin 1992; Utkin et al. 1999 for details).

We also assume that the control parameters $k_i \in \mathbb{R}$ and the linear part of the control (9.5) are bounded as

$$\|(\tilde{C} B)^{-1} \tilde{C} A\| \leq \alpha \text{ and } 0 < k_i \leq \beta_i, i = 1, 2, \ldots, m, \tag{9.6}$$

where the positive numbers α and β_i are given and $\| \cdot \|$ is the spectral matrix norm defined as

$$\|H\| = \sqrt{\lambda_{\max}(H^T H)}.$$

Remark 9.1. Using the prediction method, a **linear** stabilizing controller can also be designed based on AEM. The next section of this chapter presents the corresponding constructions for time-varying unknown delay.

9.1.3   Unavoidable Stabilization Error

Using the formula for the general solution of the system (9.1), we obtain

$$x(t + h) = e^{Ah} x(t) + \int_t^{t+h} e^{(t+h-s)A} B u(s - h) ds + \int_t^{t+h} e^{(t+h-s)A} D f(s) ds$$

$$= e^{Ah} x(t) + \int_{-h}^0 e^{-\theta A} B u(t + \theta) d\theta + \int_0^h e^{(h-\theta)A} D f(t + \theta) d\theta.$$

Hence, we easily derive the equality

$$x(t + h) = y(t) + \int_0^h e^{(h-\theta)A} Df(t + \theta)d\theta, \tag{9.7}$$

which describes the dependence of the original system state $x(t)$ on the predictor variable $y(t)$ and the external disturbances $f(t)$. The integral term on the right-hand side of (9.7) obviously does not depend on the control inputs and the predictor variables, but it is a linear functional of $f(t)$, which describes *unknown* system disturbances. So the corresponding term defines an *unavoidable stabilization error* of the system (9.1) produced by the prediction technique. Therefore, minimization of the attractive set for the original system (9.1) can be provided by the design of the appropriate controller for the prediction system (9.4).

9.1.4   *Minimal Invariant Ellipsoid for the Prediction System*

The method of linear matrix inequalities (LMIs) is the usual technique for describing an invariant ellipsoid (Gonzalez-Garcia, Polyakov, & Poznyak 2011; Polyak and Topunov 2008). It is also effective for the design of sliding mode controls (see Chap. 4 of Davila & Poznyak 2011 and Polyakov & Poznyak 2011).

Since $\text{rank}(B) = m$, the matrix B can be partitioned (perhaps after reordering the state vector components) as

$$B = \begin{pmatrix} B_1 \\ B_2 \end{pmatrix},$$

where $B_1 \in \mathbb{R}^{(n-m)\times m}$, $B_2 \in \mathbb{R}^{m\times m}$ with

$$\det(B_2) \neq 0.$$

In this case, the nondegenerate coordinate transformation

$$\begin{pmatrix} y_1 \\ y_2 \end{pmatrix} = Gy, \text{ where } G = \begin{pmatrix} I_{n-m} & -B_1 B_2^{-1} \\ 0 & B_2^{-1} \end{pmatrix},$$

reduces the system (9.4) to regular form (see, for example, Utkin et al. 1999):

$$\begin{cases} \dot{y}_1 = A_{11}y_1 + A_{12}y_2 + D_1 f, \\ \dot{y}_2 = A_{21}y_1 + A_{22}y_2 + u(t) + D_2 f. \end{cases} \tag{9.8}$$

Here

$$y_1 \in \mathbb{R}^{n-m}, \, y_2 \in \mathbb{R}^m,$$

are blocks of the system state vector,

$$A_{11} \in \mathbb{R}^{(n-m)\times(n-m)}, \; A_{12} \in \mathbb{R}^{(n-m)\times m}, \; A_{21} \in \mathbb{R}^{m\times(n-m)}, \; A_{22} \in \mathbb{R}^{m\times m},$$

are blocks of the system matrix

$$\begin{pmatrix} A_{11} & A_{12} \\ A_{21} & A_{22} \end{pmatrix} = GAG^{-1},$$

and

$$\begin{pmatrix} D_1 \\ D_2 \end{pmatrix} = Ge^{Ah}D, \quad D_1 \in \mathbb{R}^{(n-m)\times k}, \; D_2 \in \mathbb{R}^{m\times k}.$$

The classical method of sliding surface design (Utkin 1992) recommends that one select the switching (sliding) surface in the form

$$Cx_1 + x_2 = 0.$$

The lemma below gives necessary and sufficient conditions for the existence of a sliding mode on this surface.

Lemma 9.1. *The hyperplane*

$$\tilde{C}y := (C \vdots I_m)Gy = 0$$

is the sliding manifold of the system (9.4), (9.2) with control (9.5)–(9.6) if and only if the control parameters $k_i \in \mathbb{R}$ and the matrix $C \in \mathbb{R}^{m\times(n-m)}$ satisfy the following system of matrix inequalities:

$$\begin{pmatrix} \alpha^2 I_n & A^T G^T (C \vdots I_m)^T \\ (C \vdots I_m)GA & I_m \end{pmatrix} \geq 0, \tag{9.9}$$

$$\begin{pmatrix} Q_f & D^T e^{A^T h} G^T (c_i \vdots e_i)^T \\ (c_i \vdots e_i)Ge^{Ah}D & k_i^2 \end{pmatrix} \geq 0, \quad i = 1, 2, \ldots, m, \tag{9.10}$$

$$0 < k_i \leq \beta_i, \tag{9.11}$$

where $c_i^T \in \mathbb{R}^{n-m}$, c_i is the ith row of the matrix C, and

$$e_i^T \in \mathbb{R}^m, \; e_i = (0,\ldots,0,1,0,\ldots,0)$$

is the identity row vector, $i = 1,2,\ldots,m$.

Proof. Let us consider the differential equation describing the dynamics of the sliding variable $s = \tilde{C}y$:

$$\dot{s}(t) = \tilde{C}Ay(t) + \tilde{C}Bu(t) + \tilde{C}e^{Ah}Df(t) =$$

$$\tilde{C}B(-K\mathrm{Sign}\,(s\,(t))) + \tilde{C}e^{Ah}Df(t).$$

Taking into account

$$\tilde{C}B = (C \vdots I_m)GB = I_m,$$

we obtain

$$\dot{s}(t) = -K\mathrm{Sign}\,(s\,(t)) + \tilde{C}e^{Ah}Df(t).$$

Obviously, the finite-time stability criterion for the last system has the form

$$k_i > |(c_i \vdots e_i)Ge^{Ah}Df(t)|,$$

or equivalently,

$$\frac{1}{k_i^2}f^T(t)D^T e^{A^T h}G^T(c_i \vdots e_i)^T(c_i \vdots e_i)Ge^{Ah}Df(t) < 1, \tag{9.12}$$

where $i = 1,2,\ldots,m$. By the S-procedure (see, for example, Poznyak 2008), the inequality (9.2) implies (9.12) if and only if

$$\frac{1}{k_i^2}D^T e^{A^T h}G^T(c_i \vdots e_i)^T(c_i \vdots e_i)Ge^{Ah}D \le Q_f.$$

The second inequality from (9.6) can be equivalently rewritten in the form

$$A^T G^T(C \vdots I_m)^T(C \vdots I_m)GA \le \alpha^2 I_n.$$

Finally, applying the Schur complement (see, for example, Poznyak 2008) to the obtained inequalities completes the proof.   ∎

When the sliding mode appears on the hyperplane

$$\tilde{C} y = 0,$$

the sliding motion equation takes the form

$$\dot{y}_1 = (A_{11} - A_{12}C)y_1 + D_1 f(t) \tag{9.13}$$

and

$$\tilde{C} y = C y_1 + y_2 = 0.$$

The system (9.13) is linear, and the system disturbances are bounded by (9.2), so in the case of a stable matrix $A_{11} - A_{12}C$, *the minimal attractive invariant set of the system (9.13), (9.2) is some ellipsoid* (Polyak and Topunov 2008).

The proposition given below explains how to find the minimal invariant ellipsoid for the prediction system (9.4) taking into account Eq. (9.13) and the property $\tilde{C} y = 0$.

Lemma 9.2. *Let the tuple $(\tau, C, P, Z, k_1, \ldots, k_m)$ be a solution of the optimization problem*

$$\mathrm{tr}(Z) \to \min \tag{9.14}$$

subject to (9.9)–(9.11) and

$$\begin{pmatrix} P(A_{11} - A_{12}C)^T + (A_{11} - A_{12}C)P + \tau P & D_1 \\ D_1^T & -\tau Q_f \end{pmatrix} < 0, \tag{9.15}$$

$$\begin{pmatrix} P & \begin{pmatrix} P & -PC^T \end{pmatrix} \\ \begin{pmatrix} P \\ -CP \end{pmatrix} & GZG^T \end{pmatrix} \geq 0, \tag{9.16}$$

$$\tau \geq 0, \ P > 0, \ Z \geq 0, \tag{9.17}$$

$$\tau \in \mathbb{R}, \ P \in \mathbb{R}^{(n-m)\times(n-m)}, \ Z \in \mathbb{R}^{n\times n}.$$

Then the ellipsoid

$$\xi(Z) = \{y \in \mathbb{R}^n : yy^T \leq Z\}$$

is a minimal invariant ellipsoid of the system (9.4), (9.2) with controller (9.5) satisfying the condition (9.6).[1]

Proof. Consider the quadratic function

$$V(y_1) = y_1^T P^{-1} y_1, \quad P > 0.$$

The total derivative of this function along the trajectories of the system (9.13) has the form

$$\dot{V} = y_1^T ([A_{11}\text{-}A_{12}C]^T P^{-1} + P^{-1}[A_{11}\text{-}A_{12}C]) y_1 + 2 f^T D_1^T P^{-1} y_1 =$$

$$\begin{pmatrix} y_1 \\ f \end{pmatrix}^T \begin{pmatrix} (A_{11}\text{-}A_{12}C)^T P^{-1} + P^{-1}(A_{11}\text{-}A_{12}C) & P^{-1}D_1 \\ D_1^T P^{-1} & 0 \end{pmatrix} \begin{pmatrix} y_1 \\ f \end{pmatrix}.$$

Obviously, the ellipsoid $\varepsilon(P)$ is state-invariant if and only if the inequality

$$\dot{V} < 0$$

holds for every

$$y_1 \in \mathbb{R}^{n-m} : y_1^T P^{-1} y_1 \geq 1$$

and for every uncertain vector function f satisfying (9.2). Define

$$A_0 := \begin{pmatrix} (A_{11} - A_{12}C)^T P^{-1} + P^{-1}(A_{11} - A_{12}C) & P^{-1}D_1 \\ D_1^T P^{-1} & 0 \end{pmatrix},$$

$$A_1 := \begin{pmatrix} -P^{-1} & 0 \\ 0 & 0 \end{pmatrix}, \quad A_2 := \begin{pmatrix} 0 & 0 \\ 0 & Q_f \end{pmatrix},$$

$$f_i(z) := z^T A_i z \text{ for } z \in R^{n-m+k},$$

and rewrite the required condition as the implication

$$\left. \begin{matrix} f_1(z) < -1 \\ f_2(z) < 1 \end{matrix} \right\} \Rightarrow f_0(z) < 0.$$

By the S-procedure (see, for example, Poznyak 2008), the last implication holds if and only if there exist $\tau_1, \tau_2 \geq 0$ such that

---

[1] If $Z > 0$, then using the Schur complement, the inequality $x x^T \leq Z$ can be equivalently rewritten in the standard form $x^T Z^{-1} x \leq 1$.

$$A_0 < \tau_1 A_1 + \tau_2 A_2 \text{ and } 0 \geq -\tau_{.1} + \tau_2$$

Hence we obtain the following inequalities:

$$\begin{pmatrix} (A_{11} - A_{12}C)^T P^{-1} + P^{-1}(A_{11} - A_{12}C) + \tau_1 P^{-1} & P^{-1}D_1 \\ D_1^T P^{-1} & -\tau_2 Q_f \end{pmatrix} < 0, \tag{9.18}$$

$$\tau_1 \geq \tau_2 \geq 0.$$

These inequalities together with (9.9)–(9.11) define an invariant ellipsoid $\xi(P)$ for the subsystem (9.13). Observe that applying an equivalent transformation

$$T = \text{diag}(P, I_k)$$

to the first inequality (9.18) gives (9.15). While the system trajectory slides on the surface $\tilde{C}y = 0$, we have

$$Cy_1 + y_2 = 0 \text{ and } y = G^{-1}\begin{pmatrix} I_{n-m} \\ -C \end{pmatrix} y_1.$$

So the inclusion $y_1 \in \xi(P)$ implies $y \in \xi(Z)$ with

$$Z \geq G^{-1}\begin{pmatrix} I_{n-m} \\ -C \end{pmatrix} P \begin{pmatrix} I_{n-m} \\ -C \end{pmatrix}^T G^{-T} =$$

$$\tag{9.19}$$

$$G^{-1}\begin{pmatrix} P \\ -CP \end{pmatrix} P^{-1} \begin{pmatrix} P \\ -CP \end{pmatrix}^T G^{-T},$$

i.e., $\xi(Z)$ is an invariant ellipsoid of the prediction system (9.4). Applying the Schur complement (see, for example, Poznyak 2008) to (9.19), we obtain (9.15).

The final remark about the fact that the selection $\tau_1 = \tau_2$ extends (Nazin, Polyak, & Topunov 2007; Polyak and Topunov 2008) the feasibility set (with respect to C and P) of the inequality (9.18) completes the proof. ∎

The invariant ellipsoid $\xi(Z)$ of the system (9.4) with sliding mode controller is always degenerate (some ellipsoid's semiaxes are equal to zero), since it belongs to the sliding surface $\tilde{C}y = 0$.

Lemma 9.2 gives a theoretical answer to the question how to design the sliding mode controller (9.5) for the prediction system (9.4), which decreases the effects of the unmatched system disturbances in the sense of the minimal invariant ellipsoid.

The obtained optimization problem (9.14), (9.9)–(9.11), (9.15)–(9.17) has a linear functional with constraints in the form of bilinear matrix inequalities (BMIs). Such a problem can be resolved locally using, for example, the PenBMI toolbox of MATLAB. Below, we demonstrate a method of reduction of this system of BMIs to a parametric family of LMIs. The reduced optimization problem can be solved using standard MATLAB LMI solvers. Unfortunately, this approach restricts the feasible set of the original problem, allowing only suboptimal solutions.

9.1.5 Minimal Attractive Ellipsoid of the Original System

Let us denote by S_w the set of all possible values of the unavoidable stabilization error

$$w(t) := \int_0^h e^{(h-\theta)A} Df(t+\theta)d\theta, \qquad (9.20)$$

which is bounded due to (9.2). Then the formula (9.7) gives the attractive set

$$S_x = S_x(Z)$$

for the original system (9.1) in the form

$$S_x(Z) = \xi(Z) \oplus S_w,$$

where $\xi(Z)$ is an invariant ellipsoid (see Lemma 9.2) for the prediction system (9.4), and the symbol $\oplus$ denotes the Minkowski (geometric) sum of the sets (see, for example, Kurzhanski & Valyi 1997).

Note that the set S_x is not invariant for the system (9.1), since $x(0) \in S_x$ does not imply $y(0) \in \xi(Z)$. Therefore, the presented modification of IEM for time-delay systems allows us to find only an attractive ellipsoid for the original system (9.1).

The unknown function $f(t)$ is bounded for each time moment and belongs to the ellipsoid $\xi(Q_f^{-1})$. Regardless of the fact that for every t and θ, the value $e^{(h-\theta)A} Df(t+\theta)$ belongs to the same ellipsoid, the sets S_w and S_x may not be elliptic. We therefore consider the optimization criterion for the attractive set $S_x(Z)$ based on a minimal ellipsoidal estimate of the set $S_x(Z)$.

Let us introduce the following minimality criterion of the set $S_x(Z)$:

$$\mathrm{tr}Q(Z) \to \min, \qquad (9.21)$$

where $Q(Z)$ is the configuration matrix of the minimal (with respect to trace criteria) ellipsoid estimating the set $S_x(Z)$ (see Fig. 9.1), which is calculated using the so-called *ellipsoidal calculus* technique (see Kurzhanski & Valyi 1997 or the proof of Theorem 9.1 for details).

According to this criterion, the solution of the optimization problem (9.21) subject to (9.9)–(9.11), (9.15)–(9.17) defines an ellipsoid that we call the *minimal attractive ellipsoid* of the original system (9.1).

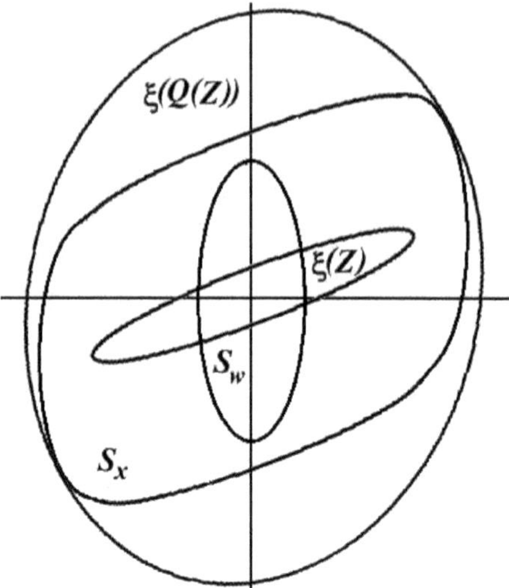

Fig. 9.1 The ellipsoidal estimate of the attractive set S_x

Theorem 9.1 given below proofs that the *minimal (with respect to trace criteria) invariant ellipsoid for the prediction system is in agreement with the minimal attractive ellipsoid for the original system.*

Theorem 9.1. *Let the tuple $(\tau, C, P, Z, k_1, \ldots, k_m)$ be a solution of the optimization problem (9.14), (9.9)–(9.11), (9.15)–(9.17), and let $X_f(t)$ be a nontrivial solution of the matrix differential equation*

$$
\begin{cases}
\dot{X}_f(t) = \pi(t)X_f(t) + \pi^{-1}(t)e^{(h-t)A}DQ_f^{-1}D^T e^{(h-t)A^T}, \\[2mm]
X_f(0) = 0,
\end{cases}
\tag{9.22}
$$

$$
\pi(t) = \sqrt{\frac{\operatorname{tr}(e^{(h-t)A}DQ_f^{-1}D^T e^{(h-t)A^T})}{\operatorname{tr}(X_f(t))}}.
$$

Then

(1) the ellipsoid $\xi(P_x)$ with configuration matrix

$$
P_x = \left(1 + \sqrt{\frac{\operatorname{tr}(X_f(h))}{\operatorname{tr}(Z)}}\right) Z + \left(1 + \sqrt{\frac{\operatorname{tr}(Z)}{\operatorname{tr}(X_f(h))}}\right) X_f(h)
\tag{9.23}
$$

is a minimal attractive ellipsoid of the original system (9.1);

(2) the controller (9.5) with $\tilde{C} := (C \,\vdots\, I_m)G$ and control parameters k_i is the optimal predictor-based sliding mode controller for the system (9.1), (9.2)

satisfying (9.6), which guarantees the asymptotic convergence of each system trajectory to the minimal attractive ellipsoid $\xi(P_x)$.

Proof.

I. Since for each time moment $t \geq 0$, we have

$$f(t) \in \xi(Q_f^{-1}),$$

the function

$$v(t, \theta) := e^{(h-\theta)A} Df(t + \theta)$$

belongs to the ellipsoid $\xi(Q_v(\theta))$ with configuration matrix

$$Q_v(\theta) = e^{(h-\theta)A} DQ_f^{-1} D^T e^{(h-\theta)A^T \theta}$$

for each $t \geq h$ and $\theta \in [0, h]$. The ellipsoidal calculus technique (see, for example, Kurzhanski & Valyi 1997) allows one to find the set S_w by the formula

$$S_w = \int_0^h \xi(Q_v(\theta))d\theta = \bigcap_{\forall p(\cdot)} \xi(X_f(h, p(\cdot))),$$

where $X_f(\tau, p(\cdot)) \in \mathbb{R}^{n \times n}$ is defined by

$$X_f(\tau, p(\cdot)) = \left(\int_0^\tau p(\theta)d\theta \right) \int_0^\tau p^{-1}(\theta) e^{(h-\theta)A} DQ_f^{-1} D^T e^{(h-\theta)A^T} d\theta,$$

and $p(\cdot)$ is an arbitrary positive scalar measurable function.

It is easy to check that the matrix

$$X_f(\tau) := X_f(\tau, p(\cdot))$$

satisfies the differential equation

$$\dot{X}_f(\tau) = \pi(\tau)X_f(\tau) + \pi^{-1}(\tau) e^{(h-\tau)A} DQ_f^{-1} D^T e^{(h-\tau)A^T}, \quad \tau \in [0, h],$$

where

$$\pi(\tau) = p(\tau) \left(\int_0^\tau p(\theta)d\theta \right)^{-1}$$

and

$$X_f(0) = 0.$$

Finally, observe that we will have the minimal (with respect to trace criteria) estimating ellipsoid

$$\xi(X_f(h)) \supseteq S_w$$

for the case

$$\pi(\tau) = \pi^*(\tau) := \sqrt{\frac{\mathrm{tr}(e^{(h-\tau)A} D Q_f^{-1} D^T e^{(h-\tau)A^T})}{\mathrm{tr}(Q(\tau))}} \tag{9.24}$$

(see Lemma 2.7.4 from Kurzhanski & Valyi 1997 for details).

II. Let $\lambda_{\min}(p(\cdot))$ and $\lambda_{\max}(p(\cdot))$ be the minimal and the maximal roots of the equation

$$\det[Z - \lambda X_f(h, p(\cdot))] = 0$$

respectively. Then (Kurzhanski & Valyi 1997)

$$\xi(Z) \oplus \xi(X_f(h, p(\cdot))) = \bigcap_{\forall q \in [\lambda_{\min}(p(\cdot)), \lambda_{\max}(p(\cdot))]} \xi(Q(q, p(\cdot))),$$

where $\xi(Q(q, p(\cdot)))$, is a parameterized family of estimating ellipsoids with configuration matrices

$$Q(q, p(\cdot)) = (1 + q^{-1})Z + (1 + q)X_f(h, p(\cdot))$$

and

$$q \in [\lambda_{\min}(p(\cdot)), \lambda_{\max}(p(\cdot))].$$

In this case,

$$S_x \subseteq \xi(Q(q, p(\cdot)))$$

and

$$S_x = \bigcap_{\forall p(\cdot)} \bigcap_{\forall q \in [\lambda_{\min}(p(\cdot)), \lambda_{\max}(p(\cdot))]} \xi(Q(q, p(\cdot))).$$

The minimal ellipsoid estimating the attractive set S_x belongs to the family of ellipsoids $\xi(Q(q, p(\cdot)))$ (Kurzhanski & Valyi 1997). To find it, let us introduce the function

$$g(q, p(\cdot)) = \mathrm{tr}(Q(q, p(\cdot)))$$

and calculate its derivative with respect to q:

$$\frac{dg}{dq} = \mathrm{tr}(Z)\frac{d}{dq}(1 + 1/q) + \mathrm{tr}(X_f(h, p(\cdot)))\frac{d}{dq}(1 + q) =$$

$$-\frac{\mathrm{tr}(Z)}{q^2} + \mathrm{tr}(X_f(h, p(\cdot))).$$

Hence, we obtain that for

$$q^* := q^*(p(\cdot)) = \sqrt{\frac{\mathrm{tr}(Z)}{\mathrm{tr}(X_f(h, p(\cdot)))}},$$

the ellipsoid $\xi(Q(q^*, p(\cdot)))$ is the minimal ellipsoid estimating the sum

$$\xi(Z) \oplus \xi(X(h, p(\cdot))).$$

Finally, taking into account the monotonicity property of the function

$$\mathrm{tr}(Q(q^*, p(\cdot))) = \left(\sqrt{\mathrm{tr}(Z)} + \sqrt{\mathrm{tr}(X_f(h, p(\cdot)))} \right)^2$$

with respect to $\mathrm{tr}(Z)$ and $\mathrm{tr}(X_f(h, p(\cdot)))$, we conclude that the formula (9.23) defines the minimal (with respect to the trace criteria) attractive ellipsoid for the original system (9.1).   ∎

Theorem 9.1 helps in the design of optimal sliding controllers of the form (9.5) with the sliding manifold of the predefined structure

$$\tilde{C}y := (C \vdots I_m)Gy = 0.$$

Consideration of other forms of sliding mode controller requires completely new constructions.

The formula (9.23) of Theorem 9.1 presents the attractive domain estimate in the form of a minimal attractive ellipsoid, which is explicitly generated by the minimal invariant ellipsoid of the prediction system. Observe that the selection of some alternative optimization criterion (for example, the criterion

$$\det(Q(Z)) \to \min,$$

minimizes the ellipsoid's volume) may violate this property, since the proof of Theorem 9.1 is substantially based the on linearity of the trace criterion.

9.1.6   Computational Aspects

The optimal control design procedure declared in Theorem 9.1 requires resolving
the optimization problem with bilinear constraints. As mentioned above, the optimal
solution of this problem can be *locally* found using the MATLAB Toolbox PenBMI.
To apply the PenBMI solver, we have to obtain an initial (starting) point belonging to
the feasible set of our optimization problem. For this purpose, we reduce the matrix
inequalities (9.9)–(9.11), (9.15)–(9.17) to the form of a two-parametric family of
LMIs.

Theorem 9.2.   *Let*

$$\gamma = \min_i\{\beta_i^2\},$$

and suppose that for some fixed numbers $\delta, \tau \in \mathbb{R}_+$ *the following system of matrix
inequalities is feasible:*

$$\begin{pmatrix} \dfrac{\alpha^2}{\gamma} I_n & A^T G^T \\ GA & \begin{pmatrix} \delta P & 0 \\ 0 & L \end{pmatrix} \end{pmatrix} \geq 0,$$

$$\begin{pmatrix} Q_f & D^T e^{A^T h} G^T \\ G e^{Ah} D & \begin{pmatrix} \delta P & 0 \\ 0 & L \end{pmatrix} \end{pmatrix} \geq 0,$$
(9.25)

$$\begin{pmatrix} PA_{11}^T + A_{11}P - Y^T A_{12}^T - A_{12}Y + \tau P & D_1 \\ D_1^T & -\tau Q_f \end{pmatrix} < 0,$$
(9.26)

$$\begin{pmatrix} \begin{pmatrix} P \\ -Y \end{pmatrix} & \begin{pmatrix} P & -Y^T \end{pmatrix} \\ & GZG^T \end{pmatrix} \geq 0, \quad \begin{pmatrix} \frac{1}{\delta}P & Y^T \\ Y & \gamma I_m - L \end{pmatrix} \geq 0,$$
(9.27)

$$P > 0,\ L > 0,\ Z > 0,\ X \in \mathbb{R}^{(n-m)\times(n-m)},$$

$$Y \in \mathbb{R}^{m\times(n-m)},\ L \in \mathbb{R}^{m\times m},\ Z \in \mathbb{R}^{n\times n}.$$

Then the corresponding

$$Z, \tau, C := YP^{-1}$$

and

$$k_i = \beta_i, i = 1, 2, \ldots, m,$$

satisfy the BMI system (9.9)–(9.11), (9.15)–(9.17).

Proof. Setting $Y := CP$ allows us to rewrite the inequalities (9.15) and (9.16) as (9.26) and (9.27), respectively.

Since

$$(C \vdots I_m)^T (C \vdots I_m) \geq (c_i \vdots e_i)^T (c_i \vdots e_i),$$

the inequality

$$Q_f \geq \frac{1}{\gamma} D^T e^{A^T h} G^T (C \vdots I_m)^T (C \vdots I_m) G e^{Ah} D \tag{9.28}$$

implies (9.10). For ever $C, X > 0$ and $\gamma > 0$, one can find a number $\delta > 0$ and matrix $L \in \mathbb{R}^{m \times m} : L > 0$ such that the upper estimate

$$\operatorname{diag}(\delta^{-1} X^{-1}, L^{-1}) \geq \frac{1}{\gamma}(C \vdots I_m)^T (C \vdots I_m) \tag{9.29}$$

holds. Consider δ and L as the new variables.

Using the transformation $T_n = \operatorname{diag}(X, I_n)$ for (9.29), we obtain

$$T_n^T \operatorname{diag}(\delta^{-1} X^{-1}, L^{-1}) T_n = \operatorname{diag}(\delta^{-1} X, L^{-1}) \geq$$

$$\frac{1}{\gamma} T_n^T (C \vdots I_m)^T (C \vdots I_m) T_n = \frac{1}{\gamma}(Y \vdots I_m)^T (Y \vdots I_m).$$

Hence, the Schur complement allows us to derive

$$\begin{pmatrix} \delta^{-1} X & 0 & Y^T \\ 0 & L^{-1} & I_m \\ Y & I_m & \gamma I_m \end{pmatrix} \geq 0.$$

Applying the equivalent transformation

$$T = \begin{pmatrix} I_{n-m} & 0 & 0 \\ 0 & 0 & I_m \\ 0 & I_m & 0 \end{pmatrix}$$

to the last inequality

$$T^T \begin{pmatrix} \delta X & 0 & Y^T \\ 0 & L^{-1} & I_m \\ Y & I_m & \gamma I_m \end{pmatrix} T = \begin{pmatrix} \delta X & Y^T & 0 \\ Y & \gamma I_m & I_m \\ 0 & I_m & L^{-1} \end{pmatrix} \geq 0$$

as well as the Schur complement, we finally obtain that the inequality (9.29) is equivalent to

$$\begin{pmatrix} \delta X & Y^T \\ Y & \gamma I_m - L \end{pmatrix} \geq 0.$$

We finally conclude that the inequalities (9.25) and the second inequality from (9.27) imply (9.28) and (9.9).   ∎

Theorem 9.2 also allows us to give a procedure for finding a suboptimal solution of our original optimization problem.

For fixed parameters τ and δ, the system of matrix inequalities (9.25)–(9.27) becomes a linear form. In this case, the optimization problem (9.14) subject to (9.25)–(9.27) can be easily solved using, for example, the MATLAB toolbox SeDuMi with the YALMIP interface. If we denote the corresponding minimal value of $\mathrm{tr}(Z)$ by $g(\tau, \delta)$ and search for the minimum of the function $g(\tau, \delta)$ using a derivative-free method, we obtain a suboptimal solution (quasiminimal invariant ellipsoid) of our original optimization problem (9.14) subject to (9.9)–(9.11), (9.15)–(9.17).

To obtain an estimate of the domain of attraction [see formula (9.23)], we have to find a solution of the matrix differential equation (9.22). For simplicity, we initially consider the equation for $q(t) = \mathrm{tr}(X_f(t))$,

$$\dot{q}(t) = 2\sqrt{\mathrm{tr}(e^{(h-t)A} D Q_f^{-1} D^T e^{(h-t)A^T})}\sqrt{q(t)}, \quad q(0) = 0,$$

for which a nontrivial solution can easily be found analytically:

$$q(t) = \left(\int_0^t \sqrt{\mathrm{tr}(e^{(h-\tau)A} D Q_f^{-1} D^T e^{(h-\tau)A^T})}\, d\tau \right)^2 > 0 \text{ for } t > 0.$$

In this case, the function $\pi(t)$ from the formula (9.22) takes the form

$$\pi(t) = \frac{\sqrt{\mathrm{tr}(e^{(h-t)A} D Q_f^{-1} D^T e^{(h-t)A^T})}}{\displaystyle\int_0^t \sqrt{\mathrm{tr}(e^{(h-\tau)A} D Q_f^{-1} D^T e^{(h-\tau)A^T})}\, d\tau} > 0 \text{ for } t > 0,$$

and Eq. (9.22) can be solved numerically using, for example, the implicit Euler method.

9.1.7 *Numerical Example*

Consider the following benchmark example:

$$A = \begin{pmatrix} 1 & 2 & 0.4 \\ -1.5 & 0.7 & 2 \\ 0.5 & -0.6 & 1 \end{pmatrix}, \quad B = \begin{pmatrix} 0.3 \\ 0 \\ 1.2 \end{pmatrix},$$

$$D = \begin{pmatrix} 1 & 0 \\ 1 & -1 \\ 0 & 1 \end{pmatrix}, \quad Q_f = \begin{pmatrix} 130 & 15 \\ 15 & 400 \end{pmatrix},$$

$$h = 0.5, \alpha = 5, \beta = 1.$$

A numerical procedure for resolving the optimization problem (9.14), (9.25)–(9.27) gives

$$\tau^0 = 0.5135, \ \delta^0 = 6.8804.$$

The corresponding matrix of the quasiminimal invariant ellipsoid for the predictor variable has the form

$$Z = \begin{pmatrix} 0.0854 & -0.0924 & 0.0504 \\ -0.0924 & 0.1384 & -0.1394 \\ 0.0504 & -0.1394 & 0.2173 \end{pmatrix},$$

and the matrix of the switching surface is given by

$$\tilde{C} = (1.0341, \ 1.2697, \ 0.5748).$$

Calculation of the ellipsoidal estimate of the unavoidable error gives the following formula for the configuration matrix:

$$X_z(h) = \begin{pmatrix} 0.0062 & 0.0020 & -0.0001 \\ 0.0020 & 0.0013 & -0.0004 \\ -0.0001 & -0.0004 & 0.0012 \end{pmatrix}.$$

Finally, the matrix of the minimal estimating attractive ellipsoid for the original system has the following form:

$$P_x = \begin{pmatrix} 0.1476 & -0.0894 & 0.0568 \\ -0.0894 & 0.1680 & -0.1620 \\ 0.0568 & -0.1620 & 0.2577 \end{pmatrix}.$$

The disturbance function $f(t)$ has been chosen in the form

$$f(t) = \begin{pmatrix} 0.0028\cos(0.6t) - 0.0879\sin(0.6t) \\ 0.0499\cos(0.6t) + 0.0049\sin(0.6t) \end{pmatrix}$$

and

$$u(t) = 0 \text{ for } t \in [-h, 0].$$

Figure 9.2 shows the corresponding quasiminimal attractive ellipsoid and phase trajectory of the original system for the case

$$x(0) = (0.05, -0.2, 0.15)^T.$$

Figure 9.3 presents system trajectories begun outside the ellipsoid with

$$x(0) = (0, -2, 2)^T,$$

and Fig. 9.4 shows the evolution of the control input.

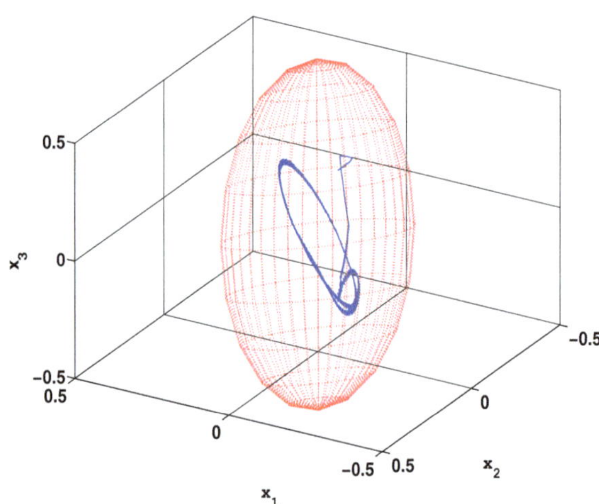

Fig. 9.2 Attractive ellipsoid and phase trajectory

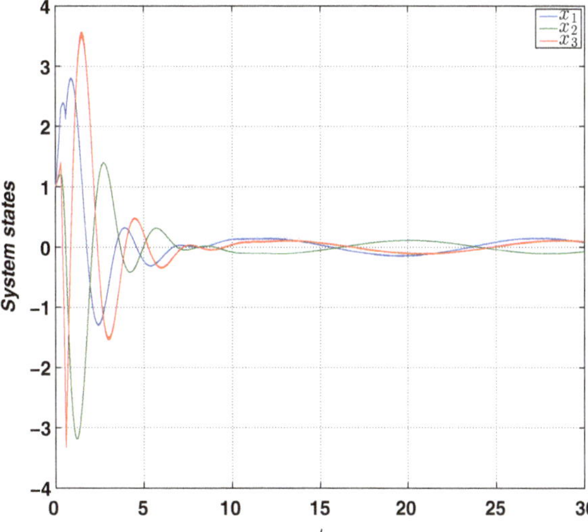

Fig. 9.3 Evolution of system states

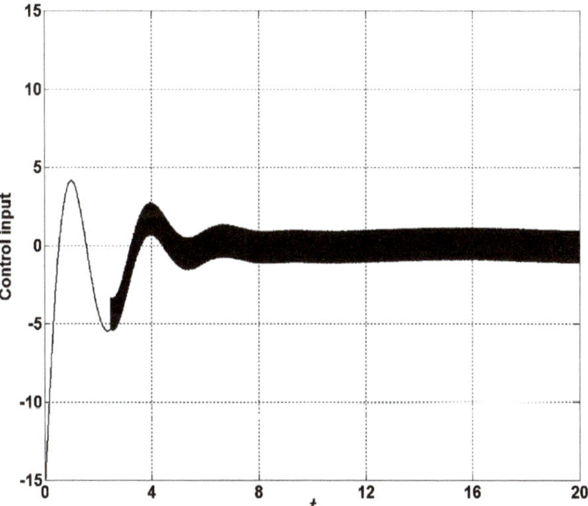

Fig. 9.4 Evolution of control input

9.2 Control of Systems with Unknown Input Delay

9.2.1 Introduction

Models with time delays frequently appear in networked systems (see, for example, (Kruszewski, Jiang, Fridman, Richard, & Toguyeni 2012), chemical (Stephanopoulos 1984)) and biological processes (Flower & Mackey 2002), and the automobile (Choi & Hedrick 1996) and aerospace industries (Yefremov, Polyakov, & Strygin 2006). Real-life applications require control algorithms that are robust with respect to exogenous disturbances, system uncertainties, and measurement noise. Control problems for systems with *known* delays are studied for both linear and nonlinear systems, state, input, and output delays (see, for example, Artstein 1982; Bekiaris-Liberis & Krstic 2012; Krstic 2009; Mazenc, Niculescu, & Krstic 2012; Richard 2003; Polyakov 2012; Watanabe 1986 and references therein). However, an important performance index of control systems with time delays is the robustness with respect to unknown and time-varying delay. Such analysis for full-state feedback control algorithms has been done for systems with linear (see, for example, Fridman, Seuret, & Richard 2004) and relay feedbacks (Polyakov 2010) and first-order sliding mode control algorithms (Han, Fridman, & Spurgeon 2012).

There has been little research devoted to output control design for systems with time-varying and unknown delay (Choi & Lim 2010; Polyakov, Efimov, Perruquetti, & Richard 2013). And even then, robustness properties of time-delay control systems have been studied only with respect to uncertainties in time delay.

This chapter treats the problem of output control design for linear systems with unknown and time-varying input delay, bounded exogenous disturbances, and bounded deterministic measurement noise. Two approaches can be considered to tackling this problem.

The first is based on finite-time (or fixed-time) observer design for systems with unknown control input (Seuret, Floquet, Richard & Spurgeon 2007; Zheng, Barbot, Boutat, Floquet, & Richard 2011). Theoretically, such an observer guarantees that after a finite (or fixed) period of time, the observed states will coincide with the real ones. This property ensures that a separation principle holds, and any existing full-state control can be applied. Unfortunately, in practice, the exact convergence of the observer cannot be guaranteed, due to noise, inaccuracy of digital realization of the continuous-time observer, etc. So in fact, additional robustness analysis of the closed-loop system is required.

The second approach does not assume that the separation principle is satisfied. The stability and robustness analysis in this case has to be done for the whole closed-loop system, including observer and controller parts. This section presents an output-based control design algorithm for nonlinear (quasi-Lipschitz) systems with time-varying and unknown input delay.

9.2.2 Problem Statement

Consider an input delay control system of the form

$$\dot{x} = Ax + Bu(t - h(t)) + Df(t, x),$$

$$y = Cx + Eg(t),$$

(9.30)

where

- $x \in \mathbb{R}^n$ is the system state,
- $u \in \mathbb{R}^m$ is the vector of control inputs,
- $y \in \mathbb{R}^k$ is the measured output,
- $A \in \mathbb{R}^{n \times n}$, $B \in \mathbb{R}^{n \times m}$, $C \in \mathbb{R}^{k \times n}$, $D \in \mathbb{R}^{n \times r}$, and $E \in \mathbb{R}^{k \times p}$ are known matrices.

The input delay $h(t)$ is assumed to be unknown, locally measurable, and bounded by

$$0 \leq \underline{h} \leq h(t) \leq \bar{h}, \tag{9.31}$$

where the numbers $\underline{h}$ and $\bar{h}$ are given.

The continuous function $f : \mathbb{R}^{n+1} \rightarrow \mathbb{R}^r$ describes bounded exogenous disturbances and system uncertainties such that

$$f^T(t, x) Q_f f(t, x) \leq 1 \quad \forall t \in \mathbb{R} \text{ and } \forall x \in \mathbb{R}^n, \tag{9.32}$$

where $Q_f \in \mathbb{R}^{r \times r}$ is a given positive definite matrix.

The locally measurable function $g : \mathbb{R} \rightarrow \mathbb{R}^p$ describes bounded deterministic measurement noises

$$g^T(t) Q_g g(t) \leq 1 \quad \forall t \in \mathbb{R}_+, \tag{9.33}$$

where $Q_g \in \mathbb{R}^{p \times p}$ is a given positive definite matrix.

The system (9.30) is studied with the initial conditions

$$x(0) = x_0,$$
$$u(t) = v(t) \text{ for } t \in [-\bar{h}, 0),$$

where $v(t)$ is some continuous bounded function.

Assumption 1. The pair (A, B) is controllable, and the pair (A, C) is observable.

Assumption 2. The information on the control signal $u(t)$ on the time interval $[t - \bar{h}, t)$ can be stored and used for control design proposes.

The main goal of this chapter is to present a control algorithm that

- stabilizes the states of the system (9.30) at the origin;
- minimizes the negative effects of exogenous disturbances and uncertainties of time delay.

9.2.3   *Attractive Ellipsoid Method for Time-Delay Systems*

The classical optimal control concept addresses the problem of the minimization of some functional (for example, quadratic) subject to all possible trajectories of the system with admissible controls. Taking system uncertainties and disturbances into account requires the "robustification" of the optimal control schemes. The optimality concept presented in the paper (Nazin et al. 2007) considers the stabilization problem for linear *disturbed* control systems. It introduces a criterion based on the minimal attractive ellipsoid of the system, which characterizes the influence of disturbances to the closed-loop system. The optimal attractive ellipsoid method (AEM) control minimizes (in some sense) the effects of disturbances (i.e., suppresses them). This concept is definitely very close to H^∞-control approaches. Similar control ideas were presented in the paper (Usoro, Schweppe, Gould, & Wormley 1982) in the early 1980s. However, at that time, the authors were unable to propose effective computational schemes for tuning the control parameters. Today, control design using attractive ellipsoid methods admits LMI formalization (Boyd, Ghaoui, Feron, & Balakrishnan 1994; Nazin et al. 2007), and a scheme for optimal adjustment of control parameters can be represented as a semidefinite programming problem (Gonzalez-Garcia et al. 2011; Nazin et al. 2007; Polyakov & Poznyak 2011).

9.2.4   *Predictor-Based Output Feedback Design*

Introduce a Luenberger observer of the form

$$\dot{\tilde{x}} = (A + \mathcal{L}C)\tilde{x} + Bu(t - \underline{h}) - \mathcal{L}y(t), \tag{9.34}$$

where the matrix $\mathcal{L} \in \mathbb{R}^{n \times k}$ is to be defined.

Consider also the error equation

$$\dot{e} = (A + \mathcal{L}C)e + B(u(t - \underline{h}) - u(t - h(t))) - Df - \mathcal{L}Eg, \tag{9.35}$$

where $e = \tilde{x} - x$.

Select a control of the form

$$u(t) = \mathcal{K}z(t), \tag{9.36}$$

where the matrix of the control $\mathcal{K} \in \mathbb{R}^{m \times n}$ must be designed, and

$$z(t) = e^{A\underline{h}}\tilde{x}(t) + \int_{-\underline{h}}^{0} e^{-A\theta} Bu(t + \theta)d\theta \tag{9.37}$$

is the predictor variable.

In this case, the closed-loop system can be rewritten in the form

$$
\begin{cases}
\dot{e} = (A + \mathcal{L}C)e + B\mathcal{K} \displaystyle\int_{t-h(t)}^{t-\underline{h}} \dot{z}(\tau)d\tau - Df - \mathcal{L}Eg, \\[2mm]
\dot{z} = (A + B\mathcal{K})z + e^{A\underline{h}}\mathcal{L}Ce.
\end{cases}
\tag{9.38}
$$

Theorem 9.3. *If the tuple* $(\alpha, \tau_1, \tau_2, \mathcal{X}, \mathcal{Z}, \mathcal{L}, \mathcal{Y})$ *satisfies the matrix inequality*

$$
W =
$$
$$
\begin{pmatrix}
\Pi_1 & (e^{A\underline{h}}\mathcal{L}C\,\mathcal{Z})^T & (e^{A\underline{h}}\mathcal{L}C\,\mathcal{Z})^T & B\mathcal{Y} & D & \mathcal{L}E \\
e^{A\underline{h}}\mathcal{L}C\,\mathcal{Z} & \Pi_2 & A\mathcal{X}+B\mathcal{Y} & 0 & 0 & 0 \\
e^{A\underline{h}}\mathcal{L}C\,\mathcal{Z} & (A\mathcal{X}+B\mathcal{Y})^T & \mathcal{R}-\frac{1}{\Delta h}\mathcal{X} & 0 & 0 & 0 \\
\mathcal{Y}^T B^T & 0 & 0 & -\frac{1}{e^{\alpha\tilde{h}}}\mathcal{R} & 0 & 0 \\
D^T & 0 & 0 & 0 & -\tau_1 Q_f & 0 \\
E^T\mathcal{L}^T & 0 & 0 & 0 & 0 & -\tau_2 Q_f
\end{pmatrix}
\leq 0,
\tag{9.39}
$$

$$
\Pi_1 = A\mathcal{Z} + \mathcal{Z}A^T + \mathcal{L}C\mathcal{Z} + \mathcal{Z}C^T\mathcal{L}^T + \alpha\mathcal{Z}, \quad \mathcal{Z} > 0,
\tag{9.40}
$$

$$
\Pi_2 = A\mathcal{X} + \mathcal{X}A^T + B\mathcal{Y} + \mathcal{Y}^T B^T + \alpha\mathcal{X}, \quad \mathcal{X} > 0,
\tag{9.41}
$$

$$
\begin{aligned}
& \alpha, \tau_1, \tau_2 \in \mathbb{R}_+ : \alpha \geq \tau_1 + \tau_2, \\
& \mathcal{Z}, \mathcal{X}, \mathcal{R} \in \mathbb{R}^{n\times n}, \mathcal{Y} \in \mathbb{R}^{m\times n}, \mathcal{L} \in \mathbb{R}^{n\times k}, \mathcal{R} > 0,
\end{aligned}
\tag{9.42}
$$

then

$$
\varepsilon(\mathcal{Z}, \mathcal{X}) := \{e \in \mathbb{R}^n, z \in \mathbb{R}^n : e^T \mathcal{Z}^{-1} e + z^T \mathcal{X}^{-1} z \leq 1\}
\tag{9.43}
$$

is an exponentially attractive ellipsoid of the system (9.38) with

$$
\mathcal{K} = \mathcal{Y}\mathcal{X}^{-1}.
$$

Proof.

I. Consider the Lyapunov–Krasovskii functional of the form

$$
V(t, e(t), z(t), \dot{z}(\cdot)) = e^T(t)\mathcal{Q}e(t) + z^T(t)\mathcal{P}z(t) + V_1(t, \dot{z}(\cdot)) + V_2(t, \dot{z}(\cdot)),
$$
$$
V_1(t, \dot{z}(\cdot)) = (\Delta h)^2 \int_{t-\underline{h}}^{t} e^{\gamma(s-t+\underline{h})}\dot{z}^T(s)\tilde{\mathcal{R}}\dot{z}(s)ds,
$$
$$
V_2(t, \dot{z}(\cdot)) = \Delta h \int_{-\Delta h}^{0} \int_{t-\underline{h}+\theta}^{t-\underline{h}} e^{\mu(s-t+\underline{h})}\dot{z}^T(s)\tilde{\mathcal{R}}\dot{z}(s)ds\, d\theta,
$$

where $\gamma, \mu \in \mathbb{R}_+$, $\mathcal{Q}, \mathcal{P}, \tilde{\mathcal{R}} \in \mathbb{R}^{2n \times 2n}$, $\mathcal{P} > 0, \mathcal{Q} > 0, \tilde{\mathcal{R}} > 0$. It can be established that the time derivative of the functional $V(t, e(t), z(t), \dot{z}(\cdot))$ is the following:

$$\dot{V}(t, e(t), z(t), \dot{z}(\cdot)) = 2e^T(t)\mathcal{Q}\dot{e}(t) + 2z^T(t)\mathcal{P}\dot{z}(t) - \gamma V_1(t, \dot{z}(\cdot))$$

$$+(\Delta h)^2 e^{\gamma \underline{h}} \dot{z}^T(t)\tilde{\mathcal{R}}\dot{z}(t) - \mu V_2(t, \dot{z}(\cdot)) - \Delta h \int_{t-\bar{h}}^{t-\underline{h}} e^{\mu(s-t+\underline{h})} \dot{z}^T(s)\tilde{\mathcal{R}}\dot{z}(s)\,ds.$$

On the one hand, we have

$$e^{-\mu \Delta h} \int_{t-\Delta h}^{t-\underline{h}} \dot{z}^T(s)\tilde{\mathcal{R}}\dot{z}(s)\,ds \leq \int_{t-\Delta h}^{t-\underline{h}} e^{\mu(s-t+\underline{h})} \dot{z}^T(s)\tilde{\mathcal{R}}\dot{z}(s)\,ds.$$

On the other hand, taking into account Jensen's inequality,

$$\left(\int_a^b \dot{z}(s)\,ds\right)^T \tilde{\mathcal{R}} \left(\int_a^b \dot{z}(s)\,ds\right) \leq (b-a) \int_a^b \dot{z}^T(s)\tilde{\mathcal{R}}\dot{z}(s)\,ds,$$

we derive

$$e^{-\mu \Delta h} \left(\int_{t-h(t)}^{t-\underline{h}} \dot{z}(s)\,ds\right)^T \tilde{\mathcal{R}} \left(\int_{t-h(t)}^{t-\underline{h}} \dot{z}(s)\,ds\right) \leq \Delta h \int_{t-\bar{h}}^{t-\underline{h}} e^{\mu(s-t+\underline{h})} \dot{z}^T(s)\tilde{\mathcal{R}}\dot{z}(s)\,ds.$$

Hence, the obtained time derivative of the functional V can be estimated as follows:

$$\dot{V}(t, e(t), z(t), \dot{z}(\cdot)) \leq -\alpha e^T(t)\mathcal{Q}e(t) - \beta z^T(t)\mathcal{P}z(t)$$
$$-\gamma V_1(t, \dot{z}(\cdot)) - \mu V_2(t, \dot{z}(\cdot)) + 2e^T(t)\mathcal{Q}\dot{e}(t) + g^T W_1 q,$$

where $\alpha, \beta \in \mathbb{R}_+$ and

$$g = \begin{pmatrix} e(t) \\ z(t) \\ \dot{z}(t) \\ \int_{t-h(t)}^{t-\underline{h}} \dot{z}(s)\,ds \\ f \\ g \end{pmatrix}, \quad W_1 = \begin{pmatrix} \alpha Q & 0 & 0 & 0 & 0 & 0 \\ 0 & \beta P & P & 0 & 0 & 0 \\ 0 & P & (\Delta h)^2 e^{\gamma \underline{h}} \tilde{\mathcal{R}} & 0 & 0 & 0 \\ 0 & 0 & 0 & -e^{-\mu \Delta h} \tilde{\mathcal{R}} & 0 & 0 \\ 0 & 0 & 0 & 0 & 0 & 0 \\ 0 & 0 & 0 & 0 & 0 & 0 \end{pmatrix}.$$

II. Employing the method of Lagrange multipliers (or the descriptor approach Fridman 2006), we consider the equality

$$0 = 2e^T(t)\mathcal{Q}\left((A + \mathcal{L}C)e(t) + B\mathcal{K}\int_{t-h(t)}^{t-\bar{h}}\dot{z}(s)ds - Df - \mathcal{L}Eg - \dot{e}(t)\right)$$
$$+ 2(\mathcal{P}z(t) + \Delta h\mathcal{P}\dot{z}(t))^T\left((A + B\mathcal{K})z(t) + e^{A\bar{h}}\mathcal{L}Ce(t) - \dot{z}(t)\right),$$

which obviously holds for every solution $(e(t), z(t))$ of the system (9.38) if $t > \bar{h}$. This equality can be rewritten in the form

$$0 = g^T W_2 g + \tau_1 f^T Q_f f + \tau_2 g^T Q_g g, \qquad \tau_1, \tau_2 \in \mathbb{R}_+,$$

where

$$W_2 = \begin{pmatrix} \tilde{\Pi}_{11} & \tilde{\Pi}_{12} & \tilde{\Pi}_{13} & \mathcal{Q}B\mathcal{K} & -\mathcal{Q}D & \mathcal{L}E \\ \tilde{\Pi}_{12}^T & \tilde{\Pi}_{22} & \tilde{\Pi}_{23} & 0 & 0 & 0 \\ \tilde{\Pi}_{13}^T & \tilde{\Pi}_{23} & -\Delta h\mathcal{P} & 0 & 0 & 0 \\ \mathcal{K}^T B^T \mathcal{Q} & 0 & 0 & 0 & 0 & 0 \\ -D^T \mathcal{Q} & 0 & 0 & 0 & -\tau_1 Q_f & 0 \\ E^T \mathcal{L}\mathcal{Q} & 0 & 0 & 0 & 0 & -\tau_2 Q_g \end{pmatrix},$$

where

$$\tilde{\Pi}_{11} = \mathcal{Q}(A + \mathcal{L}C) + (A + \mathcal{L}C)^T \mathcal{Q}, \tilde{\Pi}_{12} = C^T \mathcal{L}^T e^{A^T \bar{h}}\mathcal{P},$$

$$\tilde{\Pi}_{22} = \mathcal{P}(A + B\mathcal{K}) + (A + B\mathcal{K})^T \mathcal{P}, \tilde{\Pi}_{13} = \Delta h C^T \mathcal{L}^T e^{A^T \bar{h}}\mathcal{P},$$

$$\tilde{\Pi}_{23} = \Delta h \mathcal{P}(A + B\mathcal{K}) - \mathcal{P}.$$

Hence, if we take into account the conditions (9.32) and (9.33), the time derivative of the functional V calculated along the trajectories of the system (9.38) can be estimated as

$$\dot{V}(t, e(t), z(t), \dot{z}(\cdot)) \leq -rV(t, e(t), z(t), \dot{z}(\cdot))$$

$$+ g^T(W_1 + W_2)q + \tau_1 + \tau_2,$$

where

$$r = \min\{\alpha, \beta, \gamma, \mu\}.$$

If

$$W_1 + W_2 \leq 0 \text{ and } (\tau_1 + \tau_2)/r \leq 1,$$

then we have

$$V(t, e(t), z(t), \dot{z}(\cdot)) \leq 1 + e^{-rt} V(0, e(0), z(0), \dot{z}(\cdot)),$$

and the structure of the functional V implies that the ellipsoidal set

$$\varepsilon(\mathcal{P}, \mathcal{Q}) := \{(e^T, z^T)^T \in \mathbb{R}^{2n} : e^T \mathcal{Q} e + z^T \mathcal{P} z \leq 1\}$$

is exponentially attractive.

Finally, for $\mu = \gamma = \beta = \alpha$ and

$$\mathcal{Z} = \mathcal{Q}^{-1}, \; \mathcal{X} = \mathcal{P}^{-1}, \; \tilde{\mathcal{R}} = e^{-\gamma h}\mathcal{P}\mathcal{R}\mathcal{P}, \; \mathcal{Y} = \mathcal{K}\mathcal{P}^{-1},$$

we have

$$G^T(W_1 + W_2)G = W,$$

where

$$G = \mathrm{diag}\{\mathcal{Q}^{-1}, \mathcal{P}^{-1}, \mathcal{P}^{-1}, \mathcal{P}^{-1}, -I_n, -I_n.$$

Therefore, the feasibility of the system of matrix inequalities (9.39)–(9.42) implies the exponential attractivity of the ellipsoid (9.43).   ∎

The next lemma treats the question of feasibility of the system of matrix inequalities (9.39)–(9.42).

Lemma 9.3. *Under Assumptions 1 and 2, the system of matrix inequalities (9.39)–(9.42) is feasible at least for sufficiently small Δh.*

Proof.

I. Define $\mathcal{R} = \dfrac{0.5}{\Delta h}\mathcal{X} > 0$. In this case, for $\Delta h \to 0$, using the Schur complement, it can be easily shown that the feasibility of the system (9.39)–(9.42) is defined only by the feasibility of the following matrix inequality:

$$\begin{pmatrix} \Pi_1 & (e^{A h}\mathcal{L}C\,\mathcal{Z})^T & D & \mathcal{L}E \\ e^{A h}\mathcal{L}C\,\mathcal{Z} & \Pi_2 & 0 & 0 \\ D^T & 0 & -\tau_1 Q_f & 0 \\ E^T\mathcal{L}^T & 0 & 0 & -\tau_2 Q_g \end{pmatrix} < 0, \qquad (9.44)$$

$$\mathcal{Z} > 0, \mathcal{X} > 0, \mathcal{L} \in \mathbb{R}^{n \times k}, \mathcal{Y} \in \mathbb{R}^{m \times n}, \alpha, \tau_1, \tau_2 \in \mathbb{R}_+ : \alpha \geq \tau_1 + \tau_2,$$

where Π_1 and Π_2 are defined by (9.48) and (9.49), respectively.

II. Observe that the controllability of the pair (A, B) and the observability of the pair (A, C) imply that $(A + 0.5\alpha I_n, B)$ is also controllable and $(A + 0.5\alpha I_n, C)$ is also observable.

Let $\alpha, \tau_1, \tau_2 \in \mathbb{R}_+$ be arbitrary numbers such that

$$\alpha \geq \tau_1 + \tau_2.$$

The controllability of the pair $(A + 0.5\alpha I_n, B)$ implies that the LMI

$$\Pi_2(\alpha, \mathcal{X}, \mathcal{Y}) < 0$$

is feasible, i.e.,

$$\exists \delta_0 \in \mathbb{R}_+, \ \mathcal{X}, \mathcal{Y} : \Pi_2(\alpha, \mathcal{X}_0, \mathcal{Y}_0) \leq -\delta_0 I_n.$$

In this case, for every $\delta \in \mathbb{R}_+$, there exist $\mathcal{X}$ and $\mathcal{Y}$ such that

$$\Pi_2(\alpha, \mathcal{X}, \mathcal{Y}) \leq -\delta I_n.$$

Indeed, it is sufficient to select

$$\mathcal{X} = \delta/\delta_0 \mathcal{X}_0 \text{ and } \mathcal{Y} = \delta/\delta_0 \mathcal{Y}_0.$$

Hence, using the Schur complement, we derive that feasibility of the inequality (9.44) is restricted only by feasibility of the following matrix inequality:

$$\begin{pmatrix} (A + 0.5\alpha I_n + LC)Z + Z(A + 0.5\alpha I_n + LC)^T & D & LE \\ D^T & -\tau_1 Q_f & 0 \\ E^T L^T & 0 & -\tau_2 Q_g \end{pmatrix} < 0. \tag{9.45}$$

The observability of the pair $(A + 0.5\alpha I_n, C)$ implies existence of $\mathcal{Z}_0$, $\mathcal{L}_0$, and $\mu_0 \in \mathbb{R}_+$ such that

$$(A + 0.5\alpha I_n + \mathcal{L}_0 C)\mathcal{Z}_0 + \mathcal{Z}_0(A + 0.5\alpha I_n + \mathcal{L}_0 C)^T < -\mu_0 I_n.$$

For arbitrary $\mu \in \mathbb{R}_+$, defining

$$Z = \mu/\mu_0 \mathcal{Z}_0$$

guarantees

$$(A + 0.5\alpha I_n + \mathcal{L}_0 C)\mathcal{Z} + \mathcal{Z}(A + 0.5\alpha I_n + \mathcal{L}_0 C)^T \leq -\mu I_n.$$

Applying the Schur complement to (9.45) for $L = L_0$ and $Z = \mu/\mu_0 Z_0$, we derive

$$(A + 0.5\alpha I_n + \mathcal{L}_0 C)\mathcal{Z} + \mathcal{Z}(A + 0.5\alpha I_n + \mathcal{L}_0 C)^T$$

$$+ \frac{1}{\tau_1} D Q_f^{-1} D^T + \frac{1}{\tau_1} \mathcal{L}_0 E Q_f^{-1} E^T \mathcal{L}_0^T \leq$$

$$-\mu I_n + \frac{1}{\tau_1} D Q_f^{-1} D^T + \frac{1}{\tau_2} \mathcal{L}_0 E Q_g^{-1} E^T \mathcal{L}_0^T < 0$$

for

$$\mu > \lambda_{\max}(\frac{1}{\tau_1} D Q_f^{-1} D^T + \frac{1}{\tau_2} \mathcal{L}_0 E Q_g^{-1} E^T \mathcal{L}_0^T). \qquad \blacksquare$$

9.2.5 Adjustment of Control Parameters: Computational Aspects

The predictor variable z estimates the future state of the observer variable $\tilde{x}(t+\underline{h})$, and the vector e describes the observation error $e = \tilde{x} - x$, so in order to increase control precision, we may minimize the size of the attractive ellipsoid of the system (9.38), i.e., we need to solve the following optimization problem:

$$\mathrm{tr}\{\mathcal{Z}\} + \mathrm{tr}\{\mathcal{X}\} \to \min$$

$$\text{s.t. (9.39)–(9.42).} \tag{9.46}$$

The optimization problem (9.46), (9.39)–(9.42) has its linear cost functional and constraints represented in the form of a bilinear matrix inequality (BMI). Such a problem can be solved by a MATLAB BMI solver, for example, PENBMI. At the same time, the constraints can be restricted in order to obtain an SDP problem.

Lemma 9.4. *If for some* $\alpha, \beta \in \mathbb{R}_+$ *the following LMI system*

$$\left(\begin{array}{cc} -\mathcal{S} & \begin{bmatrix} \mathcal{F}C \\ \mathcal{F}C \end{bmatrix}^T \\ \begin{bmatrix} \mathcal{F}C \\ \mathcal{F}C \end{bmatrix} & -\mathcal{V} \end{array} \right) \leq 0, \mathcal{S} > 0, \mathcal{V} > 0, \tag{9.47}$$

$$\left(\begin{array}{cc} -\frac{\beta}{e^{\alpha \bar{h}}} I_n & B\mathcal{Y} \\ \mathcal{Y}^T B^T & -\mathcal{R} \end{array} \right) \leq 0, \mathcal{R} > 0, \mathcal{Q} > 0, \mathcal{X} > 0, \mathcal{V} > 0, \tag{9.48}$$

$$\begin{pmatrix} QA + A^T Q + FC + C^T F + \alpha Q + S & Q & QD & FE \\ Q & -\frac{1}{\beta} I_n & 0 & 0 \\ D^T Q & 0 & -\tau_1 Q_f & 0 \\ E^T F^T & 0 & 0 & -\tau_2 Q_g \end{pmatrix} \le 0,$$

$$(9.49)$$

$$\begin{pmatrix} -2\begin{bmatrix} \mathcal{Q} & 0 \\ 0 & \frac{1}{\Delta h}\mathcal{Q} \end{bmatrix} + \mathcal{V} & \begin{bmatrix} e^{A\bar{h}} & 0 \\ 0 & e^{A\bar{h}} \end{bmatrix} \\ \begin{bmatrix} e^{A^T \bar{h}} & 0 \\ 0 & e^{A^T \bar{h}} \end{bmatrix} & \begin{bmatrix} \Pi_2 & \sqrt{\Delta h}(A\mathcal{X}+B\mathcal{Y}) \\ \sqrt{\Delta h}(A\mathcal{X}+B\mathcal{Y})^T & \Delta h \mathcal{R} - \mathcal{X} \end{bmatrix} \end{pmatrix} \le 0, \quad (9.50)$$

$$\mathcal{Q}, \mathcal{X}, \mathcal{S} \in \mathbb{R}^{n\times n}, \mathcal{Y} \in \mathbb{R}^{m\times n}, \mathcal{F} \in \mathbb{R}^{n\times k}, \mathcal{V} \in \mathbb{R}^{2n\times 2n}, \alpha \ge \tau_1 + \tau_2 \qquad (9.51)$$

is feasible, then the tuple $(\alpha, \mathcal{X}, \mathcal{Z}, \mathcal{L}, \mathcal{Y})$, *where* $\mathcal{X} = \mathcal{Q}^{-1}$ *and* $\mathcal{L} = \mathcal{Q}^{-1}\mathcal{F}$, *satisfies* (9.39)–(9.42).

Proof. Define

$$\Theta = \begin{pmatrix} e^{-A\bar{h}} & 0 \\ 0 & e^{-A\bar{h}} \end{pmatrix} \begin{pmatrix} \Pi_2 & A\mathcal{X}+B\mathcal{Y} \\ (A\mathcal{X}+B\mathcal{Y})^T & \mathcal{R} - \frac{1}{\Delta h}\mathcal{X} \end{pmatrix} \begin{pmatrix} e^{-A^T \bar{h}} & 0 \\ 0 & e^{-A^T \bar{h}} \end{pmatrix},$$

$$\mathcal{Q} = \mathcal{Z}^{-1}, \ \mathcal{F} = \mathcal{Q}\mathcal{L}.$$

Using the Schur complement, the inequality (9.39) can be equivalently rewritten in the form

$$\tilde{W} =$$

$$\begin{pmatrix} \Pi_1 + \frac{1}{\alpha} D Q_f^{-1} D^T + e^{\alpha \bar{h}} B \mathcal{Y} \mathcal{R}^{-1} \mathcal{Y}^T B^T & \begin{bmatrix} e^{A\bar{h}} \mathcal{L} C \mathcal{Z} \\ e^{A\bar{h}} \mathcal{L} C \mathcal{Z} \end{bmatrix}^T \\ \begin{bmatrix} e^{A\bar{h}} \mathcal{L} C \mathcal{Z} \\ e^{A\bar{h}} \mathcal{L} C \mathcal{Z} \end{bmatrix} & \begin{bmatrix} \Pi_2 & A\mathcal{X}+B\mathcal{Y} \\ (A\mathcal{X}+B\mathcal{Y})^T & \mathcal{R} - \frac{1}{\Delta h}\mathcal{X} \end{bmatrix} \end{pmatrix} \le 0.$$

Applying the equivalent transformation

$$T^T \tilde{W} T \le 0,$$

where $T = \mathrm{diag}\{\mathcal{Q}, e^{-A^T \bar{h}} \mathcal{Q}, e^{-A^T \bar{h}} \mathcal{Q}\}$, we obtain

$$\begin{pmatrix} \mathcal{Q}(\Pi_1 + \frac{1}{\alpha} D Q_f^{-1} D^T + e^{\alpha \bar{h}} B \mathcal{Y} \mathcal{R}^{-1} \mathcal{Y}^T B^T)\mathcal{Q} & \begin{bmatrix} \mathcal{F}C \\ \mathcal{F}C \end{bmatrix}^T \\ \begin{bmatrix} \mathcal{F}C \\ \mathcal{F}C \end{bmatrix} & \Phi\Theta\Phi \end{pmatrix} \le 0, \qquad (9.52)$$

where

$$\Phi = \mathrm{diag}\{\mathcal{Q}, \mathcal{Q}\}.$$

Introduce the new variables $\beta \in \mathbb{R}_+$, $\mathcal{S} \in \mathbb{R}^{n \times n}$, $\mathcal{S} > 0$, and $\mathcal{V} \in \mathbb{R}^{2n \times 2n}$, $\mathcal{V} > 0$ such that

$$e^{\alpha \bar{h}} B \mathcal{Y} \mathcal{R}^{-1} \mathcal{Y}^T B^T \leq \beta I_n, \tag{9.53}$$

$$\mathcal{Q}(\Pi_1 + \frac{1}{\tau_1} D Q_f^{-1} D^T + \frac{1}{\tau_2} \mathcal{L} E Q_g^{-1} E^T \mathcal{L}^T + e^{\alpha \bar{h}} B \mathcal{Y} \mathcal{R}^{-1} \mathcal{Y}^T B^T) \mathcal{Q} \leq -\mathcal{S}, \tag{9.54}$$

and

$$\Phi \Theta \Phi \leq -\mathcal{V}. \tag{9.55}$$

I. Using the Schur complement, the inequality (9.53) can be rewritten in the form (9.48).

II. Taking into account the inequality (9.53), we derive that the inequality

$$\mathcal{Q}A + A^T \mathcal{Q} + \mathcal{F}C + C^T \mathcal{F} + \alpha \mathcal{Q}+$$

$$\frac{1}{\tau_1} \mathcal{Q} D Q_f^{-1} D^T \mathcal{Q} + \frac{1}{\tau_2} \mathcal{F} E Q_g^{-1} E^T \mathcal{F}^T + \beta \mathcal{Q}^2 \leq -\mathcal{S}$$

implies (9.54). Applying again the Schur complement, the last inequality becomes (9.49).

III. Our next considerations use the so-called Λ-inequality (the matrix Young inequality) (Poznyak 2008):

$$X^T Y + Y^T X \leq X^T \Lambda X + Y^T \Lambda^{-1} Y$$

$$\forall X, Y \in \mathbb{R}^{n \times k}, \ \Lambda \in \mathbb{R}^{n \times n}, \ \Lambda > 0.$$

Since the feasibility of (9.47) and (9.49) requires $\Theta < 0$, it follows that

$$E \begin{pmatrix} I_n & 0 \\ 0 & -\frac{1}{\sqrt{\Delta h}} I_n \end{pmatrix} + \begin{pmatrix} I_n & 0 \\ 0 & -\frac{1}{\sqrt{\Delta h}} I_n \end{pmatrix} E \leq$$

$$E(-\Theta)E + \begin{pmatrix} I_n & 0 \\ 0 & -\frac{1}{\sqrt{\Delta h}} I_n \end{pmatrix} (-\Theta^{-1}) \begin{pmatrix} I_n & 0 \\ 0 & -\frac{1}{\sqrt{\Delta h}} I_n \end{pmatrix}.$$

Hence, the inequality

$$-2\begin{pmatrix} \mathcal{Q} & 0 \\ 0 & -\frac{1}{\sqrt{\Delta h}}\mathcal{Q} \end{pmatrix} - \begin{pmatrix} I_n & 0 \\ 0 & -\frac{1}{\sqrt{\Delta h}}I_n \end{pmatrix}\Theta^{-1}\begin{pmatrix} I_n & 0 \\ 0 & -\frac{1}{\sqrt{\Delta h}}I_n \end{pmatrix} + \mathcal{V} \le 0,$$

which is equivalent to (9.42), implies (9.55).

Therefore, the inequalities (9.47)–(9.51) imply (9.47)–(9.50).   ∎

This lemma allows us to organize the procedure for tuning of controller and observer parameters based on semidefinite programming techniques. Indeed, for fixed $\alpha, \beta \in \mathbb{R}_+$, the system of matrix inequalities (9.47)–(9.51) assumes LMI form. So in order to minimize the attractive ellipsoid (9.51), we need to solve the following optimization problem:

$$\operatorname{tr}\{\mathcal{Q}^{-1}\} + \operatorname{tr}\{\mathcal{X}\} \to \min$$
$$\text{s.t. } (9.47)\text{–}(9.51).$$

The cost functional of the problem in nonlinear. Fortunately, this optimization problem is equivalent to the following SDP problem:

$$\operatorname{tr}\{\mathcal{H}\} + \operatorname{tr}\{\mathcal{X}\} \to \min$$
$$\text{s.t. } (9.47)\text{–}(9.51) \tag{9.56}$$

and

$$\begin{pmatrix} \mathcal{H} & I_n \\ I_n & \mathcal{Q} \end{pmatrix} \ge 0, \ \mathcal{H} > 0, \ \mathcal{H} \in \mathbb{R}^{n \times n}. \tag{9.57}$$

This equivalence obviously follows from the inequality

$$\mathcal{H} \ge \mathcal{Q}^{-1}$$

and the Schur complement.

For a given $\alpha, \beta \in \mathbb{R}_+$, let us denote by $J(\alpha, \beta)$ the solution of the optimization problem (9.56). The corresponding solution can be found using any SDP solver of MATLAB (for example, `SeDuMi`). In order to minimize the function $J(\alpha, \beta)$, some derivative-free method can be used (for example, the procedure `fminsearch` of MATLAB). Since Lemma 9.4 does not prove equivalence between conditions (9.47)–(9.42) and (9.47)–(9.57), this optimization scheme may give only a *suboptimal solution*. However, it may provide a good first approximation to the AEM optimal feedback.

9.2.6   Numerical Example

Consider the system (9.30) with parameters

$$
A = \begin{pmatrix} -0.1 & 1 & 0 \\ -0.1 & 0 & -0.4 \\ 0 & -0.1 & 0.2 \end{pmatrix}, \quad B = \begin{pmatrix} 1 \\ 0 \\ 1 \end{pmatrix},
$$

$$
C = \begin{pmatrix} 1 & 0 & 0 \\ 0 & 0 & 1 \end{pmatrix}, \quad D = \begin{pmatrix} 1 \\ 0.5 \\ 0 \end{pmatrix}, \quad E = \begin{pmatrix} 1 & 0 \\ 0 & 1 \end{pmatrix},
$$

$$
Q_f = 10^4 \quad \text{and} \quad Q_g = 10^4 \begin{pmatrix} 2 & -0.3 \\ -0.3 & 1.5 \end{pmatrix},
$$

$$
\underline{h} = 0.35, \quad \overline{h} = 0.5.
$$

The matrix A is unstable, $\lambda_1 = 0.273$, $\lambda_{2,3} = -0.0869 \pm 0.2830i$. The selected matrices Q_f and Q_g correspond to disturbances and noises of order $\mathcal{O}(10^{-2})$.

Using the optimization procedure (9.56) for $\alpha = 0.05$ and $\beta = 0.004$, we obtain the following suboptimal solution:

$$
K_{subopt} = \begin{pmatrix} -0.0232 & 0.0943 & -0.4889 \end{pmatrix},
$$

$$
L_{subopt} = \begin{pmatrix} -1.4276 & -0.4486 \\ -0.5563 & 0.1173 \\ 0.3131 & -0.6700 \end{pmatrix}.
$$

The numerical simulation results for the obtained output feedback control application are depicted in Figs. 9.5, 9.6, 9.7, and 9.8. They were carried out for

$$
h(t) = 0.35 + 0.15 \sin^2(t),
$$

$$
f(t, x) = 0.01 \cos(t),
$$

$$
g(t) = \begin{pmatrix} 0.0036 \sin(3t) - 0.0062 \cos(3t) \\ 0.0078 \sin(3t) + 0.0029 \cos(3t) \end{pmatrix},
$$

$x_0 = (0.5, 0, -0.1)^T$, and $u(t) = 0$ for $t \in [-\overline{h}, 0]$.

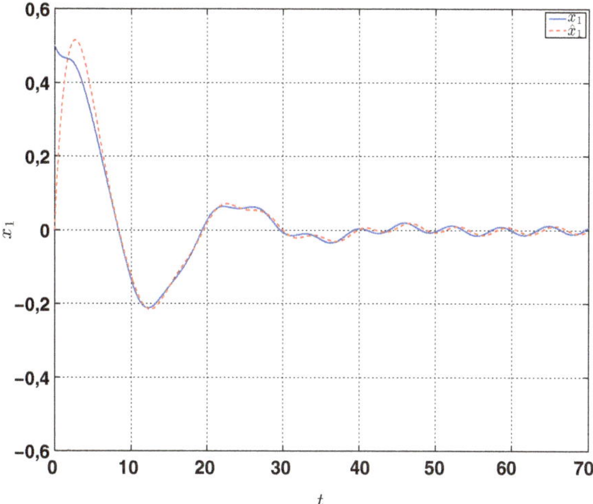

Fig. 9.5 Evolution of real and observed states x_1

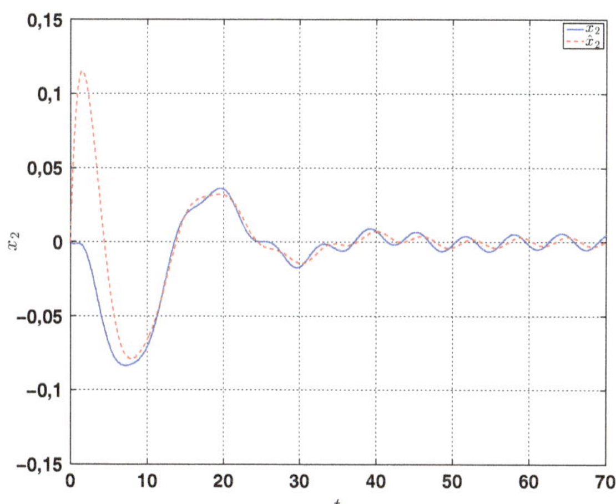

Fig. 9.6 Evolution of real and observed states x_2

9.3 Conclusion

In this chapter, a new approach to sliding mode control design for time-delay systems was introduced. It combines the classical predictor-based control algorithm with the attractive ellipsoid method, minimizing the effects of system disturbances. Corresponding attractive domain estimates were given. It was also shown that the minimal invariant ellipsoid of the prediction system corresponds to the minimal attractive ellipsoid for the original system. A numerical simulation is used to support the obtained theoretical results.

As we have shown in this chapter, the attractive ellipsoid method is applicable to a wide class of nonlinear systems containing a delay in the current state variable. Here we selected the control action as a full-state predictor-based linear control in

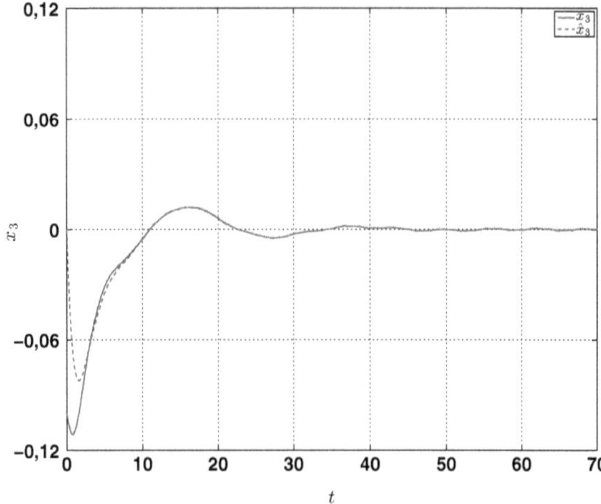

Fig. 9.7 Evolution of real and observed states x_3

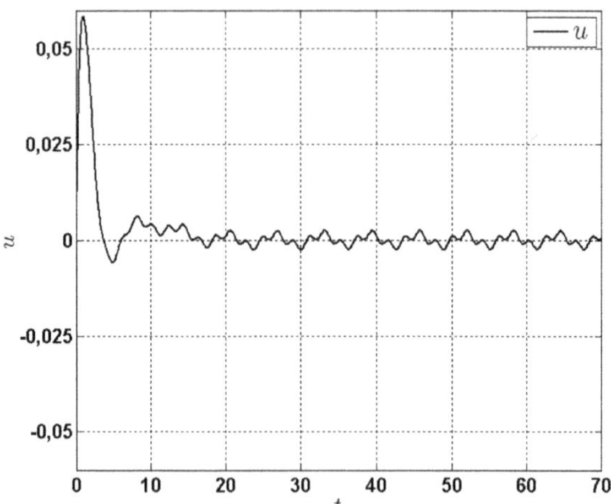

Fig. 9.8 Control signal u

spite of the fact the system to be controlled has nonlinear dynamics. We presented a method for numerical calculation of these parameters providing the "smallest" zone convergence for controlled trajectories. An illustrative example demonstrated the effectiveness of the suggested approach (Fig. 9.9).

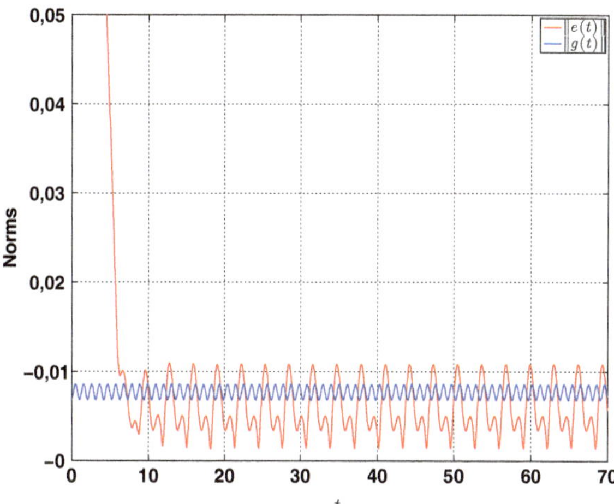

Fig. 9.9 The norms of the vectors of measurement noises $g(t)$ and observation errors $e(t)$

Chapter 10
Robust Control of Switched Systems

Abstract This chapter deals with the problem of robust feedback design for a class of switched systems in the presence of bounded model uncertainties as well as external perturbations. Only the output of the system is supposed to be available for a designer. We consider nonlinear dynamic models under arbitrary switching mechanisms assuming that sample-switching times are known. Online state estimates are obtained by the use of a Luenberger-like observer using only current inputs and general information on the class of model uncertainties. The stabilization issue is solved in the sense of practical stability, and it is carried out by a linear (with respect to a current state estimate) feedback switching controller subject to an average dwell time scheme. We apply the newly elaborated (extended) version of the conventional attractive ellipsoid method for this purpose. Numerically implementable sufficient conditions for the practical stability of systems are derived using bilinear matrix inequalities. The effectiveness of the proposed method is illustrated by an example of a continuous stirred tank reactor in which only the temperature (not the concentration) is available during the process.

Keywords Switched systems • Attractive ellipsoids • Output-based control

This chapter deals with the problem of robust feedback design for a class of switched systems in the presence of bounded model uncertainties as well as external perturbations. Only the output of the system is supposed to be available for a designer. We consider nonlinear dynamic models under arbitrary switching mechanisms assuming that sample-switching times are known. Online state estimates are obtained by the use of a Luenberger-like observer using only current inputs and general information on the class of model uncertainties. The stabilization issue is solved in the sense of practical stability, and it is carried out by a linear (with respect to a current state estimate) feedback switching controller subject to an average dwell time scheme. We apply the newly elaborated (extended) version of the conventional attractive ellipsoid method for this purpose. Numerically implementable sufficient conditions for the practical stability of systems are derived using bilinear matrix inequalities (BMIs). The effectiveness of the proposed method is illustrated by an example of a continuous stirred tank reactor in which only the temperature (not the concentration) is available during the process.

© Springer International Publishing Switzerland 2014

A. Poznyak et al., *Attractive Ellipsoids in Robust Control*, Systems & Control:
Foundations & Applications, DOI 10.1007/978-3-319-09210-2_10

10.1 Introduction

In recent years, switched model systems have received growing interest because of numerous engineering applications that can be modeled as switched systems—see, for example, Antsaklis (2000), Barkhordari-Yazdi and Jahed-Motlagh (2009), Boskovic and Mehra (2000), Kruszewski, Jiang, Fridman, Richard, and Toguyeni (2012), and Shorten (1996)—where models from industrial electronics, aircraft control, automotive control, networked systems, and chemical processes are treated.

By a *switched system* we mean a dynamical system that contains a set of continuous-time subsystems and a rule that determines the switching between them (see Liberzon 2003; Lin & Antsaklis 2009; Lunze & Lamnabhi-Lagarrigue 2009). In particular, such systems are suitable for describing dynamical systems located at a lower level of processes subject to upper-level logical dynamics or supervisor.

Many contributions regarding stability analysis of switched systems have been presented in the framework of Lyapunov stability (see Branicky 1998; DeCarlo et al. 2000; Liberzon & Morse 1999; Lin & Antsaklis 2009; Shorten & Cairbre 2001; Wicks et al. 1994 and references therein). The large number of existing results in the automatic control literature concerning the stability (Lyapunov stability) of these systems can be divided between two topics: the first deals with conditions that guarantee stability of a switched system under arbitrary switching rules, whereas the second deals with the issue of design a switching signal that makes a system stable (Liberzon & Morse 1999).

A significant problem inherent in all practical systems is the presence of perturbations or uncertainties. This issue has been an active area of research for many years (see Duncan & Schweppe 1971; Doyle 1983; Barabanov & Granichin 1984; Dahleh et al. 1988). Here we study the robust stabilization of a specific family of nonlinear switched systems with bounded uncertainties. Roughly speaking, the nonlinear uncertainty effects in the given dynamic models are modeled by the quasi-Lipschitz right-hand sides of the corresponding state equations. The stabilization issue is in the sense of practical stability, and the basic notions of this idea can be found in Ben-Abdallah 2009; Corless 1990, and Lakshmikantham, Leela, and Martynyuk (1990). We are interested in effective algorithms for an appropriate robust control design based on output feedback that guarantees the "practical stability" of the resulting closed-loop system. The robust control approach to be discussed in our contribution is based on three fundamental ideas:

- the well-known invariant ellipsoid approach (Azhmyakov 2011; Davila & Poznyak 2011; Gonzalez-Garcia, Polyakov, & Poznyak 2009; Kurzhanski & Veliov 1994; Kurzhanski & Varaiya 2006; Polyak et al. 2004; Poznyak et al. 2011);
- the multiple Lyapunov functions approach (Branicky 1998; Daafouz, Riedinger, & Iung 2002; Liberzon 2003); and
- the dwell time scheme approach (Hespanha & Morse 1999; Zhai, Hu, Yasuda, & Michel 2001).

We restrict our attention here to a class of linear output feedback control laws that stabilize the system for a class of switching signals. We are interested in a synchronously switching output-feedback control law, that is, one for which the switchings in the plant and the controller are synchronous. The switching signal is characterized by a dwell time scheme. We also assume that it is unknown a priori but that its instantaneous value is available at every time instant. The rate of convergence of the trajectories achieved by feedback controls is exponential. The robust control design strategy proposed in the present work is based on the "attractive ellipsoid" approach involving the theory of asymptotically stable invariant sets of dynamical systems interpreted here as "practical stability." Under some structural assumptions, it is possible to determine an attractive set (Lyapunov stable set) constructively. Here, this set is chosen in the form of an (attractive) ellipsoid in the given state space of the system. From the numerical point of view, the control synthesis problem is reduced to an auxiliary (relaxed) BMI-constrained optimization problem. The resulting ellipsoidal set possesses some optimal (minimal) properties and is constructively used in the main feedback-type control design procedure.

10.1.1 Some Preliminaries

Consider a nonlinear dynamical system

$$\dot{z}(t) = g(t, z(t), u(z(t))), \ \forall t \geq 0, \ z(0) = z_0, \tag{10.1}$$

where $z(t) \in \mathbb{R}^n$ is the state vector, $u(t) \in \mathbb{R}^m$ is the control input, and

$$g : \mathbb{R}_+ \times \mathbb{R}^n \times \mathbb{R}^m \mapsto \mathbb{R}^n$$

is a continuous function. Assume that for every admissible control $u(\cdot)$, the resulting system (10.1) has a solution $z(\cdot)$.

Here we study a family of admissible linear (with respect to state estimates) feedback controls. We now consider the closed-loop realization of (10.1) and the classical concepts from the set stability of dynamical systems (see, e.g., Blanchini & Miami 2008). A set $\mathcal{D}$ in the state space of the dynamical system (10.1) is said to be *(positively) invariant* if every trajectory that begins in this set remains inside the set at all future time instants. Denote by $\Omega(z(0))$ the (positive) limit set of (10.1) (the set of all positive limit points). We refer to the following conventional set stability concept (see Haddad & Chellaboina 2008): a compact invariant set $\mathcal{D} \subset \mathbb{R}^n$ is said to be *asymptotically Lyapunov stable* for a nonlinear dynamical system (in our case for the closed-loop system (10.1)) if

$$\Omega(z(0)) \in \mathcal{D}$$

and

- (*attractivity property*): there exists $\delta_1 > 0$ such that $\mathrm{dist}\{z_0, \mathcal{D}\} < \delta_1$ implies

$$\lim_{t \to \infty} \mathrm{dist}\{z(t), \mathcal{D}\} = 0,$$

where

$$\mathrm{dist}(\rho, \mathcal{D}) = \inf_{x \in \mathcal{D}} \|\rho - x\|;$$

- (*Lyapunov stability*): for all $\epsilon > 0$ there exists $\delta_2 > 0$ such that $\mathrm{dist}\{z_0, \mathcal{D}\} < \delta_2$ implies $\mathrm{dist}(z(t), \mathcal{D}) < \epsilon$ for all $t \geq 0$.

Below, we follow the concept of the attractive ellipsoid method introduced and discussed in previous chapters.

10.1.2   Problem Formulation

Let us analyze a specific class of (10.1), namely an affine system of switched nature with *quasi-Lipschitz right-hand sides*

$$\begin{aligned}
\dot{x}(t) &= f_{\sigma(t)}(x(t)) + B_{\sigma(t)}u(t) + v(t), \\
x(0) &= x_0, \quad \sigma(0) = \sigma_0, \\
y(t) &= C_{\sigma(t)}x(t) + \xi_y(t),
\end{aligned} \tag{10.2}$$

where $x(t)$, $x_0 \in \mathbb{R}^n$, $t \in \mathbb{R}_+$ are the state vector and the initial state vector, respectively. The control input is denoted by $u(t) \in \mathbb{R}^m$. The vector $y(t) \in \mathbb{R}^q$ describes the system output, and $v(t)$, $\xi_y(t)$ are bounded perturbations associated with the variables $x(t)$ and $y(t)$, respectively. Moreover,

$$\{f_i(\cdot)\}, \; i = 1, \ldots, M \in \mathbb{N}$$

is a family of quasi-Lipschitz functions $f_i : \mathbb{R}^n \to \mathbb{R}^n$ (see Definition 2.1). By

$$\{B_i, C_i\}, \; B_i \in \mathbb{R}^{n \times m}, \; C_i \in \mathbb{R}^{q \times n}$$

we denote the family of matrices of the given system. The time-dependent switching mechanism in (10.2) is determined by a piecewise-constant function

$$\sigma(t) \in \mathcal{I} = \{1, \ldots, M\}, \; t \in \mathbb{R}_+. \tag{10.3}$$

This function indicates the currently active subsystem given by the selected $f_{\sigma(t)}(\cdot)$, $B_{\sigma(t)}$, and $C_{\sigma(t)}$. The nonlinearities $f_i(x)$ belong to the class $\mathcal{C}(A_i, c_{1i}, c_{2i})$ of the quasi-Lispchitz functions, namely,

$$\mathcal{C}\left(A, c_{1i}, c_{2i}\right) := \{f_i : \mathbb{R}^n \mapsto \mathbb{R}^n |,$$

$$\|f_i(x) - A_i x\|^2 \le c_{1i} + c_{2i}\|x\|^2, \quad \forall x \in \mathbb{R}^n\}, \tag{10.4}$$

where $i = 1, \ldots, M$, $A_i \in \mathbb{R}^{n \times n}$ are suitable matrices, $c_{1i} > 0$ and $c_{2i} > 0$. We also introduce the following technical assumptions, hypothesis (**A**):

- perturbations acting on the dynamics are bounded, that is,

$$\|v(t)\|^2 \le \bar{v};$$

- the output noise is also assumed to be bounded, that is,

$$\|\xi_y(t)\|^2 \le \bar{\xi}_y;$$

- the pairs (A_i, B_i) and (A_i, C_i) are controllable/observable, respectively, for every $i = 1, \ldots, M$; and
- the *Zeno behavior* (infinite switchings in a finite time) in (10.3) is assumed to be excluded.

The dynamic transitions between the given subsystems (initially enumerated by i) occur at the given switching times t_k, where $k \in \mathbb{Z}$. We introduce some concepts concerning switched systems when the dwell time approach is used.

Definition 10.1. A constant $\tau_d > 0$, such that $t_k - t_{k-1} \ge \tau_d$, is called the **dwell time** because $\sigma(\cdot)$ dwells on each of its values for at least τ_d units of time.

Definition 10.2 (Hespanha & Morse 1999; Zhai et al. 2001). For a switching signal $\sigma(\cdot)$ and any $T_2 > T_1 \ge 0$, let $N(T_1, T_2)$ be the switching number of $\sigma(t)$ over the interval $[T_1, T_2)$. If

$$N(T_1, T_2) \le N_0 + \frac{T_2 - T_1}{\tau_{\mathrm{av}}} \tag{10.5}$$

holds for $N_0 \ge 1$, $\tau_{\mathrm{av}} > 0$, then τ_{av} is called the **average dwell time**, and N_0 the **chatter bound**.

Let us also define the auxiliary variable

$$\xi_x(x(t), t) := f_{\sigma(t)}(x(t)) - A_{\sigma(t)}x(t) + v(t), \; t \in \mathbb{R}_+$$

and rewrite (10.2) in a *quasilinear format* as

$$\begin{aligned} \dot{x}(t) &= A_{\sigma(t)}x(t) + B_{\sigma(t)}u(t) + \xi_x(x(t), t), \\ x(0) &= x_o, \; \sigma(0) = \sigma_0, \\ y(t) &= C_{\sigma(t)}x(t) + \xi_y(t). \end{aligned} \tag{10.6}$$

A control system of type (10.2) is usually associated with a set $\mathcal{U}$ of feasible control functions $u\,(\cdot)$. Here we deal with a class of linear control strategies based on the output feedback. To realize this strategy, we suggest using a suitable observation scheme. We apply the standard Luenberger observer governed by the following ODE:

$$\dot{\hat{x}}\,(t) = A_{\sigma(t)}\hat{x}\,(t) + B_{\sigma(t)}u\,(t) + L_{\sigma(t)}\,(y\,(t) - C\hat{x}\,(t))$$

$$\forall t \geq 0,\ \hat{x}\,(0) = \hat{x}_0,$$

(10.7)

where $L \in \mathbb{R}^{n \times q}$ is the matrix of the observer. Application of (10.7) to the switched system (10.2) gives rise to the explicit structure of the control function, namely

$$u\,(t) = -K_{\sigma(t)}\hat{x}\,(t)\,.$$

(10.8)

A control of type (10.8) is characterized by suitable gain matrices $K_i \in \mathbb{R}^{m \times n}$, $i = 1,\ldots,M$. For the estimation error vector

$$e\,(t) = x\,(t) - \hat{x}\,(t)\,,$$

we have $\forall t \geq 0$

$$\dot{e}\,(t) = \left(A_{\sigma(t)} - L_{\sigma(t)}C_{\sigma(t)}\right)e\,(t) +$$

$$\left(\xi_x\,(t) - L_{\sigma(t)}\xi_y\,(t)\right).$$

(10.9)

From (10.6)–(10.9), we can write the resulting closed-loop system in the following compact form:

$$\dot{z}\,(t) = \tilde{A}_{\sigma(t)}z\,(t) + F_{\sigma(t)}\psi\,(t)\,,\ z\,(0) = (x_0, e_0)^\mathsf{T}\,,$$

(10.10)

where

$$z\,(t) = [\hat{x}^\mathsf{T}\,(t)\,, e^\mathsf{T}\,(t)]^\mathsf{T}\,,\ \xi\,(t) = \left[\xi_x^\mathsf{T}\,(t)\,, \xi_y^\mathsf{T}\,(t)\right]^\mathsf{T}$$

and

$$\tilde{A}_{\sigma(t)} := \begin{bmatrix} A_{\sigma(t)} - B_{\sigma(t)}K_{\sigma(t)} & L_{\sigma(t)}C_{\sigma(t)} \\ 0 & A_{\sigma(t)} - L_{\sigma(t)}C_{\sigma(t)} \end{bmatrix},$$

$$F_{\sigma(t)} := \begin{bmatrix} 0 & L_{\sigma(t)} \\ I_{n \times n} & -L_{\sigma(t)} \end{bmatrix},$$

with $I_{n \times n} \in \mathbb{R}^{n \times n}$ as the n-unitary matrix.

Note that using the quasi-Lispchitz property, the assumptions in (**A**), and the Λ-inequality, we may conclude that the variable $\xi(t)$ satisfies

$$\|\xi(t)\|^2 = \|\xi_x(t)\|^2 + \|\xi_y(t)\|^2 \le$$

$$\bar{\xi}_y + \|f_i(x(t)) - A_i x(t) + v(t)\|^2 \le$$

$$\bar{\xi}_y + v^{\mathsf{T}}(t)\left(I + \Lambda^{-1}\right) v(t) +$$

$$(f_i(x(t)) - A_i x(t))^{\mathsf{T}}(I + \Lambda)(f_i(x(t)) - A_i x(t)).$$

Selecting $\Lambda = \rho I$, $\rho > 0$, we get

$$\|\xi(t)\|^2 \le \bar{\xi}_y + (\rho + 1)\|f_i(x(t)) - A_i x(t)\|^2 +$$
$$(\rho^{-1} + 1)\|v(t)\|^2 \le \tilde{c}_{1i} + \tilde{c}_{2i}\|x(t)\|^2$$
$$\tilde{c}_{1i} = \bar{\xi}_y + (\rho + 1)c_{1i} + (\rho^{-1} + 1)\bar{v},\ \tilde{c}_{2i} = (\rho + 1)c_{2i}.$$

Since

$$x = \hat{x} + e = Gz,\ G = \begin{bmatrix} I_n & I_n \end{bmatrix},$$

one may conclude that

$$\|\xi(t)\|^2 \le \tilde{c}_{1i} + \tilde{c}_{2i}\|Gz(t)\|^2. \tag{10.11}$$

Our aim is to design an output feedback control and determine suitable dynamic controller and observer matrices $\{K_i, L_i\}$ such that each invariant asymptotically stable ellipsoid

$$\mathcal{E}(P_i) := \{z \in \mathbb{R}^{2n} \mid z^T P_i^{-1} z \le 1\}$$

has "minimal size" (in this case, the *trace* of the ellipsoid's shape matrix P_i). This minimality requirement can be easily formalized by

$$\begin{aligned} \text{minimize tr}\,(P_i) \\ \text{subject to} \\ P_i^{\mathsf{T}} = P_i > 0, \\ P_i \in \Gamma_1(z(\cdot)),\ K_i,\ L_i \in \Gamma_2(z(\cdot)) \quad (i = 1, \dots, M). \end{aligned} \tag{10.12}$$

Here

- $\Gamma_1(z(\cdot))$ is a set of symmetric and positive definite $2n \times 2n$ matrices P_i $(i = 1, \dots, M)$ that guarantee the property from Definition 2.1 for $\mathcal{E}(0, P_i)$.
- $\Gamma_2(z(\cdot))$ is a subset of the space of K_i, L_i stabilizing matrices.

These subsets formally describe the admissible feedback K_i, observer L_i gains, and ellipsoidal matrices P_i. We assume that the basic optimization problem (10.12) has at least one optimal solution, denoted by $\{\hat{P}_i, \hat{K}_i, \hat{L}_i\}$ $(i = 1, \ldots, M)$.

This minimization problem formulated above guarantees the minimal "size" of the invariant ellipsoid $\mathcal{E}(P_i)$ under construction. Note that the set $\Gamma_1 \otimes \Gamma_2$ in (10.12) is a set of restrictions. Evidently, the main problem is to give a concrete constructive characterization of the given set of restrictions $\Gamma_1 \otimes \Gamma_2$.

An ellipsoid determined by an "optimal" family $\{\hat{P}_i\}$, where $i = 1, \ldots, M$, shall be called a *minimal attractive ellipsoid* for system (10.10). It is clear that the control strategy given by the corresponding sets $\{\hat{K}_i\}$, $\{L_i\}$ of optimal gain matrices will possess some robustness properties with respect to the above-mentioned attractive ellipsoidal region. A constructive solution of (10.12) constitutes our main generic approach to robust output feedback control design for the class of uncertain control systems of type (10.2).

10.2 Application of the Attractive Ellipsoid Method

The aim of this section is a constructive characterization of the set of restrictions $\Gamma_1\left(z\left(\cdot\right)\right) \otimes \Gamma_2\left(z\left(\cdot\right)\right)$ in (10.12). We use some specific facts for this purpose and finally obtain these restrictions in the form of BMIs. We introduce specific functions that possess a similarity with a Lyapunov function used in the stability of dynamical systems:

$$V_i(z) = z^{\mathsf{T}} P_i^{-1} z, \ i = 1, \ldots, M. \tag{10.13}$$

Here $P_i \in \mathbb{R}^{2n \times 2n}$ are positive definite symmetric matrices of the following structure:

$$P_i = \begin{bmatrix} P_{1i} & 0 \\ 0 & P_{2i} \end{bmatrix},$$

where $P_{i1} \in \mathbb{R}^{n \times n}$, $P_{2i} \in \mathbb{R}^{n \times n}$. Let us first compute the derivative of the function $V_i(z)$ along the trajectories of the corresponding subsystems in (10.10):

$$\dot{V}_i\left(z\left(t\right)\right) = z^{\mathsf{T}}\left(t\right) P_i^{-1}\left[\tilde{A}_i z\left(t\right) + F_i \xi\left(t\right)\right] + \left[\tilde{A}_i z\left(t\right) + F_i \xi\left(t\right)\right]^{\mathsf{T}} P_i^{-1} z\left(t\right)$$

$$= \eta^{\mathsf{T}}\left(t\right) \begin{bmatrix} P_i^{-1}\tilde{A}_i + \tilde{A}_i^{\mathsf{T}} P_i^{-1} & P_i^{-1} F_i \\ F_i^{\mathsf{T}} P_i^{-1} & 0 \end{bmatrix} \eta\left(t\right),$$

$$\tag{10.14}$$

where $\eta\left(t\right) = \left[z^{\mathsf{T}}\left(t\right), \xi^{\mathsf{T}}\left(t\right)\right]^{\mathsf{T}}$. From (10.11) and for constants $\alpha_i > 0$ and $\beta := \max_i \{\tilde{c}_{1i}\}$, we obtain

$$\dot{V}_i\left(z\left(t\right)\right) + \alpha_i\, V\left(z\left(t\right)\right) - \beta =$$

$$\eta^{\mathsf{T}}\left(t\right)\begin{bmatrix} P_i^{-1}\tilde{A}_i + \tilde{A}_i^{\mathsf{T}} P_i^{-1} + \alpha_i P_i^{-1} + \tilde{c}_{2i}\, G^{\mathsf{T}}G & P_i^{-1} F_i \\ F_i^{\mathsf{T}} P_i^{-1} & -I \end{bmatrix}\eta\left(t\right)$$

$$+\ \|\xi\|^2 - \beta - \tilde{c}_{2i}\, z^{\mathsf{T}}\left(t\right) G^{\mathsf{T}}G z\left(t\right).$$

By (10.11), it follows that

$$\|\xi\|^2 - \tilde{c}_{2i}\, z^{\mathsf{T}}\left(t\right) G^{\mathsf{T}}G z\left(t\right) \leq \beta,$$

and if additionally,

$$W_i\left(P_i, K_i, L_i, \alpha_i\right) = \begin{bmatrix} w_{11i} & P_i^{-1} F_i \\ F_i^{\mathsf{T}} P_i^{-1} & -I \end{bmatrix} \leq 0,$$

$$w_{11i} = P_i^{-1}\tilde{A}_i + \tilde{A}_i^{\mathsf{T}} P_i^{-1} + \alpha_i P_i^{-1} + \tilde{c}_{2i}\, G^{\mathsf{T}}G,$$

(10.15)

then one may conclude that the function $V_i\left(z(t)\right)$ satisfies the differential inequality

$$\dot{V}_i\left(z(t)\right) \leq -\alpha_i\, V_i\left(z(t)\right) + \beta, \quad \sigma(t) = i. \tag{10.16}$$

We may now formulate the following result.

Lemma 10.1. *Let the function $V_i\left(z(t)\right)$ be continuous together with its derivative on the half-open interval $\tau_r = [t_{r-1}, t_r)$ such that $\sigma\left(t\right) = i$, $t \in \tau_r$, and*

$$W_i\left(P_i, K_i, L_i, \alpha_i\right) \leq 0.$$

Then

$$V_i\left(z\left(t\right)\right) \leq V_i\left(z\left(t_{r-1}\right)\right) \exp\left(-\alpha_i\left(t - t_{r-1}\right)\right)$$

$$+ \frac{\beta}{\alpha_i}\left(1 - \exp\left(-\alpha_i\left(t - t_{k-1}\right)\right)\right). \tag{10.17}$$

Note that if the system remains in mode i for a large period of time, then an upper bound like $\dfrac{\beta}{\alpha_i}$ can be achieved.

10.2.1   Practical Stability

It is well known that the stability of each subsystem (10.2) does not guarantee the stability of the complete switched system. We refer to Liberzon (2003) for examples.

In this section, we study conditions that imply the practical stability of the switched systems under consideration.

Recall that we say that the system (10.2) is *practically stable* if there exists an attractive ellipsoidal set of the prescribed form associated with the dynamics of the system. Considering the ellipsoidal sets as attractive, we may associate the property of the practical stability with the state vector $z(t)$ satisfying

$$\limsup_{t \to \infty} z^{\mathsf{T}}(t) Q_{\sigma(t)} z(t) \leq 1$$

under the matrix constraints

$$Q_i \geq Q_{prescr} > 0, \quad i = 1, \dots, M$$

for a given matrix $Q_{prescr} \in \mathbb{R}^{2n \times 2n}$.

Let us now analyze the practical stability property when the switched system is under a dwell-time scheme of commutation. We define the piecewise continuous "Lyapunov-like function"

$$\mathbf{V}(t) = V_{\sigma(t)}(z(t)) = z^{\mathsf{T}}(t) P_{\sigma(t)}^{-1} z(t) \tag{10.18}$$

for the switched system (10.10), where each $V_i(z)$ is as in (10.17). Also suppose that there exists a constant $\mu > 1$ such that

$$V_i(z) \leq \mu V_j(z), \qquad \forall z \in \mathbb{R}^{2n}, \quad \forall i, j \in \mathcal{I}. \tag{10.19}$$

This last property is satisfied, for example, with

$$\mu = \sup_{r,l \in \mathcal{I}} \frac{\lambda_{\max}\left(P_r^{-1}\right)}{\lambda_{\min}\left(P_l^{-1}\right)},$$

where $\lambda_{\max}(P)$ $(\lambda_{\min}(P))$ denotes the largest (smallest) eigenvalue of the positive definite symmetric matrix P. Therefore, in the switching times, we have

$$\mathbf{V}(t_k) \leq \mu \lim_{t \to t_k - 0} V_{\sigma(t)}(z(t)) = \mu \mathbf{V}(t_k^-), \quad k \in \mathbb{Z}. \tag{10.20}$$

Let $N(t_0, t)$ be the number of switchings of $\sigma(\cdot)$ in the open interval (t_0, t), such that

$$0 \leq t_0 < t_1 \cdots < t_{N(t_0,t)} < t < t_{N(t_0,t)+1} = T.$$

Define

$$\overline{\alpha}_k := \alpha_{\sigma(t)} \in \mathcal{I}, \, t \in \tau_k.$$

By (10.17), for an arbitrary switching signal $\sigma(\cdot)$ such that $\tau_k \geq \tau_{av}$, we have (10.20) and

$$\mathbf{V}(t_N) \leq \mu \mathbf{V}(t_N^-) \leq \mu \exp\left(-\bar{\alpha}(t_N - t_{N-1})\right) \mathbf{V}(t_{N-1})$$

$$+ \frac{\beta}{\bar{\alpha}_N} \mu \left[1 - \exp\left(-\bar{\alpha}(t_N - t_{N-1})\right)\right]$$

$$\leq \mu^2 \exp\left(-\sum_{k=0}^{1} \bar{\alpha}_{N-k}\bar{\tau}_{N-k}\right) \mathbf{V}(t_{N-2}) + \frac{\beta}{\bar{\alpha}_N} \mu \left[1 - \exp\left(-\bar{\alpha}_N \bar{\tau}_N\right)\right]$$

$$+ \frac{\beta}{\bar{\alpha}_{N-1}} \mu^2 \left[1 - \exp\left(-\bar{\alpha}_{N-1}\bar{\tau}_{N-1}\right)\right] \exp\left(-\bar{\alpha}_N \bar{\tau}_N\right)$$

$$\leq \mu^3 \exp\left(-\sum_{k=0}^{2} \bar{\alpha}_{N-k}\bar{\tau}_{N-k}\right) \mathbf{V}(t_{N-3}) + \frac{\beta}{\bar{\alpha}_N} \mu \left[1 - \exp\left(-\bar{\alpha}_N \bar{\tau}_N\right)\right]$$

$$+ \frac{\beta}{\bar{\alpha}_{N-1}} \mu^2 \left[1 - \exp\left(-\bar{\alpha}_{N-1}\bar{\tau}_{N-1}\right)\right] \exp\left(-\bar{\alpha}_N \bar{\tau}_N\right)$$

$$+ \frac{\beta}{\bar{\alpha}_{N-2}} \mu^3 \left[1 - \exp\left(-\bar{\alpha}_{N-2}\bar{\tau}_{N-2}\right)\right] \exp\left(-\bar{\alpha}_N \bar{\tau}_N - \bar{\alpha}_{N-1}\bar{\tau}_{N-1}\right) \leq \cdots$$

$$\leq \mu^{N(t_0,t)} \exp\left(-\sum_{k=0}^{N(t_0,t)-1} \bar{\alpha}_{N-k}\bar{\tau}_{N-k}\right) \mathbf{V}(t_0) + \frac{\beta}{\bar{\alpha}_N} \mu \left[1 - \exp\left(-\bar{\alpha}_N \bar{\tau}_N\right)\right]$$

$$+ \beta \sum_{k=1}^{N(t_0,t)-1} \frac{\mu^{k+1}}{\bar{\alpha}_{N-k}} \left[1 - \exp\left(-\bar{\alpha}_{N-k}\bar{\tau}_{N-k}\right)\right] \exp\left(-\sum_{l=0}^{k-1} \bar{\alpha}_{N-l}\bar{\tau}_{N-l}\right).$$

Here $\bar{\tau}_k$ is the length of the interval τ_k. The last inequality implies

$$\mathbf{V}(t_N) \leq \mu^{N(t_0,t)} \exp\left(-\alpha_{\min} \sum_{k=0}^{N(t_0,t)-1} \bar{\tau}_{N-k}\right) \mathbf{V}(t_0) +$$

$$\frac{\beta}{\alpha_{\min}} \mu \left(1 - \exp\left(-\alpha_{\max}\bar{\tau}_N\right)\right) + \tag{10.21}$$

$$\frac{\beta}{\alpha_{\min}} \sum_{k=0}^{N(t_0,t)-1} \mu^{k+1} \left(1 - \exp\left(-\alpha_{\max}\bar{\tau}_N\right)\right) \exp\left(-\alpha_{\min} \sum_{l=0}^{k-1} \bar{\tau}_{N-l}\right).$$

For the first term of (10.21), we have

$$\mu^{N(t_0,t)} \exp\left(-\alpha_{\min} \sum_{k=0}^{N(t_0,t)-1} \bar{\tau}_{N-k}\right) =$$

$$\mu^{N(t_0,t)} \exp\left(-\alpha_{\min}(t_N - t_0)\right) = \exp\left(N \log \mu - \alpha_{\min}(t_N - t_0)\right).$$

To guarantee a decay rate γ_1, we must have that

$$N(t_N^+, t_0) \log \mu - \alpha_{\min}(t_N - t_0) \leq \gamma_0 - \gamma_1(t_N - t_0),$$

where $\gamma_0 > 0$, $\gamma_1 > 0$, or equivalently,

$$N(t_N^+, t_0) \leq \frac{\gamma_0}{\log \mu} + \frac{(\alpha_{\min} - \gamma_1)(t_N - t_0)}{\log \mu}. \tag{10.22}$$

Inequality (10.22) has the form

$$N(t_N^+, t_0) \leq N_0 + \frac{t_N - t_0}{\tau_{av}}$$

with

$$N_0 = \frac{\gamma_0}{\log \mu} \quad \text{and} \quad \tau_{av} = \frac{\log \mu}{\alpha_{\min} - \gamma_1}$$

subject to

$$0 < \gamma_1 < \alpha_{\min}.$$

For the other two terms, we observe that

$$1 - \exp\left(-\alpha_{\max} \bar{\tau}_{N-k}\right) < 1, \quad k = 0, 1, 2, \ldots, N(t_0, t) - 1.$$

Considering $\bar{\tau}_{N-l} \geq \tau_{av}$, we get

$$\frac{\beta}{\alpha_{\min}} \mu \left[1 - \exp\left(-\alpha_{\max} \bar{\tau}_N\right)\right] +$$

$$\frac{\beta}{\alpha_{\min}} \sum_{k=1}^{N(t_0,t)-1} \mu^{k+1} \left(1 - \exp\left(-\alpha_{\max} \bar{\tau}_{N-k}\right)\right) \exp\left(-\alpha_{\min} \sum_{l=0}^{k-1} \bar{\tau}_{N-l}\right)$$

$$\leq \frac{\beta}{\alpha_{\min}} \mu \left[1 + \sum_{k=1}^{N(t_0,t)-1} \mu^k \exp\left(-\alpha_{\min} \sum_{l=0}^{k-1} \bar{\tau}_{N-l}\right)\right]$$

$$\leq \frac{\beta}{\alpha_{\min}} \mu \sum_{k=0}^{N(t_0,t)-1} \mu^k \exp\left(-\alpha_{\min} k \tau_{av}\right)$$

$$\leq \frac{\beta}{\alpha_{\min}} \mu \sum_{k=0}^{N(t_0,t)-1} \exp\left(k(\log \mu - \alpha_{\min} \tau_{av})\right).$$

Choose

$$\gamma_1 = \frac{1}{2}\alpha_{\min},$$

which implies that

$$\tau_{av} = \frac{2\log\mu}{\alpha_{\min}}.$$

Then we may conclude that

$$\frac{\beta}{\alpha_{\min}}\mu \sum_{k=0}^{N(t_0,t)-1} \exp\left(-k\log\mu\right) = \frac{\beta}{\alpha_{\min}}\mu \sum_{k=0}^{N(t_0,t)-1} \left(\frac{1}{\mu}\right)^k,$$

implying

$$\frac{\beta}{\alpha_{\min}}\mu \sum_{k=0}^{N(t_0,t)-1}\left(\frac{1}{\mu}\right)^k = \frac{\beta}{\alpha_{\min}}\mu \left[\frac{1-\left(\frac{1}{\mu}\right)^{N(t_0,t)}}{1-\frac{1}{\mu}}\right]$$

$$= \frac{\beta}{\alpha_{\min}}\mu^2 \left[\frac{1-\left(\frac{1}{\mu}\right)^{N(t_0,t)}}{\mu-1}\right].$$

Therefore,

$$\mathbf{V}(t_N) \le \exp\left(\gamma_0 - \gamma_1\left(t_N - t_0\right)\right)\mathbf{V}(t_0)$$

$$+\frac{\beta}{\alpha_{\min}}\mu^2 \left[\frac{1-\left(\frac{1}{\mu}\right)^{N(t_0,t)}}{\mu-1}\right], \tag{10.23}$$

and hence

$$\limsup_{N(t_0,t)\to\infty} \mathbf{V}(t_N) \le \frac{\beta}{\alpha_{\min}}\left(\frac{\mu^2}{\mu-1}\right). \tag{10.24}$$

We formalize the result of the foregoing in the following theorem.

Theorem 10.1. *Assume that all subsystems in (10.9) satisfy the assumptions of Lemma 10.1. Let* $\mathbf{V}(t) = z^\mathsf{T}(t)\, P_{\sigma(t)}^{-1} z(t)$ *be a piecewise continuous function, and suppose that there exists a constant* $\mu > 1$ *such that*

$$V_i(z) \le \mu V_j(z), \quad \forall i, j \in \mathcal{I}.$$

Then for all positive constants γ_0 and

$$\gamma_1 = \alpha_{\min} \left(\frac{r-1}{r} \right), \quad r = 2, 3, 4, \dots,$$

($\alpha_{\min} := \min\limits_{i} \alpha_i$, α_i are decay parameters in Lemma 10.1), there exists a finite constant

$$\tau_{\mathrm{av}} = \frac{\log \mu}{\alpha_{\min} - \gamma_1}$$

such that $\mathbf{V}(t)$ is a storage function for the switched system satisfying

$$\mathbf{V}(t) \leq \exp\left(\gamma_0 - \gamma_1 (t - t_0)\right) \mathbf{V}(t_0)$$

$$+ \frac{\beta}{\alpha_{\min}} \mu^r \left(\frac{1 - \left(\frac{1}{\mu^{r-1}}\right)^{N(t_0,t)}}{\mu^{r-1} - 1} \right),$$

with decay rate γ_1 and for any average dwell time $\tau_d \geq \tau_{\mathrm{av}}$. Moreover,

$$\limsup_{t \to \infty} \mathbf{V}(t) \leq \frac{\beta}{\alpha_{\min}} \left(\frac{\mu^r}{\mu^{r-1} - 1} \right). \tag{10.25}$$

Remark 10.1. The bound (10.24) corresponds to the choice of decay rate $\gamma_1 = \frac{1}{2}\alpha_{\min}$. It is clear that there is a tradeoff in the choice of the bound

$$\epsilon = \frac{\beta}{\alpha_{\min}} \left(\frac{\mu^r}{\mu^{r-1} - 1} \right)$$

and the dwell time τ_d.

10.2.2   Intersection of Ellipsoids

As it has been mentioned before for that each half-open time-interval $t \in [\tau_{r-1}, \tau_r)$, we have

$$\frac{d}{dt} V_{q(t)}(t) \leq -\alpha_{q(t)} V_{q(t)}(t) + \beta,$$

$$V_{q(t)}(t) = x^{\mathsf{T}}(t) P_{q(t)}^{-1} x(t).$$

We introduce the new functions

$$
\mathcal{G}_{q(t)}(t) = \left(\left[\sqrt{\frac{\alpha_{q(t)}}{\beta} V_{q(t)}(t)} - 1 \right]_+ \right)^2,
\tag{10.26}
$$

where $[\cdot]_+$ is defined as

$$
[z]_+ = \begin{cases} z, & z \geq 0, \\ 0, & z < 0. \end{cases}
\tag{10.27}
$$

Observe that this function is not differentiable at the point $z = 0$, but $\left([z]_+\right)^2$ is differentiable everywhere.

Next, consider the following Lyapunov-like function defined on the trajectories for our switched system where the designated feedback is also switched synchronously:

$$
G(t) := \sum_{r=1}^{\infty} \chi_r(t) \, \mathcal{G}_{q(t)}(t),
\tag{10.28}
$$

where the characteristic function $\chi_r(t)$ of the interval $t \in [\tau_{r-1}, \tau_r)$ is defined as

$$
\chi_r(t) := \begin{cases} 1 & \text{if } t \in [\tau_{r-1}, \tau_r), \\ 0 & \text{if } t \notin [\tau_{r-1}, \tau_r). \end{cases}
$$

Note that

$$
\sum_{r=1}^{\infty} \chi_r(t) = 1.
$$

Suppose also that each structure appears during the process infinitely many times (*ergodicity property*), that is, for all $i = 1, \ldots, M$, we have

$$
\int_{t=0}^{\infty} \chi_{q(t)=i} \, dt = \sum_{r=1}^{\infty} \int_{t=t_{r-1}}^{t_r} \chi_{q(t)=i} \, dt
\tag{10.29}
$$

$$
= \sum_{r=1}^{\infty} \chi_{q(t_{r-1})=i} \, (t_r - t_{r-1}) = \infty.
$$

Theorem 10.2. *On the trajectories of the system (10.2) with switched structure closed by the switched feedback (10.8) satisfying*

$$\mathcal{G}_{q(\tau_{r-1})}(\tau_r) = \left(\left[\sqrt{x^{\mathsf{T}}(\tau_r)\,\frac{\alpha_{q(\tau_{r-1})}}{\beta}\,P^{-1}_{q(\tau_{r-1})}x(\tau_r)} - 1\right]_+\right)^2$$
$$\geq \left(\left[\sqrt{x^{\mathsf{T}}(\tau_r)\,\frac{\alpha_{q(\tau_r)}}{\beta}\,P^{-1}_{q(\tau_r)}x(\tau_r)} - 1\right]_+\right)^2 = \mathcal{G}_{q(\tau_r)}(\tau_r),$$

(10.30)

we have the following:

(1) There exists a "dominating process" $\tilde{G}(t)$ satisfying

$$G(t) \leq \tilde{G}(t) \quad \text{for all } t \geq 0, \; G(0) = \tilde{G}(0),$$

$$G(s) = \tilde{G}(s) \quad \text{for all } s \in T_0 := \{s : G(s) = 0\},$$

(10.31)

$$\frac{d}{dt}\tilde{G}(t) < 0 \quad \text{for all } t \;\; \text{such that } \mathcal{G}_r(t) > 0,$$

which means that $G(t)$ is a monotonically nonincreasing function.

(2)

$$\lim_{t\to\infty} G(t) = 0.$$

(10.32)

Proof. (1) Recall (Gel'fand & Shilov 1968) that the "generalized" derivative of the Heaviside function

$$\theta(t) := \begin{cases} 1 & \text{if} \quad t \geq 0, \\ 0 & \text{if} \quad t < 0, \end{cases}$$

is the Dirac delta function, namely

$$\delta(t) = \theta'(t),$$

possessing the property

$$\int_{t=-\infty}^{\infty} \delta(t)\, f(t)\, dt = f(0)$$

for every function $f(t)$ right-continuous at the origin. In view of the fact that

$$\chi_r(t) = \theta(t - \tau_{r-1}) - \theta(t - \tau_r),$$

differentiation of (10.28) leads to

$$\frac{d}{dt}G\left(t\right) = \sum_{r=1}^{\infty} \frac{d}{dt}\left[\chi_r\left(t\right)\mathcal{G}_{q(t)}\left(t\right)\right] = \sum_{r=1}^{\infty} \chi_r\left(t\right)\frac{d}{dt}\mathcal{G}_{q(\tau_{r-1})}\left(t\right)$$

$$+ \sum_{r=1}^{\infty}\left[\delta\left(t - \tau_{r-1}\right) - \delta\left(t - \tau_r\right)\right]\mathcal{G}_{q(t)}\left(t\right).$$

This is a singularly perturbed differential equation that in the equivalent integral form can be represented as follows:

$$G\left(t\right) - G(0) = \int_{s=0}^{t}\sum_{r=1}^{\infty}\left[\delta\left(s - \tau_{r-1}\right) - \delta\left(s - \tau_r\right)\right]\mathcal{G}_{q(s)}\left(s\right) ds$$

$$+ \int_{s=0}^{t}\left[\sum_{r=1}^{\infty}\chi_r\left(s\right)\frac{d}{ds}\mathcal{G}_{(t_{r-1})}\left(s\right)\right]ds = \sum_{r=1}^{\infty}\left[\mathcal{G}_{q(\tau_{r-1})}\left(\tau_{r-1}\right) - \mathcal{G}_{q(\tau_r)}\left(\tau_r\right)\right]$$

$$+ \int_{s=0}^{t}\sum_{r=1}^{\infty}\chi_r\left(s\right)\left[\sqrt{\frac{\alpha_{q(\tau_{r-1})}}{\beta}V_{q(\tau_{r-1})}(s)} - 1\right]_{+}\frac{\frac{\alpha_{q(\tau_{r-1})}}{\beta}\left(\frac{d}{ds}V_{q(\tau_{r-1})}(s)\right)}{\sqrt{\frac{\alpha_{q(\tau_{r-1})}}{\beta}V_{q(\tau_{r-1})}(s)}}ds$$

$$\leq \mathcal{G}_{q(\tau_0)}\left(\tau_0\right) + \left[-\mathcal{G}_{q(\tau_1)}\left(\tau_1\right) + \mathcal{G}_{q(\tau_1)}\left(\tau_1\right)\right]$$

$$+ \left[-\mathcal{G}_{q(\tau_2)}\left(\tau_2\right) + \mathcal{G}_{q(\tau_2)}\left(\tau_2\right)\right] + \ldots + \left[-\mathcal{G}_{q(\tau_r)}\left(\tau_r\right) + \mathcal{G}_{q(\tau_r)}\left(\tau_r\right)\right] + \ldots$$

$$\int_{s=0}^{t}\sum_{r=1}^{\infty}\chi_r\left(s\right)\left[\sqrt{\frac{\alpha_{q(\tau_{r-1})}}{\beta}V_{q(\tau_{r-1})}(s)} - 1\right]_{+}\frac{\frac{\alpha_{q(\tau_{r-1})}}{\beta}\left(-\alpha_{q(\tau_{r-1})}V_{q(\tau_{r-1})}(s)+\beta\right)}{\sqrt{\frac{\alpha_{q(\tau_{r-1})}}{\beta}V_{q(\tau_{r-1})}(s)}}ds.$$

Taking into account the "monotonicity condition" (10.30), we get

$$G\left(t\right) - G(0) \leq \mathcal{G}_{q(\tau_0)}\left(\tau_0\right) + I(t),$$

where

$$I(t) := \int_{s=0}^{t}\sum_{r=1}^{\infty}\chi_r\left(s\right)\left[\sqrt{\frac{\alpha_{q(\tau_{r-1})}}{\beta}V_{q(\tau_{r-1})}(s)} - 1\right]_{+}\frac{\frac{\alpha_{q(\tau_{r-1})}}{\beta}\left(-\alpha_{q(\tau_{r-1})}V_{q(\tau_{r-1})}(s)+\beta\right)}{\sqrt{\frac{\alpha_{q(\tau_{r-1})}}{\beta}V_{q(\tau_{r-1})}(s)}}ds$$

$$= -\int_{s=0}^{t}\sum_{r=1}^{\infty}\alpha_{q(\tau_{r-1})}\chi_r\left(s\right)\left[\sqrt{\frac{\alpha_{q(\tau_{r-1})}}{\beta}V_{q(\tau_{r-1})}(s)} - 1\right]_{+}^{2}\frac{\left(\sqrt{\frac{\alpha_{q(\tau_{r-1})}}{\beta}V_{q(\tau_{r-1})}(s)}+1\right)}{\sqrt{\frac{\alpha_{q(\tau_{r-1})}}{\beta}V_{q(\tau_{r-1})}(s)}}ds$$

$$= -\int_{s=0}^{t}\sum_{r=1}^{\infty}\alpha_{q(\tau_{r-1})}\chi_r\left(s\right)\mathcal{G}_{q(\tau_{r-1})}\left(s\right)\frac{\left(\sqrt{\frac{\alpha_{q(\tau_{r-1})}}{\beta}V_{q(\tau_{r-1})}(s)}+1\right)}{\sqrt{\frac{\alpha_{q(\tau_{r-1})}}{\beta}V_{q(\tau_{r-1})}(s)}}ds \leq 0.$$

$$(10.33)$$

We now introduce the so-called dominating process $\tilde{G}(t)$, for which

$$\tilde{G}(t) - \tilde{G}(0) = \mathcal{G}_{q(\tau_0)}(\tau_0) + I(t).$$

Obviously, if $G(0) = \tilde{G}(0)$, then

$$G(t) \leq \tilde{G}(t).$$

Differentiation of the last identity implies

$$\frac{d}{dt}\tilde{G}(t) = -\sum_{r=1}^{\infty} \alpha_{q(\tau_{r-1})} \chi_r(t) \mathcal{G}_{q(\tau_{r-1})}(t) \mathcal{H}(q(\tau_{r-1}), t),$$

where

$$\mathcal{H}(q(\tau_{r-1}), t) = \frac{\left(\sqrt{\frac{\alpha_{q(\tau_{r-1})}}{\beta} V_{q(\tau_{r-1})}(t)} + 1\right)}{\sqrt{\frac{\alpha_{q(\tau_{r-1})}}{\beta} V_{q(\tau_{r-1})}(t)}}.$$

The right-hand side of the last expression is strictly negative because of the inequality

$$\mathcal{G}_{q(\tau_{r-1})}(t) > 0.$$

This completes the proof of assertion (1).

(2) Since $\tilde{G}(t)$ is a nonnegative monotonically nonincreasing function, by Weierstrass's theorem, it follows that $\tilde{G}(t)$ has a limit, that is, that the limit

$$\lim_{t\to\infty} \tilde{G}(t) = \tilde{G}^*$$

exists. From the previous relations, it follows that

$$0 \leq \tilde{G}(t) + |I(t)| = \mathcal{G}_{q(\tau_0)}(\tau_0) + \tilde{G}(0) = \text{const}.$$

Taking $t \to \infty$, we obtain

$$\limsup_{t\to\infty} |I(t)| < \infty.$$

The convergence of the integral $|I(\infty)|$ means that there exists a time sequence $\{s_k\}_{k=1,2,\ldots}$ such that

$$\sum_{r=1}^{\infty} \alpha_{q(\tau_{r-1})} \chi_r(s_k) \mathcal{G}_{q(\tau_{r-1})}(s_k) \mathcal{H}(q(\tau_{r-1}), s_k) \xrightarrow[k\to\infty]{} 0,$$

which means that

$$\sum_{r=1}^{\infty} \chi_r(s_k) \mathcal{G}_{q(\tau_{r-1})}(s_k) = G(s_k) \xrightarrow[k \to \infty]{} 0, \tag{10.34}$$

since

$$\alpha_{q(\tau_{r-1})} \mathcal{H}(q(\tau_{r-1}), s_k) \geq \alpha_{q(\tau_{r-1})} \geq \min_{i=1,\dots,M} \alpha_i > 0.$$

Hence, from the continuity of $\tilde{G}(s)$, we have

$$\tilde{G}(s_k) \xrightarrow[k \to \infty]{} 0,$$

which implies $\tilde{G}^* = 0$, and consequently,

$$G(t) \xrightarrow[t \to \infty]{} 0.$$

This completes the proof of the theorem. ∎

Remark 10.2. In view of the identity

$$\chi_r(s) \chi_{q(\tau_{r-1})=i} = \chi_{q(s)=i},$$

it follows that

$$\int_{s=0}^{\infty} \sum_{r=1}^{\infty} \chi_r(s) \mathcal{G}_{q(\tau_{r-1})}(s)\,ds = \int_{s=0}^{\infty} \sum_{r=1}^{\infty} \chi_r(s) \left[\sum_{i=1}^{M} \chi_{q(\tau_{r-1})=i} \mathcal{G}_i(s) \right] ds$$

$$= \sum_{i=1}^{M} \int_{s=0}^{\infty} \sum_{r=1}^{\infty} \chi_r(s) \chi_{q(\tau_{r-1})=i} \mathcal{G}_i(s)\,ds =$$

$$\sum_{i=1}^{M} \int_{s=0}^{\infty} \sum_{r=1}^{\infty} \chi_r(s) \chi_{q(\tau_{r-1})=i} \mathcal{G}_i(s)\,ds = \sum_{i=1}^{M} \int_{s=0}^{\infty} \chi_{q(s)=i} \mathcal{G}_i(s)\,ds < \infty,$$

implying that for all $i = 1, \dots, M$,

$$\int_{s=0}^{\infty} \chi_{q(s)=i} \mathcal{G}_i(s)\,ds < \infty.$$

But taking into account the assumption (10.29), we get

$$\mathcal{G}_i(s_k) = \left[\sqrt{\frac{\alpha_i}{\beta}} V_i(s_k) - 1 \right]_+^2 \xrightarrow[k \to \infty]{} 0.$$

This means that every admissible trajectory of the considered controlled switched system converges to the intersection of individual ellipsoids, namely

$$z(t) \underset{t \to \infty}{\to} \bigcap_{i=1}^{M} \mathcal{E}\left(\frac{\beta}{\alpha_i} P_i\right).$$

Corollary 10.1. *Assume that the optimization problem*

$$minimize \sum_{i=1}^{M} \frac{\beta}{\alpha_i} \mathrm{tr}(P_i) + \mu$$
$$subject\ to$$
$$W_i\,(P_i, K_i, L_i, \alpha_i) \leq 0, \quad \forall i, j = \{1, \ldots, M\},$$
$$P_i^{-1} \leq \mu P_j^{-1}, \ i \neq j, \ P_i^{-1} \geq \frac{\beta}{\alpha_i} Q_{presc},$$
$$P_i > 0, \ \alpha_i > 0, \ \mu > 1,$$

(10.35)

has the optimal solution

$$\hat{\Upsilon} := \left(\hat{P}_i, \hat{K}_i, \hat{L}_i, \hat{\alpha}_i, \hat{\mu}\right)_{i=1,\ldots,M}.$$

Then the intersection of the ellipsoids $\mathcal{E}\left(\dfrac{\beta}{\alpha_i}\hat{P}_i\right)$ *is the minimal attractive set for the system (10.2).*

10.2.3   Bilinear Matrix Inequality Representation

The matrix constraints in problem (10.35) are given by BMIs. We can solve the optimization problem using some standard and advanced computational tools; namely, we use the PENBMI MATLAB package for this purpose. But at this point, it is still difficult to solve the optimization problem with bilinear matrix restrictions as outlined in (10.35). In order to overcome this issue, we give some relaxations of the constraints in the optimization problem.

Note that the main constraint in the above optimization problem is given by (10.15). From (10.14), using the Λ-matrix inequality, we get

$$z^\mathsf{T}(t)\, P_i^{-1} F_i \xi(t) + \xi^\mathsf{T}(t)\, F_i^\mathsf{T} P_i^{-1} z(t) \leq$$

$$z^\mathsf{T}(t)\, P_i^{-1} \Lambda P_i^{-1} z(t) + \xi^\mathsf{T}(t)\, F_i^\mathsf{T} \Lambda^{-1} F_i \xi(t).$$

We also include two new restrictions:

$$P_i^{-1} \Lambda P_i^{-1} \leq \Pi_{1i}, \ \Pi_{1i} = \Pi_{1i}^{\mathsf{T}} > 0, \ \Pi_{1i} \in \mathbb{R}^{2n \times 2n},$$

$$F_i^{\mathsf{T}} \Lambda^{-1} F_i \leq \Pi_{2i}, \ \Pi_{2i} = \Pi_{2i}^{\mathsf{T}} > 0, \ \Pi_{2i} \in \mathbb{R}^{(n+q) \times (n+q)}.$$

Using the Schur complement, these matrix inequalities may be represented as

$$\Theta_{1i} = \begin{bmatrix} \Pi_{1i} & P_i^{-1} \\ P_i^{-1} & \Lambda^{-1} \end{bmatrix} \geq 0, \ \Theta_{2i} = \begin{bmatrix} \Pi_{2i} & F_i^{\mathsf{T}} \\ F_i & \Lambda \end{bmatrix} \geq 0.$$

Then (10.15) becomes

$$\bar{W}_i \left(P_i, K_i, L_i, \alpha_i \right) = \begin{bmatrix} \bar{w}_{11i} & 0 \\ 0 & \Pi_{2i} - I \end{bmatrix} \leq 0,$$

$$\bar{w}_{11i} = P_i^{-1} \tilde{A}_i + \tilde{A}_i^{\mathsf{T}} P_i^{-1} + \alpha_i P_i^{-1} + \tilde{c}_{2i} G^{\mathsf{T}} G + \Pi_{1i}.$$

Observe that in (10.35), the size-minimization of $\mathcal{E}(Q_i)$ is achieved by the trace minimization of P_i. Nevertheless, minimization of $\mathrm{tr}(P_i)$ is a hard nonlinear problem. Therefore, we consider an upper bound

$$P_i \leq H_i.$$

We next use the Schur complement, and we obtain the LMI

$$\Theta_{3i} = \begin{bmatrix} H_i & I_n \\ I_n & P_i^{-1} \end{bmatrix} > 0.$$

Furthermore, in (10.35), we consider the minimization of μ with the aim of reducing the dwell time, but there exists a tradeoff in the selection of μ and ϵ. Fixing β and $\alpha_{\min}$, one can see that the size of ϵ depends on $\mu^2 / (\mu - 1)$, in the case $\gamma_1 = \alpha_{\min}/2$. It is easily verified that if μ is close to 1, then ϵ is large. Therefore, μ is selected such that $\mu^2 / (\mu - 1)$ will be minimal, which is true for $\mu = 2$. Now consider a positive constant ϵ^+ such that

$$\epsilon = \frac{\beta}{\alpha_{\min}} \left(\frac{\mu^2}{\mu - 1} \right) \leq \epsilon^+.$$

This is equivalent, for $\mu = 2$, to

$$4\beta - \alpha_{\min} \epsilon^+ \leq 0,$$

and now we ask for the minimum for ϵ^+.

We next use the above constraints in a modified (relaxed) optimization problem given in the following corollary.

Corollary 10.2. *Assume that the optimization problem*

$$\text{minimize} \ \sum_{i=1}^{M} \text{tr}(H_i) + \epsilon^+$$

$$\text{subject to}$$

$$\bar{W}_i\,(P_i, K_i, L_i, \alpha_i) \leq 0, \quad \forall i, j = \{1, \ldots, M\},$$

$$\Theta_{1i} > 0, \ \Theta_{2i} > 0, \ \Theta_{3i} > 0, \ 4\beta - \alpha_i \epsilon^+ \leq 0, \tag{10.36}$$

$$P_i^{-1} \leq \mu P_j^{-1}, \ i \neq j, \ \frac{\alpha_i}{\beta} P_i^{-1} > Q_{presc},$$

$$P_i > 0, \ H_i > 0, \ \alpha_i > 0, \ \epsilon^+ > 0,$$

has optimal solution $\hat{\Upsilon} := \left(\hat{P}_i, \hat{K}_i, \hat{L}_i, \hat{H}_i, \hat{\alpha}_i, \hat{\epsilon}^+ \right)$. *Then the ellipsoid* $\mathcal{E}\,(Q_i)$ *determined by* $Q_i = \dfrac{1}{\epsilon^+} P_i^{-1}$ *is an attractive ellipsoid for the switched system (10.2).*

The resulting relaxed optimization problem provides a basis for a numerical approach to the robust output feedback control for the nonlinear systems under consideration (Fig. 10.1).

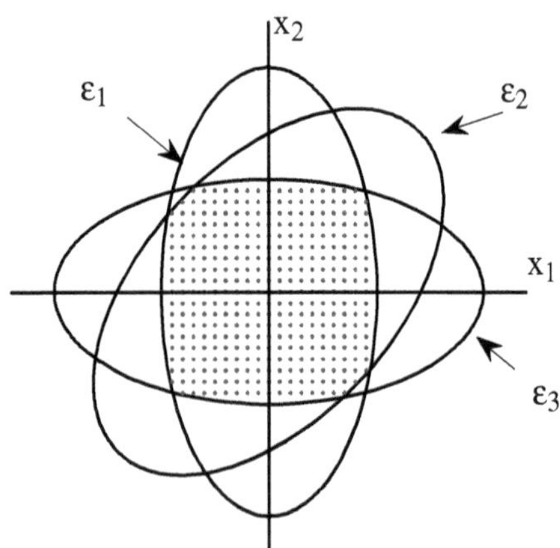

Fig. 10.1 Intersection of ellipsoids (for $M = 3$)

10.2.4 Simulation Results

Consider now a continuous stirred tank reactor (CSTR) that can be modeled as a switched system under arbitrary switching. The system consists of a constant-volume CSTR fed by an inlet stream through a valve that selects one of two different source streams, so the reactor will have two operating modes. The process is depicted in Fig. 10.2. The position of the valve is determined by a supervisor or a higher process, and therefore, the switching signal is considered arbitrary but is known at each time instant (Figs. 10.3–10.6).

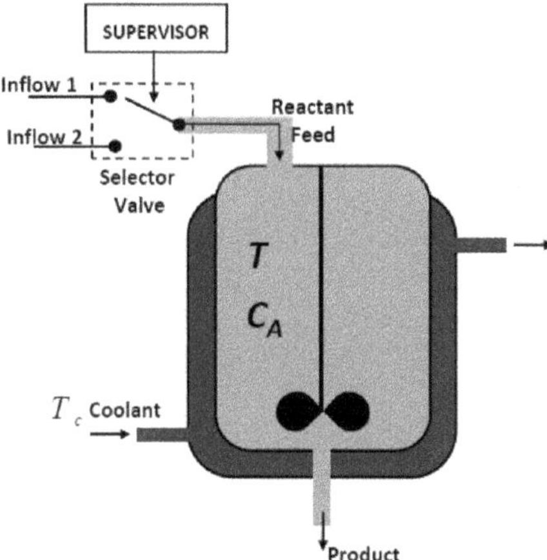

Fig. 10.2 Continuous stirred
tank reactor

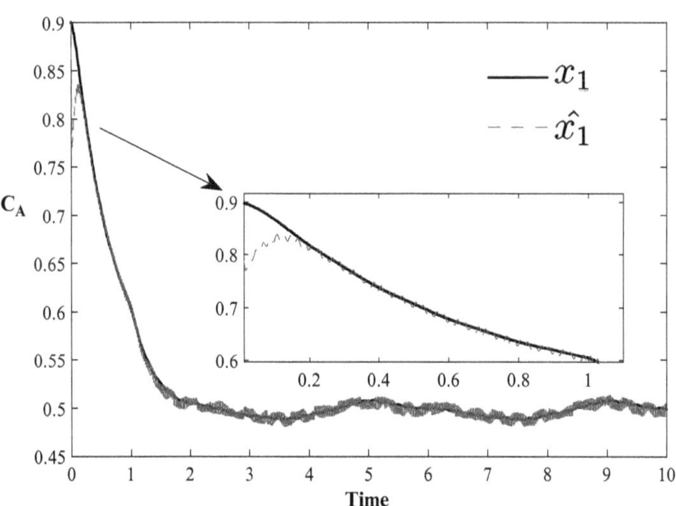

Fig. 10.3 C_A estimating process

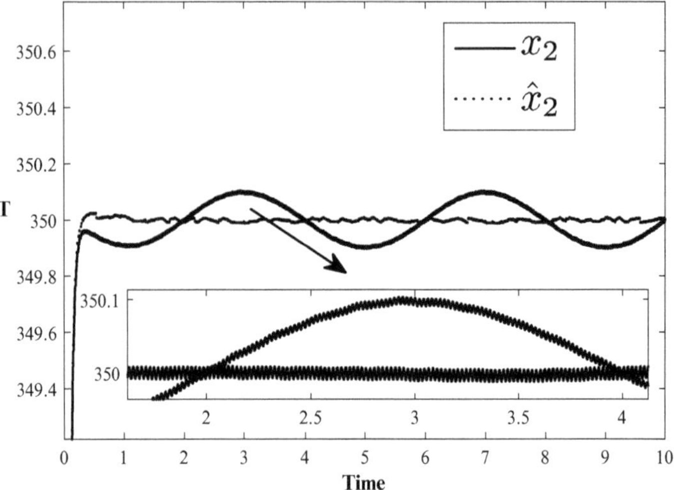

Fig. 10.4 T estimating process

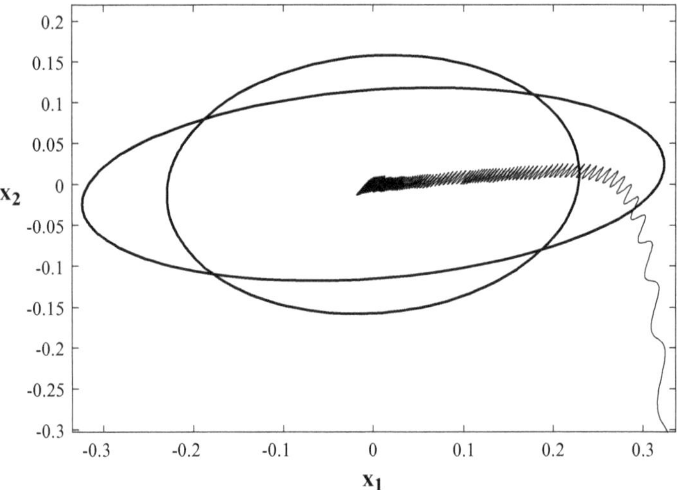

Fig. 10.5 Attractive ellipsoid in the state space

An irreversible exothermic reaction $A \rightarrow B$ occurs in the reactor, and this is cooled by a stream with a constant flow rate and a variable temperature T_c. Assuming constant liquid volume, negligible heat loss, perfect mixing, and a first-order reaction in reactant A, the CSTR at each operating mode is described by

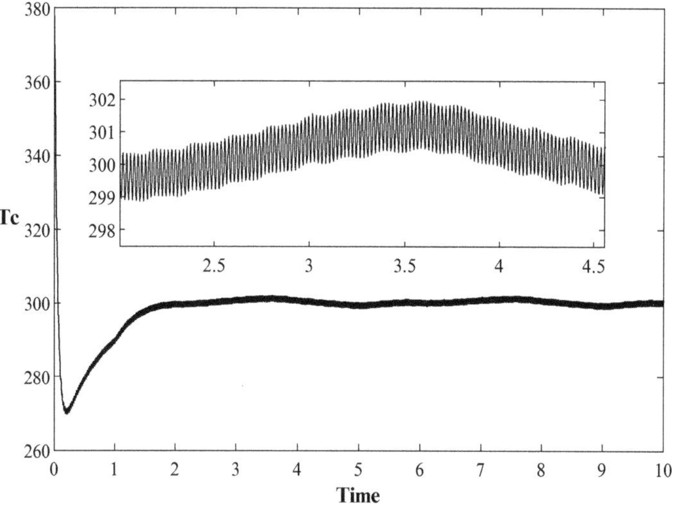

Fig. 10.6 Input and observer estimates

Table 10.1 Model parameters of CSTR

Symbol	Value	Meaning
a_0	7.2×10^{10}	Constant coefficient
a_1	-1.506×10^{13}	Constant coefficient
a_2	2.092	Constant coefficient
E/R	8750	$\dfrac{\text{activation energy}}{\text{gas constant}}$
V	100	Volume of the reactor
F_1	50	Feed flow rate 1
F_2	200	Feed flow rate 2
C_{A1}	1.5	Reactant A1 concentration
C_{A2}	0.75	Reactant A2 concentration
T_1	350	Inlet stream temperature 1
T_2	350	Inlet stream temperature 2

$$\frac{d}{dt}C_A(t) = \frac{F_{q(t)}}{V}\left(C_{Aq(t)} - C_A(t)\right) - a_0 \exp\left(-\frac{E}{RT(t)}\right) C_A(t),$$

$$\frac{d}{dt}T(t) = \frac{F_{q(t)}}{V}\left(T_{q(t)} - T(t)\right) - a_1 \exp\left(-\frac{E}{RT(t)}\right) C_A(t) + a_2(T_c(t) - T(t)).$$

The parameter values are shown in Table 10.1 (see Barkhordari-Yazdi & Jahed-Motlagh 2009). The nominal operating conditions corresponding to an unstable equilibrium point are

$$T_c^* = 300 \,\text{K}, \ C_A^* = 0.5 \,\text{mol/L}, \ \text{and} \ T^* = 350 \,\text{K}$$

for both modes. Also suppose that the values of the system states, T and C_A, are available at each time instant. The objective is to regulate C_A and T to their nominal values by manipulating T_c under an arbitrary switching of the valve position, but T is the only measured variable. We select as state variables

$$x = [C_A - C_A^*, \; T - T^*]^{\mathsf{T}}.$$

The control input is

$$u = T_c - T^*.$$

Rewriting (10.4) in the form of (10.2), we have

$$f_{1i}(x) = \tfrac{F_i}{V}\left(C_{Ai} - C_A^* - x_1\right) - a_0\left(x_1 + C_A^*\right)\exp\left(-\tfrac{E/R}{x_2+T^*}\right),$$

$$f_{2i}(x) = \tfrac{F_i}{V}\left(T_i - T^* - x_2\right) - a_1\left(x_1 + C_A^*\right)\exp\left(-\tfrac{E/R}{x_2+T^*}\right)$$
$$+ a_2\left(T_c^* - x_2 - T^*\right),$$

$$A_i = \begin{bmatrix} -\tfrac{F_i}{V} - a_0 C_e & -a_0 C_e C_A^* \tfrac{E/R}{(T^*)^2} \\ -a_1 C_e & -\tfrac{F_i}{V} - a_1 C_e C_A^* \tfrac{E/R}{(T^*)^2} - a_2 \end{bmatrix},$$

$$B_i = \begin{bmatrix} 0 \\ a_2 \end{bmatrix}, \; C_i = \begin{bmatrix} 0 & 1 \end{bmatrix}, \; C_e = \exp\left(-\tfrac{E/R}{T^*}\right).$$

Consider

$$Q_{presc} = \begin{bmatrix} 10 & 0 & 0 & 0 \\ 0 & 25 & 0 & 0 \\ 0 & 0 & 250 & -250 \\ 0 & 0 & -250 & 500 \end{bmatrix}$$

Figures 10.3–10.6 show the convergence of the real state and estimated state variables C_A and T to the desired values. The graphics show the dynamics of the systems under the initial condition

$$[C_A(t_0), T(t_0)] = [0.9, 340].$$

The gain matrices and others data obtained are as follows:

$$K_1 = \begin{bmatrix} 99.3088 & 11.7369 \end{bmatrix}, \; K_2 = \begin{bmatrix} 99.6727 & 8.0830 \end{bmatrix},$$

$$L_1 = \begin{bmatrix} 191.3623 \\ 251.7547 \end{bmatrix}, \qquad L_2 = \begin{bmatrix} 200.4465 \\ 244.5836 \end{bmatrix},$$

$$P_1 = \begin{bmatrix} 5.106 & -2.880 & 0 & 0 \\ -2.880 & 38.447 & 0 & 0 \\ 0 & 0 & 522.798 & -309.919 \\ 0 & 0 & -309.919 & 409.301 \end{bmatrix},$$

$$P_2 = \begin{bmatrix} 9.832 & -1.195 & 0 & 0 \\ -1.195 & 20.648 & 0 & 0 \\ 0 & 0 & 444.401 & -260.797 \\ 0 & 0 & -260.797 & 377.356 \end{bmatrix},$$

$$a_1 = 2.5023 \ a_2 = 3.0631 \ \epsilon = 7.8.$$

10.3   Switched Systems with Quantized and Sampled Output Feedback

10.3.1   System Description

Consider a class of switched nonlinear system described by

$$\dot{x}(t) = f_{\sigma(t)}(t, x(t)) + B_{\sigma(t)}u(t) + \upsilon_x(t),$$

$$(10.37)$$

$$x(0) = x_0, \ \sigma(0) = \sigma_0,$$

where $x(t) \in \mathbb{R}^n$, $u(t) \in \mathbb{R}^m$, and $\upsilon_x(t) \in \mathbb{R}^n$ are the state vector, control input, and exogenous disturbance at time $t \in \mathbb{R}_+$. Moreover, $\{f_i(\cdot, \cdot)\}$, $i = 1, \ldots, M \in \mathbb{N}$ is a family of quasi-Lipschitz functions $f_i : \mathbb{R}_+ \times \mathbb{R}^n \mapsto \mathbb{R}^n$. The switching signal in (10.37) is determined by a time-dependent piecewise-constant function

$$\sigma : \mathbb{R}_+ \mapsto \mathcal{I} = \{1, \ldots, M\},$$

where $\mathcal{I}$ is a set of finite index. The initial conditions are given by the pair $\{x_0, \sigma_0\} \in \mathbb{R}^n \times \mathcal{I}$. Transitions between the subsystems occur at the switching times t_r, where $r \in \mathbb{N}$, i.e.,

$$\sigma(t) = i \in \mathcal{I}, t \in [t_{r-1}, t_r).$$

A constant $\tau_d > 0$ such that $t_r - t_{r-1} \geq \tau_d$ is the *dwell time*, because $\sigma(\cdot)$ dwells on each of its values for at least τ_d units of time.

We use the following model to describe a noisy sampled and quantized output of the above switched system:

$$\bar{\bar{y}}(t) = Cx(t) + \omega_y(t), \tag{10.38a}$$

$$\bar{y}(t) = \sum_{\bar{t}_k} \bar{\bar{y}}(\bar{t}_k)\chi_{[\bar{t}_k,\bar{t}_{k+1})}(t), \tag{10.39}$$

$$y(t) = \pi(\bar{y}(t)). \tag{10.40}$$

The vector $\omega_y(t) \in R^q$ in (10.38a) is the deterministic noise. The symbol $\chi_{[\bar{t}_k,\bar{t}_{k+1})}$ in (10.39) denotes the characteristic function of the time interval $[\bar{t}_k, \bar{t}_{k+1})$, i.e.,

$$\chi_{[\bar{t}_k,\bar{t}_{k+1})}(t) := \begin{cases} 1 & \text{if } t \in [\bar{t}_k, \bar{t}_{k+1}), \\ 0 & \text{otherwise,} \end{cases} \qquad k = 0, 1, 2, \ldots.$$

Thus, $\bar{y} : \mathbb{R}_+ \to \mathbb{R}^q$ is the piecewise constant function obtained by sampling and holding $\bar{\bar{y}}$ at the discrete instants $\bar{t}_k$ (the sample times). The measurable system output at time t is $y(t) \in \mathbb{R}^q$, and it is obtained by quantizing the sampled signal $\bar{y}(t)$. Formally, let $Y \subset \mathbb{R}^q$ be a countable set of possible output values or quantization levels. Then $\pi : \mathbb{R}^q \to Y$ in (10.40) is a function such that

$$\pi(\bar{y}(t)) := \operatorname*{argmin}_{y(t) \in Y} \varrho(y(t), \bar{y}(t))$$

with

$$\varrho(y(t), \bar{y}(t)) := \|y(t) - \bar{y}(t)\|_{Q_y}^2.$$

By $\{B_i, C\}$, $B_i \in \mathbb{R}^{n \times m}$, $C \in \mathbb{R}^{q \times n}$ we denote here a family of given system matrices.

Let us now formulate our basic assumptions.

Assumptions:

1. There are known positive definite matrices $Q_x \in \mathbb{R}^{n \times n}$ and $Q_y \in \mathbb{R}^{q \times q}$ such that

$$\|\upsilon_x(t)\|_{Q_x}^2 + \|\omega_y(t)\|_{Q_y}^2 \leq 1 \ \forall t \in \mathbb{R}_+. \tag{10.41}$$

Here $\|\cdot\|_{Q_x}$ and $\|\cdot\|_{Q_y}$ are weighted norms given by Q_x and Q_y.
2. The functions f_i satisfy the quasi-Lipschitz bound

$$\|f_i(t, x) - A_i x(t)\|_{Q_x}^2 \leq \delta + \|x(t)\|_{Q_i}^2 \tag{10.42}$$

for all $(t, x) \in \mathbb{R}^+ \times \mathbb{R}^n$, where $\delta > 0$ is a scalar and $Q_i > 0$ and A_i are known $n \times n$ matrices.

3. The pairs (A_i, B_i) are controllable, and (A_i, C) are observable.
4. The sampling intervals do not need to be regular, but there exists a maximum sampling interval

$$ h := \max_k |\bar{t}_{k+1} - \bar{t}_k|. $$

5. The quantization error is bounded, i.e., the positive scalar

$$ c := \max_{\bar{y} \in \mathbb{R}^q} \| \pi(\bar{y}) - \bar{y} \|_{Q_y}^2 \qquad (10.43) $$

is finite. The "Zeno behavior" (infinite switchings in finite time) in $\sigma(t)$ is assumed to be excluded. Also, it is a natural consequence to impose a dwell-time scheme in the switching signal.

Note that (10.42) is not restrictive and comprises a large class of unknown non-linear functions (Gonzalez-Garcia et al. 2011). By defining the auxiliary function

$$ \omega_x(t) := \upsilon_x(t) + f_{\sigma(t)}(t, x(t)) - A_{\sigma(t)}x(t), $$

we can rewrite (10.37) as

$$ \dot{x}(t) = A_{\sigma(t)}x(t) + B_{\sigma(t)}u(t) + \omega_x(t). \qquad (10.44) $$

We propose a classical Luenberger observer for state estimation:

$$ \dot{\hat{x}}(t) = A_{\sigma(t)}\hat{x}(t) + B_{\sigma(t)}u(t) + L_{\sigma(t)}\left[y(t) - C\hat{x}(t)\right], \qquad (10.45) $$

where $L_i \in \mathbb{R}^{n \times q}$ are the observer gains. The control law is taken as a feedback

$$ u(t) - K_{\sigma(t)}\hat{x}(t) \qquad (10.46) $$

with $K_{\sigma(t)} \in \mathbb{R}^{m \times n}$ as the control gains.

We now introduce the estimation error vector

$$ e(t) := x(t) - \hat{x}(t) $$

and the auxiliary variable

$$ \Delta y(t) := y(t) - \bar{\bar{y}}(t). $$

It can be readily seen that $e(t)$ satisfies the dynamic equation

$$ \dot{e}(t) = (A_{\sigma(t)} - L_{\sigma(t)}C)e(t) - L_{\sigma(t)}\left[\Delta y(t) + \omega_y(t)\right] + \omega_x(t). \qquad (10.47) $$

It is possible to write the closed-loop equations (10.45) and (10.47) more compactly as

$$\dot{z}(t) = \tilde{A}_{\sigma(t)}z(t) + F_{\sigma(t)}\omega(t) + \psi(t), \tag{10.48}$$

where we have defined the vectors

$$z(t) := \begin{pmatrix} \hat{x}(t) \\ e(t) \end{pmatrix}, \quad \omega(t) := \begin{pmatrix} \omega_x(t) \\ \omega_y(t) \end{pmatrix},$$

$$\psi(t) := \begin{pmatrix} L_{\sigma(t)} \\ -L_{\sigma(t)} \end{pmatrix} \Delta y(t),$$

and the matrices

$$\tilde{A}_{\sigma(t)} = \begin{pmatrix} A_{\sigma(t)} + B_{\sigma(t)}K_{\sigma(t)} & L_{\sigma(t)}C \\ 0 & A_{\sigma(t)} - L_{\sigma(t)}C \end{pmatrix},$$

$$F_{\sigma(t)} = \begin{pmatrix} 0 & L_{\sigma(t)} \\ I & -L_{\sigma(t)} \end{pmatrix}.$$

Because of the presence of ω and ψ, the convergence of $z(t)$ at the origin as $t \to \infty$ is not to be expected. But if $K_{\sigma(t)}$ and $L_{\sigma(t)}$ are properly chosen, it is reasonable to expect $z(t)$ to converge to a "small" set containing the origin. Our problem is first to find a characterization of such a set and then to find $L_{\sigma(t)}$ and $K_{\sigma(t)}$ that minimize (in a particular sense to be defined later) its size.

To estimate the region where the states of (10.48) converge, we use the ellipsoid method and propose an extension of it to deal with the sampling and quantization of the output.

10.3.2   Lyapunov–Krasovskii-Like Functional

Considering that the sampling phenomenon involves a delay, we decided to use a Lyapunov–Krasovskii-like functional instead of a regular function. Let $C^0(\mathbb{R}, \mathbb{R}^{2n})$ be the space of all continuous functions of $\mathbb{R}$ into $\mathbb{R}^{2n}$, differentiable almost everywhere; let $R_i > 0$ and $P_i > 0$ be $2n \times 2n$ matrices, and let $\alpha_i > 0$ be a scalar. We propose the functional $V_i : \mathbb{R} \times C^0(\mathbb{R}, \mathbb{R}^{2n}) \to \mathbb{R}_+, i \in \mathcal{I}$, defined as

$$V_i(t, z(\cdot)) := z^\top(t) P_i^{-1} z(t) +$$

$$h \int_{\theta=-h}^0 \int_{s=t+\theta}^t e^{\alpha_i(s-t)} \dot{z}^\top(s) R_i \dot{z}(s) \, ds \, d\theta.$$

Our primary goal is to derive sufficient conditions for $V_i(t, z(\cdot))$ to satisfy (10.16) with $\alpha_i > 0$ and $\beta \geq 0$ when z is a solution of (10.48). Let us begin with the case that z is arbitrary. It is easy to see (repeating the analogous calculations as in previous sections) that for a given

$$z(\cdot) \in \mathcal{C}^0(\mathbb{R}, \mathbb{R}^{2n}) , \quad h, \alpha_i, b \in \mathbb{R},$$

$$P_i, R_i \in \mathbb{R}^{2n \times 2n}, \, i \in \mathcal{I},$$

such that

$$h > 0, \alpha_i > 0, P_i > 0, R_i > 0,$$

the time derivative of $V_i(t, z(\cdot))$ satisfies the differential inequality

$$\dot{V}_i(t, z(\cdot)) \leq -\alpha_i V_i(t, z(\cdot)) + b\bar{\delta} + \eta(t, z(\cdot))^\top W_i \eta(t, z(\cdot)), \tag{10.49}$$

where $\bar{\delta} := \delta + 1$ and

$$\eta(t, z(\cdot)) := \left(z^\top(t) \ \dot{z}^\top(t) \ z^\top(t) - z^\top(t_k) \ \omega^\top(t) \right)^\top,$$

$$W_i := \begin{pmatrix} \alpha_i P_i^{-1} + bQ_{zi} & P_i^{-1} & 0 & 0 \\ P_i^{-1} & h^2 R_i & 0 & 0 \\ 0 & 0 & -he^{-\alpha_i h} R_i & 0 \\ 0 & 0 & 0 & -b\bar{Q} \end{pmatrix}, \tag{10.50}$$

$$\bar{Q} := \begin{pmatrix} Q_x & 0 \\ 0 & Q_y \end{pmatrix}, \, Q_{zi} := \begin{pmatrix} I \\ I \end{pmatrix} Q_i \begin{pmatrix} I & I \end{pmatrix}.$$

Now we will refine the bound given in (10.49) by restricting $z(\cdot)$ to the set of solutions of (10.48) on the interval $[t_{r-1}, t_r)$, $r \in \mathbb{N}$. In order to do so, we follow the idea presented in Fridman 2006, which, originally devised for systems in descriptor form, consists in adding a term (the *descriptor term*) to the expression for $\dot{V}_i$. The descriptor term has to be zero for every solution z of the system. In our case, we add

$$\mathcal{D}_i(t, z(\cdot)) := 2 \left(z(t)^\top \Pi_{ai} + \dot{z}^\top(t) \Pi_{bi} \right) \times$$

$$\left(\tilde{A}_i z(t) + F_i \omega(t) + \psi(t) - \dot{z}(t) \right),$$

where Π_{ai} and Π_{bi} are in $\mathbb{R}^{2n}$. Obviously, $\mathcal{D}_i$ is zero along the solutions of (10.48).

Theorem 10.3. *Let ρ_1 be a positive scalar satisfying*

$$L_i^\top L_i \leq \rho_1 I. \tag{10.51}$$

Then for every

$$z(\cdot) \in \mathcal{C}^0(\mathbb{R}, \mathbb{R}^{2n}), \ h, \alpha_i, b, \varepsilon \in \mathbb{R},$$

$$P_i, R_i, \Pi_{ai}, \Pi_{bi} \in \mathbb{R}^{2n \times 2n}, \ i \in \mathcal{I},$$

$$h > 0, \ \alpha_i > 0, \ P_i > 0, \ R_i > 0,$$

such that z is a solution of (10.48), *the time derivative of $V_i(t, z(\cdot))$ satisfies*

$$\dot{V}_i(t, z(\cdot)) \leq -\alpha_i V_i(t, z(\cdot)) + \beta$$

$$+ \xi(t, z(\cdot))^\top \Omega_i \xi(t, z(\cdot)) \tag{10.52}$$

for all $t \in [t_{r-1}, t_r), \ r \in \mathbb{N}$, and $\sigma(t) = i$, where

$$\Omega_i := \begin{pmatrix} \omega_{11} & \omega_{12} & 0 & \Pi_{ai} F_i & \Pi_{ai} \\ * & \omega_{22} & 0 & \Pi_{bi} F_i & \Pi_{bi} \\ * & * & \omega_{33} & 0 & 0 \\ * & * & * & -b\bar{Q} & 0 \\ * & * & * & * & -\varepsilon I \end{pmatrix} \tag{10.53}$$

($*$ means the transposed symmetric element) and

$$\omega_{11} = \alpha_i P_i^{-1} + bQ_{zi} + 2\Pi_{ai} \tilde{A}_i,$$

$$\omega_{12} = P_i^{-1} - \Pi_{ai} + \Pi_{bi} \tilde{A}_i,$$

$$\omega_{22} = h^2 R_i - 2\Pi_{bi},$$

$$\omega_{33} = -he^{-\alpha_i h} R_i + \varepsilon \rho Q_c,$$

and

$$\beta := b\bar{\delta} + \varepsilon \rho (2 + c),$$

$$\rho := 2\rho_1 / \lambda_{\min}(Q_y),$$

$$\xi(t, z(\cdot)) := (z^\top(t), \dot{z}^\top(t), z^\top(t) - z^\top(t_k), \omega^\top(t), \psi^\top(t))^\top,$$

$$ \tag{10.54}$$

$$Q_c := (I \ \ I)^\top C^\top Q_y C (I \ \ I).$$

The following lemma will be needed before the proof of the theorem.

Lemma 10.2. *The uncertainty resulting from noise, sampling, and quantization is bounded as*

$$\|\psi(t)\|^2 \leq \rho\left((z(t) - z(t_k))^{\top} Q_c(z(t) - z(t_k)) + 2 + c\right). \tag{10.55}$$

This can be verified directly using equations (10.41) and (10.43).

Proof. (of Theorem 10.3): Adding the null term

$$\mathcal{D}_i(t, z(\cdot)) + \varepsilon \|\psi(t)\|^2 - \varepsilon \|\psi(t)\|^2$$

to (10.49) gives

$$\dot{V}_i(t, z(\cdot)) \leq -\alpha_i V_i(t, z(\cdot)) + b\bar{\delta} +$$

$$\varepsilon \|\psi(t)\|^2 + \eta^{\top}(t, z(\cdot)) W_i \eta(t, z(\cdot)) +$$

$$2\left(z^{\top}(t) \Pi_{ai} + \dot{z}^{\top}(t) \Pi_{bi}\right) \times \tag{10.56}$$

$$\left(\tilde{A}_i z(t) + F_i \omega(t) + \psi(t) - \dot{z}(t)\right) - \varepsilon \|\psi(t)\|^2.$$

Substituting (10.55) in (10.56) establishes

$$\dot{V}_i(t, z(\cdot)) \leq -\alpha_i V_i(t, z(\cdot)) +$$

$$\beta + \varepsilon\rho(z(t) - z^{\top}(t_k)) Q_c(z(t) - z(t_k)) +$$

$$\eta^{\top}(t, z(\cdot)) W_i \eta(t, z(\cdot)) + 2\left(z^{\top}(t) \Pi_{ai} + \dot{z}^{\top}(t) \Pi_{bi}\right) \tag{10.57}$$

$$\times \left(\tilde{A}_i z(t) + F_i \omega(t) + \psi(t) - \dot{z}(t)\right) - \varepsilon \|\psi(t)\|^2.$$

Equation (10.52) is (10.57) rewritten in a compact form.   ∎

10.3.3   On Practical Stability

The system (10.48) is said to be *practically stable* if there exists a prescribed attractive set associated with the dynamics of the system. Considering the ellipsoidal sets attractive, we may associate the property of practical stability with the state vector $z(t)$ satisfying

$$\limsup_{t \to \infty} z^{\top}(t) Q_{\sigma(t)} z(t) \leq 1$$

under the matrix constraints

$$Q_i \geq Q_0 > 0, \quad i = 1, \ldots, M,$$

for a given matrix $Q_0 \in \mathbb{R}^{2n \times 2n}$.

We derive the practical stability property subject to an average dwell-time condition for the switching signal. We use the property given in Theorem 10.3 to construct a storage function for the switched system (10.48).

Theorem 10.4. *Let* $\mathbf{V}(t) = V_{\sigma(t)}(t, z(t))$ *be a piecewise continuous function, where each* $V_i(t, z(t))$ *satisfies Theorem 10.3. Furthermore, we ask for*

$$\Omega_i < 0$$

and suppose that there exists a constant $\mu > 1$ *such that*

$$V_i(t, z) \leq \mu V_j(t, z), \quad \forall i, j \in \mathcal{I}, \ t \in \mathbb{R}_+. \tag{10.58}$$

Then for positive constants $(\gamma_0, \gamma_1, \alpha_{\min})$, *there exists a finite constant*

$$\tau_{av} = \frac{\log \mu}{\alpha_{\min} - \gamma_1}$$

such that $\mathbf{V}(t)$ *is a storage function for the switched system satisfying*

$$\mathbf{V}(t) \leq \exp(\gamma_0 - \gamma_1(t - t_0))\mathbf{V}(t_0) +$$

$$\frac{\beta}{\alpha_{\min}} \left(\frac{\mu^2}{\mu - 1} \right) (1 - \exp(-N(t_0, t) \log \mu)) \tag{10.59}$$

with $t_0 \geq 0$, *decay rate* γ_1, *and average dwell time* τ_{av}. *Moreover,*

$$\limsup_{t \to \infty} \mathbf{V}(t) \leq \frac{\beta}{\alpha_{\min}} \left(\frac{\mu^2}{\mu - 1} \right) := \kappa. \tag{10.60}$$

Proof. The property (10.58) implies the conditions

$$P_i^{-1} \leq \mu P_j^{-1}, \quad i \neq j,$$

$$e^{-\alpha_i \bar{h}} R_i \leq \mu e^{-\alpha_j \bar{h}} R_j, \quad \bar{h} \in [0, h], \ i \neq j.$$

These last conditions are satisfied, for example, with

$$\mu = \max\{\mu_P, \mu_R\},$$

$$\mu_P = \sup_{a,b\in\mathcal{I}} \lambda_{\max}(P_a^{-1})/\lambda_{\min}(P_b^{-1}),$$

$$\mu_R = \sup_{c,d\in\mathcal{I}} \lambda_{\max}(R_c)/\lambda_{\min}(R_d),$$

where $\lambda_{\max}(X)(\lambda_{\min}(X))$ denotes the largest (smallest) eigenvalue of the matrix X. Using this condition, we have that in the switching instants t_r,

$$\mathbf{V}(t_r) \leq \mu \lim_{t\to t_r^-} V_{\sigma(t)}(t, z(t)) = \mu\mathbf{V}(t_r^-), \quad r \in \mathbb{N}. \tag{10.61}$$

Consider that every $V_i(z(t))$ satisfies Theorem 10.3 and also

$$\Omega_i < 0, \forall i \in \mathcal{I}.$$

Then

$$V_i(t, z(t)) \leq V_i(t_{r-1}, z(t_{r-1})) \exp\left(-\alpha_i(t - t_{r-1})\right)$$

$$+ \frac{\beta}{\alpha_i}\left[1 - \exp\left(-\alpha_i(t - t_{r-1})\right)\right]$$

for all $t \in [t_{r-1}, t_r)$. Let $N(t_0, t)$ be the number of switchings of $\sigma(\cdot)$ in the interval $[t_0, t)$ such that

$$0 \leq t_0 < t_1 \cdots < t_{N(t_0,t)} < t < t_{N(t_0,t)+1} = T.$$

Define

$$\alpha_{\min} = \min_{i\in\mathcal{I}} \alpha_i$$

and

$$\tau_r = t_r - t_{r-1}.$$

From the foregoing inequality and (10.61), it follows that by backward iteration from t_0 to $t_{N(t_0,t)}$, we get (we omit the arguments of $N(t_0, t)$)

$$V(t_N) \leq \exp\left[N(t_0, t)\log\mu - \alpha_{\min}(t_N - t_0)\right] V(t_N) +$$

$$\frac{\beta}{\bar{\alpha}_{\min}}\mu\left[1 + \sum_{k=1}^{N(t_0,t)-1} \exp\left(k\log\mu - \alpha_{\min}\sum_{l=0}^{k-1} \tau_{N-l}\right)\right]. \tag{10.62}$$

To guarantee a decay rate γ_1, for the first term of (10.62), we must have

$$N(t_0, t) \log \mu - \alpha_{\min}(t_N - t_0) \leq \gamma_0 - \gamma_1(t_N - t_0), \qquad (10.63)$$

where

$$\gamma_0 > 0, \gamma_1 > 0.$$

This last expression is equivalent to (10.5) with

$$N_0 = \frac{\gamma_0}{\log \mu}$$

and

$$\tau_{\mathrm{av}} = \frac{\log \mu}{\alpha_{\min} - \gamma_1},$$

subject to

$$0 < \gamma_1 < \alpha_{\min}.$$

For the second term of (10.62), there is no loss of generality if we consider

$$\sum_{l=0}^{k-1} \tau_{N-l} \geq k \tau_{\mathrm{av}}.$$

So we get

$$1 + \sum_{k=1}^{N(t_0,t)-1} \exp\left(k \log \mu - \alpha_{\min} \sum_{l=0}^{k-1} \tau_{N-l} \right) \leq$$

$$\sum_{k=0}^{N(t_0,t)-1} \exp\left(k(\log \mu - \alpha_{\min} \tau_{\mathrm{av}}) \right).$$

Choosing $\gamma_1 = \frac{\alpha_{\min}}{2}$, which implies $\tau_{\mathrm{av}} = \dfrac{2 \log \mu}{\alpha_{\min}}$, this allows us to rewrite the right-hand side of the last inequality as

$$\sum_{k=0}^{N(t_0,t)-1} \exp(1/\mu)^k = \frac{1 - (1/\mu)^{N(t_0,t)}}{1 - 1/\mu}.$$

Substituting the latter into (10.62) and considering (10.63), we obtain (10.59).∎

Intersection of Ellipsoids

From the above procedure and considering

$$t_r - t_{r-1} \geq \tau_{\mathrm{av}} = \frac{2 \log \mu}{\alpha_{\min}} V_{\sigma(t)}(x(t)) \geq \kappa$$

for all

$$x \in \mathcal{X} := \left(x : \dot{V}_i(x) \geq -\alpha_i V_i(x) + \beta \right),$$

we have that

$$V_{\sigma(t_r)}(t_r) - V_{\sigma(t_{r-1})}(t_{r-1}) \leq \mu V_{\sigma(t_{r-1})}(t_r) - V_{\sigma(t_{r-1})}(t_{r-1})$$

$$\leq -V_{\sigma(t_{r-1})}(t_{r-1}) \left[\frac{\mu-1}{\mu} \right] + \mu \frac{\beta}{\alpha_{\min}} \left(1 - e^{-\alpha_{\min} \tau_r} \right)$$

$$\leq -\mu \frac{\beta}{\alpha_{\min}} e^{-\alpha_{\min} \tau_r} < 0.$$

Let t_{i_j}, $j \in \mathbb{N}$, be the switching times such that $\sigma(t_{i_j}) = i$; so the above inequality implies

$$V_i(t_{i_{j+1}}) - V_i(t_{i_j}) \leq 0.$$

Therefore, only if we suppose that each regime is active during the process infinitely many times will we have that the subsequence $V_i(x(t_{i_1}))$, $V_i(x(t_{i_2}))$, ..., is decreasing and has a limit κ. The foregoing considerations imply

$$\sum_{i=1}^{M} \left[\sqrt{V_i(t_{i_j})} - \kappa \right]^2 \xrightarrow[j \to \infty]{} 0.$$

Finally, this means that every trajectory of the switched system converges to the intersection of the individual ellipsoids, namely,

$$z(t) \xrightarrow[j \to \infty]{} \bigcap_{i=1}^{M} \mathcal{E}(0, \kappa P_i).$$

Main Result

The following corollary follows from Theorems 10.3 and 10.4.

Corollary 10.3. *Let*

$$\{\alpha_i > 0, b > 0, \varepsilon > 0, \rho_1 > 0, \mu > 1,$$

$$P_i > 0, R_i > 0, \Pi_{ai}, \Pi_{bi}, L_i, K_i\}$$

be a set of control parameters such that

$$\Omega_i \le 0, \ L_i^{\top} L_i \le \rho_1,$$

$$P_i^{-1} \le \mu P_j^{-1}, \ i, j \in \mathcal{I}, \ \frac{\alpha_i}{\beta} P_i^{-1} > Q_0, \tag{10.64}$$

with Ω_i defined by (10.53), and Q_{zi}, Q_c, $\bar{Q}$, ρ given by (10.54) and (10.50). The intersection set

$$\mathrm{Int}\mathcal{E} := \left\{ z \in \mathbb{R}^{2n} \ : \ z^{\top} P_i^{-1} z \le \kappa, \forall i \in \mathcal{I} \right\}$$

*with β given by (10.54), κ by (10.60), and for a prescribed Q_0, it is an **attractive** and **invariant** set.*

Illustrative Example

An example of a separately excited DC motor is considered. The following model describes the dynamics of the motor with switching inertia

$$J_{\sigma(t)} \frac{d\omega(t)}{dt} = c_m \phi_s(t) i_r(t) - B_m \omega(t) - \eta_1(t),$$

$$L_r \frac{d i_r(t)}{dt} = U_r(t) - R_r i_r(t) - c_m \phi_s(t) \omega(t) + \eta_2(t), \tag{10.65}$$

$$\frac{d\phi_s(t)}{dt} = U_s(t) - R_s \phi_s(t) + \eta_3(t),$$

where

- $\omega(t)$ denotes the angular velocity of the shaft,
- $i_r(t)$ is the current of the rotor circuit,
- R_r and R_s are the rotor and stator resistances, respectively,
- the rotor inductance is denoted here by L_r, and $\phi(t)$ is the stator flux,
- the parameters $J_{\sigma(t)} \in \{J_1, J, 2\}$ and B_m in the above model express the moment of inertia of the rotor and the viscous friction coefficient, respectively,

- $\eta = (\eta_1, \eta_2, \eta_3)^\top$ denotes a parametric uncertainty,
- c_m represents a constant parameter that depends on the spatial architecture of the drive,
- $U_r(t)$ and $U_s(t)$ are the rotor and stator voltages.

We choose the state variables as

$$(x_1, x_2, x_3)^\top = (\omega, i_r, \phi_s)^\top,$$

and then let us apply the conventional linearization procedure to (10.65) around a given reference point $(\Omega^{\text{ref}}, I_r^{\text{ref}}, \Phi_s^{\text{ref}})$. The resulting linearized model satisfies the quasilinear representation (10.44) with

$$A_i = \begin{pmatrix} -\dfrac{B_m}{J_i} & \dfrac{c_m \Phi_s^{\text{ref}}}{J_i} & \dfrac{c_m I_r^{\text{ref}}}{J} \\ -\dfrac{c_m \Phi_s^{\text{ref}}}{L_r} & -\dfrac{R_r}{L_r} & -\dfrac{c_m \Phi_s^{\text{ref}}}{L_r} \\ 0 & 0 & -R_s \end{pmatrix},$$

$$B_1 = B_2 = \begin{pmatrix} 0 & 0 & 1 \\ 0 & \frac{1}{L_r} & 0 \end{pmatrix}^\top, \qquad C = \begin{pmatrix} 1 & 0 & 0 \\ 0 & 1 & 0 \end{pmatrix},$$

where $i = 1, 2$, and the values of the parameters are shown in Table 10.2. We choose $h = 0.01$ and $c = 0.25$. The initial conditions are $x(0) = (1, 1, 1)^\top$. The prescribed matrix we use is a diagonal matrix $Q_0 = \text{diag}\,(4, 400, 400, 4, 400, 400)$.

The observer and the controller gains obtained using the algorithm were

$$K_1 = \begin{pmatrix} -1.0763 & -0.63941 & -0.65784 \\ 0.86956 & -0.38739 & -1.0461 \end{pmatrix},$$

$$L_1 = \begin{pmatrix} -0.5501 & -0.062781 & 1.3013 \\ 0.80277 & 1.0962 & 0.39223 \end{pmatrix}^\top,$$

$$K_2 = \begin{pmatrix} -0.48345 & -0.89499 & -0.98247 \\ 1.3188 & -0.36972 & -0.31266 \end{pmatrix},$$

Table 10.2 Parameter Values

Parameter	Value	Unit	Parameter	Value	Unit
c_m	0.03	Wb/rad	L_s	50	H
J_1	0.001	kg/m^2	B_m	0.009	Nm/rad
J_2	0.004	kg/m^2	Ω_{ref}	120	rad/s
R_r	0.5	Ohms	I_r^{ref}	0.1	A
R_s	85	Ohms	Φ_s^{ref}	15	Wb
L_r	8.9	mH			

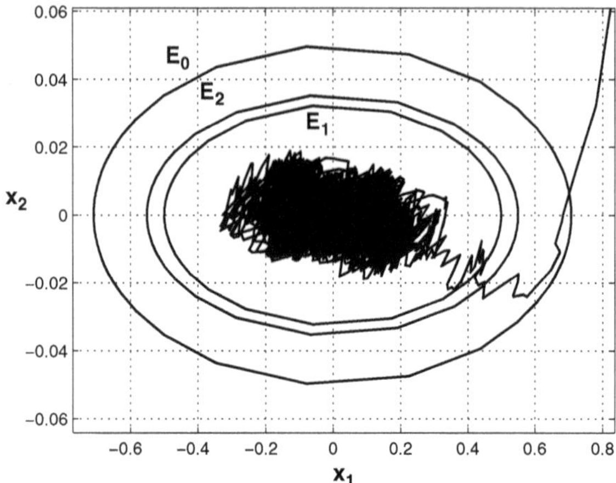

Fig. 10.7 Estimated ellipsoid and system trajectories x_1 and x_2

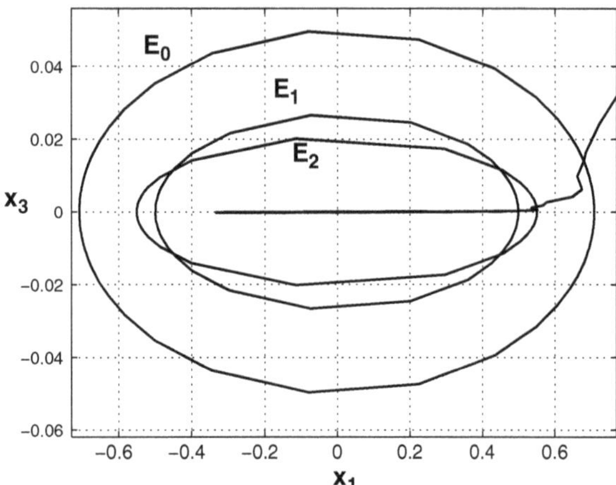

Fig. 10.8 Estimated ellipsoid and system trajectories x_1 and x_3

$$L_2 = \begin{pmatrix} -0.46817 & 0.285 & 1.3037 \\ 1.2663 & 0.53086 & 0.33868 \end{pmatrix}^{\mathsf{T}},$$

and the average dwell time was $\tau_{\mathrm{av}} = 0.126$.

The results of the system simulation are shown below. So, Figs. 10.7, 10.8, and 10.9 contain the projection (the obtained attractive ellipsoid and the system trajectory) of the three-dimensional state space on the two-dimensional subspaces (x_1, x_2), (x_1, x_3), and (x_2, x_3) respectively (Fig. 10.10).

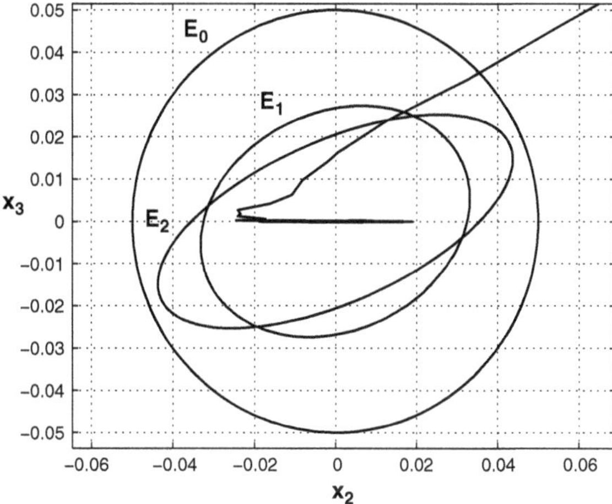

Fig. 10.9 Estimated ellipsoid and system trajectories x_2 and x_3

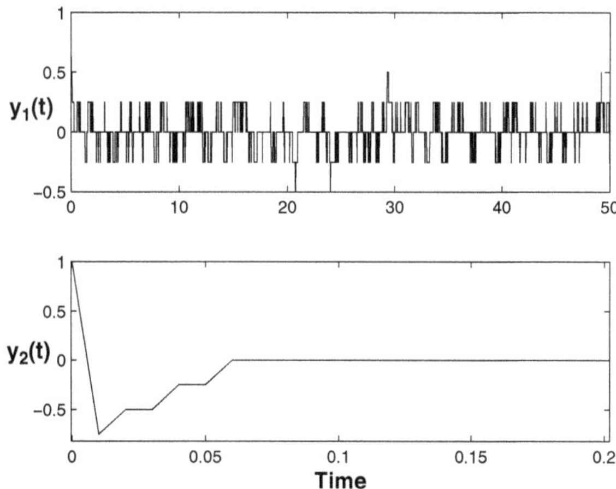

Fig. 10.10 Output of the system

10.4 Conclusions

- In this chapter, we developed a robust control design for a class of nonlinear switched systems in the presence of bounded uncertainties.
- An extension of the invariant ellipsoid method for systems with switched, sampled, and quantized outputs was also developed.
- The control design was based on a specific extension of the invariant ellipsoid method. Our approach generates an admissible linear feedback control law based on an observer that guarantees the existence and characterization of a minimal-size attractive ellipsoid for the closed-loop system (Fig. 10.11).
- These results were carried out by considering a dwell-time scheme for the switching signal.

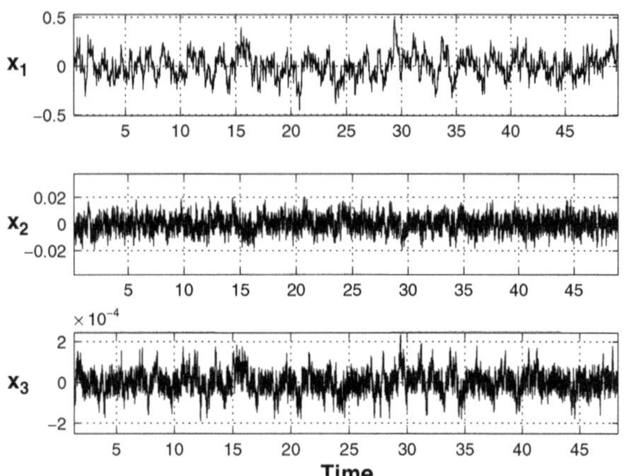

Fig. 10.11 The states of the system

- From the computational point of view, we proposed an auxiliary nonlinear minimization problem with matrix constraints. This permitted us to convert the robust controller design to an optimization problem subject to BMI constraints.
- We next obtain an attractive ellipsoid with some minimal properties that can be interpreted (under the given Q_{prescr}) as stability of the closed-loop system in a practical sense.
- Finally, the effectiveness of the proposed control strategy was demonstrated by illustrative examples.

Chapter 11
Bounded Robust Control

Abstract This chapter deals with a bounded control design for a class of nonlinear systems whose mathematical model may not be explicitly given. This class of uncertain nonlinear systems is governed by a system of ordinary differential equations with quasi-Lipschitz right-hand sides and contains external perturbations as well. The attractive ellipsoid method (AEM) allows us to describe the class of nonlinear feedbacks (containing a nonlinear projection operator, a linear state estimator, and a feedback matrix gain) guaranteeing the boundedness of all possible trajectories around the origin. To satisfy this property, some modifications of the AEM are introduced: basically, some sort of sample-time corrections of the feedback parameters are required. The optimization of feedback within this class of controllers is associated with the selection of the feedback parameters such that the trajectory converges within an ellipsoid of "minimal size." The effectiveness of the suggested approach is illustrated by its application to a flexible arm system.

Keywords Bounded control • Robust stabilization • Linear matrix inequalities

This chapter deals with a bounded control design for a class of nonlinear systems whose mathematical model may not be explicitly given. This class of uncertain nonlinear systems is governed by a system of ordinary differential equations with quasi-Lipschitz right-hand sides and contains external perturbations as well. The attractive ellipsoid method (AEM) allows us to describe the class of nonlinear feedbacks (containing a nonlinear projection operator, a linear state estimator, and a feedback matrix gain) guaranteeing the boundedness of all possible trajectories around the origin. To satisfy this property, some modifications of the AEM are introduced: basically, some sort of sample-time corrections of the feedback parameters are required. The optimization of feedback within this class of controllers is associated with the selection of the feedback parameters such that the trajectory converges within an ellipsoid of "minimal size." The effectiveness of the suggested approach is illustrated by its application to a flexible arm system.

© Springer International Publishing Switzerland 2014

267

A. Poznyak et al., *Attractive Ellipsoids in Robust Control*, Systems & Control:
Foundations & Applications, DOI 10.1007/978-3-319-09210-2_11

11.1 Introduction

Here we follow Ordaz, Alazki and Poznyak (2013) and study the workability of
the AEM when control actions (generated by a designated feedback) are bounded
by their physical nature (see, e.g., Ting Shu & Zong Li 2001; Teel 1996; Sussmann,
Sontag, & Yang, 1994) and are designed based only on sample-time output data (see
Jong & Jay 2005; Kabamba & Hara 1993; Poznyak, Azhmyakov, & Mera, 2011).
In general, such constraints are described by membership of the control action in
a given bounded convex set (compact) (Rudin 1991). The convexity property is
basically topological, depending on the connectedness of the considered subsets.
Although robust control is applied in many different branches of control theory
(linear and nonlinear control (Duncan & Schweppe 1971; Poznyak et al. 2011; Utkin
1993), adaptive control (Ioannou & Sun 1996; Narendra & Annaswamy 2005); and
others), there is little evidence of robust control designs taking into account the
boundedness (or saturation) of a set of admissible control actions.

Saturation is probably the most commonly encountered nonlinearity in control
engineering. For this reason, a projector participating in a designed nonlinear feedback plays an important role in the description of a saturated control. Thus Davila
and Poznyak (2011) and Polyakov and Poznyak (2011) construct the invariant
ellipsoid for the sliding mode control by means of linear difference inclusions.
Here we consider more general types of nonlinear bounded feedbacks that may be
corrected (adjusted) online during a control process (Ordaz & Poznyak 2012).

The basic assumptions in this chapter are as follows:

- The matrix related to the actuator in the dynamical system (i.e., the matrix B in
  the affine representation of the control action) is assumed to be known a priori.
- During the process, the states are assumed not to be directly measurable.
- Every admissible control action $u(t)$ is bounded for every $t \geq 0$ belonging to
  some bounded compact U.

As for the constraints, we assume that

- the dynamical system model is admitted to have an unknown nonlinearity from
  a given class of the quasi-Lipschitz functions;
- the external perturbations affecting the system dynamics are assumed to be
  bounded.

11.2 The Class of Uncertain Nonlinear Systems and Problem
Formulation

11.2.1 System Description

Hereinafter, we show how the combination of traditional AEM and the projection
concept are used in designing a *robust nonlinear bounded-output feedback*. Consider

a sufficiently wide class of perturbed uncertain nonlinear systems governed by the following system of ordinary differential equations:

$$\dot{x}(t) = f\,[x(t)] + Bu(t) + \zeta(t), \text{ a.e. on } \mathbb{R}_+,$$

$$y(t) = h\,[x(t)] + \zeta_y(t), \ x(0) = x_0 \in \mathbb{R}^n,$$

$$(11.1)$$

where

- $x(t) \in \mathbb{R}^n$ is the state vector at time $t \geq 0$,
- $y(t) \in \mathbb{R}^p$ is the output system at time $t \geq 0$,
- the vector functions $f : \mathbb{R}^n \to \mathbb{R}^n$ and $h : \mathbb{R}^n \to \mathbb{R}^p$ define the dynamics and output mapping of the system (11.1), respectively,
- $B \in \mathbb{R}^{n \times m}$ is the matrix realizing the actuator mapping, and
- $u(t) \in \mathbb{R}^m$ is the control input at time $t \geq 0$; $\zeta_x(t) \in \mathbb{R}^n$ and $\zeta_y(t) \in \mathbb{R}^p$ are external perturbations.

This nonlinear system can be represented in a "quasilinear" format as

$$\dot{x}(t) = Ax(t) + Bu(t) + \xi_x[x(t), t],$$

$$y(t) = Cx(t) + \xi_y[x(t), t], \ x(0) = x_0,$$

$$(11.2)$$

$$A \in \mathbb{R}^{n \times n}, \ C \in \mathbb{R}^{p \times n},$$

where

$$\xi_x[x(t), t] := \Delta f[x(t)] + \zeta_x(t),$$

$$\Delta f(x) := f(x) - Ax,$$

$$\xi_y[x(t), t] := \Delta h[x(t)] + \zeta_y(t),$$

$$\Delta h(x) := h(x) - Cx.$$

The vectors $\xi_x[x(t), t]$ and $\xi_y[x(t), t]$ characterize the *uncertain part* (or unmodeled dynamics) of the system (11.2), which contains both external perturbations $\zeta_x(t)$ and $\zeta_y(t)$, which are assumed to be bounded:

$$\sup_{t \in \mathbb{R}_+} \|\zeta_x(t)\| \leq c_4 < \infty,$$

$$(11.3)$$

$$\sup_{t \in \mathbb{R}_+} \|\zeta_y(t)\| \leq c_5 < \infty.$$

The mappings f and h may not be known exactly, but they are quasi-Lipschitz (see Definition 2.1), that is,

$$f \in \mathcal{C}(A, c_0, c_1), \quad h \in \mathcal{C}(C, c_2, c_3). \tag{11.4}$$

The constant matrices A and C as well as the constants c_k ($k = 0, 3$) are assumed to be known a priori. The memberships $f \in \mathcal{C}(A, c_1, c_2)$ and $h \in \mathcal{C}(C, c_3, c_4)$ mean exactly that the growth rates of these functions (as $\|x\| \to \infty$) are not faster than linear. In (11.4), the matrices A and C characterize the, so-called nominal linear plant contained within the $\mathcal{C}$ class; the scalars c_k, $k = \overline{0, 3}$ are nonnegative constants defining a permitted deviation of every nonlinearity from this class with respect to a nominal linear plant. Under the additional information that $f(0) = 0$ for every function $f \in \mathcal{C}(A, c_0, c_1)$, one can take $c_0 = 0$, and we shall deal with the class of Lipschitz functions commonly considered within modern control theory.

Under the conditions (11.3) and (11.4), we may conclude that

$$\left\| \xi_x[x(t), t] \right\|^2 \le d_0 + d_1 \left\| x(t) \right\|^2,$$

$$d_0 = 2\left(c_0 + c_4\right), \quad d_1 = 2c_2,$$

$$\left\| \xi_y[x(t), t] \right\|^2 \le d_2 + d_3 \left\| x(t) \right\|^2, \tag{11.5}$$

$$d_2 = 2\left(c_1 + c_5\right), \quad d_3 = 2c_3.$$

11.2.2 Basic Assumptions

We make hereinafter the following assumptions:

A1. The nonlinearity $f(x)$ in (11.1) belongs to the class $\mathcal{C}(A, c_0, c_1)$ (11.4). Certainly, knowledge of the matrix A (characterizing the "nominal linear plant") as well as two scalar parameters gives very "approximative" information about the nonlinear function f. Nevertheless, the approximate values of these class parameters can be estimated a priori based on the following consideration:

- $A \simeq \nabla_x f(x)|_{x=0}$ if the vector field $f(x)$ is differential (and hence $c_1 = 0$) at the origin;
- the parameter c_1 defines a possible upper bound of the velocity norm at the origin, i.e.,

$$\left\| f(x) \right\|_{x=x_0=0} \simeq \left\| \dot{x}(0) \right\| \le c_1;$$

- the parameter c_2 characterizes the maximum possible linear increment of the difference, i.e.,

$$\sup_{x\in\mathbb{R}^n} \|f(x) - Ax\| / \|x\| \le c_2.$$

The same interpretation can be given for the parameters of the class $\mathcal{C}(C, c_3, c_4)$.

A2. Based on the upper estimate (11.5) below, we accept that

$$\xi^\mathsf{T}\xi = \begin{pmatrix} \xi_x[x(t),t] \\ \xi_y[x(t),t] \end{pmatrix}^\mathsf{T} \begin{pmatrix} \xi_x[x(t),t] \\ \xi_y[x(t),t] \end{pmatrix}$$

$$= \xi_x^\mathsf{T}[x(t),t]\,\xi_x[x(t),t] + \xi_y^\mathsf{T}[x(t),t]\,\xi_y[x(t),t] \tag{11.6}$$

$$\le b_0 + b_1\|x(t)\|^2,$$

$$b_0 = d_0 + d_2,\ b_1 = d_1 + d_3.$$

A3. The set of all admissible control actions $\mathcal{U}$ is a convex closed bounded complete set (compact):

$$u^+ = \operatorname{diam}\mathcal{U} := \min_{p\in\mathcal{U}}\|s - p\| < \infty.$$

In that case, for every $s \in \mathbb{R}^m$, there exists a unique $p_0 \in \mathcal{U}$, called the *projection* of s to the set $\mathcal{U}$, such that

$$\|s - p\| \ge \|s - p_0\|$$

for every $p \in U$ [see Fig. 11.1 and Rudin (1991)].

In other words,

$$\|s - p_0\| = \min_{p\in\mathcal{U}}\{\|s - p\|\ \text{for every}\ s \in \mathbb{R}^m\}.$$

The control action

$$u(t) \subset \mathcal{U} \in \mathbb{R}^m \tag{11.7}$$

is obtained as a result of the application of the nonlinear operator $\pi_{\mathcal{U}}(\cdot)$ acting as

$$u(t) := \pi_{\mathcal{U}}(K_{t_i}\hat{x}(t)), \tag{11.8}$$

$$\pi_{\mathcal{U}}(s) := \{\bar{u} \in \mathcal{U} \mid \|\bar{u} - s\| \le \|u - s\|\ \forall s \in \mathbb{R}^m, u \in \mathcal{U}\},$$

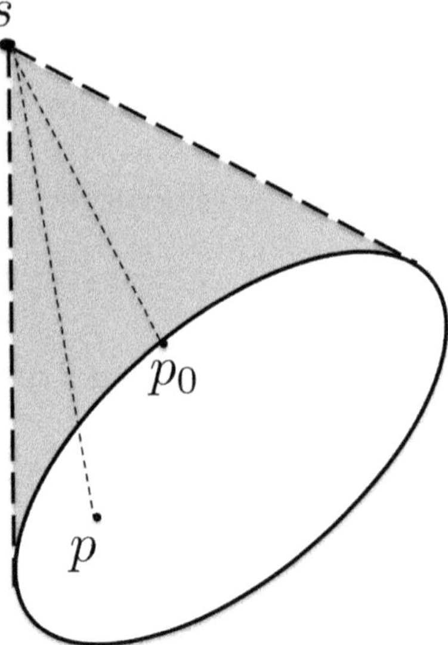

Fig. 11.1 Illustration of the projection operator

where $K_{t_i} \in \mathbb{R}^{m \times n}$ is a gain matrix that also should be designed online, so that K_{t_i} remains constant within the half-open given intervals $(t_{i-1}, t_i]$, for all $i = 1, 2, \ldots$, but subject to tuning at sample times t_i.

In (11.8), the vector $\hat{x}(t) \in \mathbb{R}^n$ is an estimate of the state $x(t)$.

A4. The state estimates $\hat{x}(t)$ are generated by the observer (for some fixed $\hat{x}(0) = \hat{x}_0$)

$$\dot{\hat{x}}(t) = A\hat{x}(t) + Bu(t) + L_{t_i}\left[y(x) - C\hat{x}(t)\right], \tag{11.9}$$

where $L_{t_i} \in \mathbb{R}^{n \times p}$ is a time-invariant gain matrix to be designated, keeping a constant value on each interval $(t_{i-1}, t_i]$.

11.2.3   Extended Dynamic Form

For the observer (11.9), in view of (11.8), we have

$$\dot{\hat{x}}(t) = \left[A + BK_{t_i}(t)\right]\hat{x}(t) + L_{t_i}\left[y(t) - C\hat{x}(t)\right] + B\Delta\pi\left[K_{t_i}\hat{x}(t)\right], \tag{11.10}$$

$$\Delta\pi[K_{t_i}\hat{x}(t)] := \pi_{\mathcal{U}}[K_{t_i}\hat{x}(t)] - K_{t_i}\hat{x}(t).$$

Define the state estimation error

$$e(t) = x(t) - \hat{x}(t), \tag{11.11}$$

for which we get

$$\dot{e}(t) = [A - L_{t_i} C]\, e(t) + \xi_x[x(t), t] - L_{t_i} \xi_y[x(t), t],$$

$$e(0) = e_0.$$

(11.12)

Combining (11.10) and (11.12) for the extended vector

$$z^\mathsf{T}(t) = [\hat{x}^\mathsf{T}(t),\ e^\mathsf{T}(t)],$$

we derive

$$\dot{z}(t) = \mathcal{A}_{t_i}\, (K_{t_i}, L_{t_i})\, z(t) + \mathcal{F}_{t_i}\, (L_{t_i})\, \xi\, [x\,(t)\,, t] + \mathcal{B}v_{t_i},$$

$$z(0) = z_0,$$

(11.13)

where

$$\mathcal{A}_{t_i}\, (K_{t_i}, L_{t_i}) := \begin{bmatrix} A + BK_{t_i} & L_{t_i} C \\ 0_{n \times n} & A - L_{t_i} C \end{bmatrix} \in \mathbb{R}^{2n \times 2n},$$

$$\mathcal{F}_{t_i}\, (L_{t_i}) := \begin{bmatrix} 0_{n \times n} & L_{t_i} \\ I_{n \times n} & -L_{t_i} \end{bmatrix} \in \mathbb{R}^{2n \times (n+p)},$$

$$\xi[x(t), t] := \begin{bmatrix} \xi_x[x(t), t] \\ \xi_y[x(t), t] \end{bmatrix} \in \mathbb{R}^{n+p},$$

$$\mathcal{B} := \begin{bmatrix} B & 0_{n \times m} \\ 0_{n \times m} & 0_{n \times m} \end{bmatrix} \in \mathbb{R}^{2n \times (n+m)},$$

$$v_{t_i} := \begin{bmatrix} \Delta \pi [K_{t_i} \hat{x}(t)] \\ 0_m \end{bmatrix} \in \mathbb{R}^{2m}.$$

11.2.4   Problem Formulation

Observe that in the presence of the unmodeled dynamics ($\xi\,[x\,(t)\,, t] \neq 0$), it is impossible to stabilize the given dynamics exactly providing asymptotic origin convergence. The boundedness of the trajectories can be guaranteed only if it is possible within the admissible feedbacks (11.8). Since every bounded trajectory can be imposed on some convex set (in this book, we select an ellipsoid), the "best designing" that one can do is to minimize the "size" of this ellipsoid by varying the gains matrices K_{t_i} and L_{t_i} using online information $\{\hat{x}\,(t)\,, u\,(t)\}_{t \geq 0}$.

Recall that a trajectory $\{x(t)\}_{t \geq 0}$ belongs asymptotically to the *attractive ellipsoid* $\mathcal{E}\,(\ell, \bar{P})$ with center at the point $x = \ell$ and corresponding matrix $\bar{P}$ if

$$\limsup_{t \to \infty} \; [x(t) - \ell]^{\mathsf{T}} \, \bar{P}^{-1} \, [x(t) - \ell] \leq 1.$$

This means that asymptotically, all trajectories of a considered system arrive at the ellipsoid set $\mathcal{E}\left(\ell, \bar{P}\right)$ referred to in this book as an *attractive ellipsoid*.

Now we are ready to formulate the problem that we are going to solve.

Problem 11.1. Based on the available information $\{y(t), \hat{x}(t), u(t)\}_{t \geq 0}$, find sequences $\{K_{t_i}\}_{i=1,2,...}$ and $\{L_{t_i}\}_{i=1,2,...}$ of the gain matrices K_{t_i}, L_{t_i} that for every plant with uncertainties from the given class $\mathcal{C}$ (11.4) guarantees the existence of an attractive ellipsoid of minimal "size" (the size $\mathcal{E}\left(0, \bar{P}\right)$ is associated with the trace of the ellipsoidal matrix $\bar{P}$):

$$\mathrm{tr}\left\{\bar{P}_{t_i}\right\} \; \to \; \min_{K_{t_i}, \, L_{t_i} \; (i=1,2,...)} . \tag{11.14}$$

The sequences $\{K_{t_i}, L_{t_i}\}_{i=1,2,...}$ of the gain matrices in (11.8) and in (11.9), realizing (11.14), guarantee the so-called *zone-stability* under bounded control signals (11.7) for every uncertain system (11.13).

11.3   Robust Bounded Output Control Synthesis

This section deals with the design of a bounded output controller that provides a robust performance for the system (11.13) under perturbations and unknown dynamics based on the AEM concept. First, let us select the feedback controller as projectional control (11.8). Observe that for the system (11.13), the gain matrix K_{t_i} can be found for each time $t \in (t_{i-1}, t_i]$. But in this case, the available information for find K_{t_i} is given by previous data, that is, at time t_i, we use information up to t_{i-1}, i.e., we use t_{i-1} data. Thus the problem in which we are interested is to find the gain matrices K_{t_i} and L_{t_i} based on data within the time interval $(t_{i-1}, t_i]$.

First, let us formulate an auxiliary result to be used below in the proof of the main result.

11.3.1   Storage Function

Proposition 11.1 (On the time-interval storage function). *If the collection*

$$P_{t_i} \in \mathbb{R}^{2n \times 2n}, \; Q_{t_i} \in \mathbb{R}^{n \times n},$$

$$K_{t_i} \in \mathbb{R}^{m \times n}, \; L_{t_i} \in \mathbb{R}^{n \times p}, \; \varepsilon_{1,i}, \; \varepsilon_{2,i}, \; \alpha_i,$$

satisfies the matrix constraints

$$
W_i := \begin{bmatrix}
\begin{array}{c} P_{t_i}^{-1} \mathcal{A}_{\alpha,t_i}\left(K_{t_i}, L_{t_i}\right) + \\ \mathcal{A}_{\alpha,t_i}^{\mathsf{T}}\left(K_{t_i}, L_{t_i}\right) P_{t_i}^{-1} + R_{t_i} \end{array} & P_{t_i}^{-1} \mathcal{F}_{t_i}\left(L_{t_i}\right) & P_{t_i}^{-1} \mathcal{B} \\
\mathcal{F}_{t_i}^{\mathsf{T}}\left(L_{t_i}\right) P_{t_i}^{-1} & -\varepsilon_{1,i} I_{(n+p)\times(n+p)} & 0_{(n+p)\times 2m} \\
\mathcal{B}^{\mathsf{T}} P_{t_i}^{-1} & 0_{2m\times(n+p)} & -\varepsilon_{2,i} I_{2m\times 2m}
\end{bmatrix} < 0, \qquad (11.15)
$$

$$
\varepsilon_{1,i} > 0, \ \varepsilon_{2,i} > 0, \ 0 < Q_{t_i} = Q_{t_i}^{\mathsf{T}},
$$

$$
K_{t_i}^{\mathsf{T}} K_{t_i} \leq Q_{t_i},
$$

$$
0 < P_{t_i} = P_{t_i}^{\mathsf{T}} = \operatorname{diag}\left[P_{A,t_i}, \ P_{A,t_i}\right],
$$

$$
\hat{\mathcal{A}}_{\alpha_i}\left(K_{t_i},, L_{t_i}\right) := \mathcal{A}_{t_i}\left(K_{t_i},, L_{t_i}\right) + \frac{\alpha_i}{2} I_{2n},
$$

$$
R_{t_i}\left(Q_{t_i}\right) := \varepsilon_{1,i} d_1 I_{2n\times 2n} + 2\varepsilon_{2,i} \operatorname{diag}(Q_{t_i}, 0_{n\times n}),
$$

then the energetic ("storage") function

$$
V_i[z(t)] := z^{\mathsf{T}}(t) P_{t_i}^{-1} z(t) \qquad (11.16)
$$

defined on $(t_{i-1}, t]$ satisfies the following differential inequality:

$$
\dot{V}_i[z(t)] \leq -\alpha_i V_i[z(t)] + \beta_i, \qquad (11.17)
$$

$$
\beta_i := \varepsilon_{1,i} d_0 + 2\varepsilon_{2,i}\left(u^+\right)^2. \qquad (11.18)
$$

Proof. The time derivative of (11.16) along the system trajectories (11.13) for the time interval $(t_{i-1}, t_i]$ is

$$
\dot{V}_i[z(t)] = 2z^{\mathsf{T}}(t) P_{t_i}^{-1} \dot{z}(t) = 2z^{\mathsf{T}}(t) P_{t_i}^{-1} \mathcal{A}_{t_i}\left(K_{t_i}, L_{t_i}\right) z(t)
$$

$$
+ 2z^{\mathsf{T}}(t) P_{t_i}^{-1} \mathcal{F}_{t_i}\left(L_{t_i}\right) \xi + 2z^{\mathsf{T}}(t) P_{t_i}^{-1} \mathcal{B} \upsilon_{t_i} =
$$

$$
\begin{bmatrix} z(t) \\ \xi \\ \upsilon_{t_i} \end{bmatrix}^{\mathsf{T}}
\begin{bmatrix}
\begin{array}{c} P_{t_i}^{-1} \mathcal{A}_{t_i}\left(K_{t_i}, L_{t_i}\right) + \\ \mathcal{A}_{t_i}^{\mathsf{T}}\left(K_{t_i}, L_{t_i}\right) P_{t_i}^{-1} \end{array} & P_{t_i}^{-1} \mathcal{F}_{t_i}\left(L_{t_i}\right) & P_{t_i}^{-1} \mathcal{B} \\
\mathcal{F}_{t_i}^{\mathsf{T}}\left(L_{t_i}\right) P_{t_i}^{-1} & 0_{(n+p)\times(n+p)} & 0_{(n+p)\times 2m} \\
\mathcal{B}^{\mathsf{T}} P_{t_i}^{-1} & 0_{2m\times(n+p)} & 0_{2m\times 2m}
\end{bmatrix}
\begin{bmatrix} z(t) \\ \xi \\ \upsilon_{t_i} \end{bmatrix}.
$$

Adding and subtracting the terms $\alpha V_i[z(t)]$, $\varepsilon_{1,i}\xi^{\mathsf{T}}\xi$ and $\varepsilon_{2,i}\upsilon_{t_i}^{\mathsf{T}}\upsilon_{t_i}$ on the right-hands side of the last equality leads to

$$
\dot{V}_i[z(t)] =
\begin{bmatrix} z(t) \\ \xi \\ \upsilon_{t_i} \end{bmatrix}^{\mathsf{T}}
\begin{bmatrix}
\begin{array}{c} P_{t_i}^{-1} \mathcal{A}_{t_i}\left(K_{t_i}, L_{t_i}\right) \\ + \mathcal{A}_{t_i}^{\mathsf{T}}\left(K_{t_i}, L_{t_i}\right) P_{t_i}^{-1} \end{array} & P_{t_i}^{-1} \mathcal{F}_{t_i}\left(L_{t_i}\right) & P_{t_i}^{-1} \mathcal{B} \\
\mathcal{F}_{t_i}^{\mathsf{T}}\left(L_{t_i}\right) P_{t_i}^{-1} & -\varepsilon_{1,i} I_{(n+p)\times(n+p)} & 0_{(n+p)\times 2m} \\
\mathcal{B}^{\mathsf{T}} P_{t_i}^{-1} & 0_{2m\times(n+p)} & -\varepsilon_{2,i} I_{2m\times 2m}
\end{bmatrix}
\begin{bmatrix} z(t) \\ \xi \\ \upsilon_{t_i} \end{bmatrix}
$$

$$
-\alpha V_i[z(t)] + \varepsilon_{1,i}\xi^{\mathsf{T}}\xi + \varepsilon_{2,i}\upsilon_{t_i}^{\mathsf{T}}\upsilon_{t_i}.
$$

Note that by the identity

$$
x(t) := \hat{x}(t) + e(t) = I_{2n\times 2n} z(t)
$$

and in view of the estimate

$$
\xi^{\mathsf{T}}\xi \le b_0 + b_1 \|x\|^2
$$

[b_0, b_1 are defined in (11.6)], we conclude that

$$
\xi^{\mathsf{T}}\xi \le b_0 + b_1 \|z(t)\|^2.
$$

Hence in view of (11.10), we have

$$
\|\Delta\pi[K_{t_i}\hat{x}(t)]\|^2 \le 2\|\pi_{\mathcal{U}}(K_{t_i}\hat{x}(t))\|^2 + 2\|K_{t_i}\hat{x}(t)\|^2
$$

$$
= 2\left(u^+\right)^2 + 2\|K_{t_i}\hat{x}(t)\|^2.
$$

Since

$$
\hat{x}(t) = Hz(t), \; H = [I_{n\times n} \; 0_{n\times n}]
$$

and in view of the conditions of this proposition, it follows that

$$
\|\Delta\pi[K_{t_i}\hat{x}(t)]\|^2 \le 2\left(u^+\right)^2 + 2z^{\mathsf{T}}(t)H^{\mathsf{T}}Q_{t_i}Hz(t).
$$

So the differential equation for $V_i[z(t)]$ given above results in the following differential inclusion:

$$
\dot{V}_i[z(t)] \le
$$

$$
\begin{bmatrix} z(t) \\ \xi \\ \upsilon_{t_i} \end{bmatrix}^{\mathsf{T}}
\begin{bmatrix}
\begin{array}{c} P_{t_i}^{-1}\mathcal{A}_{\alpha,t_i}\left(K_{t_i}, L_{t_i}\right) + \\ \mathcal{A}_{\alpha,t_i}^{\mathsf{T}}\left(K_{t_i}, L_{t_i}\right) P_{t_i}^{-1} + R_{t_i} \end{array} & P_{t_i}^{-1}\mathcal{F}_{t_i}\left(L_{t_i}\right) & P_{t_i}^{-1}\mathcal{B} \\
\mathcal{F}_{t_i}^{\mathsf{T}}\left(L_{t_i}\right) P_{t_i}^{-1} & -\varepsilon_{1,i} I_{(n+p)\times(n+p)} & 0_{(n+p)\times 2m} \\
\mathcal{B}^{\mathsf{T}} P_{t_i}^{-1} & 0_{2m\times(n+p)} & -\varepsilon_{2,i} I_{2m\times 2m}
\end{bmatrix}
\begin{bmatrix} z(t) \\ \xi \\ \upsilon_{t_i} \end{bmatrix}
$$

$$
-\alpha_i V_i[z(t)] + \beta_i,
$$

where

$$\mathcal{A}_{\alpha,t_i}\left(K_{t_i}, L_{t_i}\right) := \mathcal{A}_{t_i}\left(K_{t_i}, L_{t_i}\right) + \tfrac{\alpha_i}{2} I_{2n\times 2n},$$

$$R := \varepsilon_{1,i} d_1 I_{2n\times 2n} + 2\varepsilon_{1,i} Q_{t_i}.$$

If

$$W := \begin{bmatrix} \begin{array}{c} P_{t_i}^{-1}\mathcal{A}_{\alpha,t_i}\left(K_{t_i}, L_{t_i}\right) + \\ \mathcal{A}_{\alpha,t_i}^{\mathsf{T}}\left(K_{t_i}, L_{t_i}\right) P_{t_i}^{-1} + R_{t_i} \end{array} & P_{t_i}^{-1}\mathcal{F}_{t_i}\left(L_{t_i}\right) & P_{t_i}^{-1}\mathcal{B} \\ \mathcal{F}_{t_i}^{\mathsf{T}}\left(L_{t_i}\right) P_{t_i}^{-1} & -\varepsilon_{1,i} I_{(n+p)\times(n+p)} & 0_{(n+p)\times 2m} \\ \mathcal{B}^{\mathsf{T}} P_{t_i}^{-1} & 0_{2m\times(n+p)} & -\varepsilon_{2,i} I_{2m\times 2m} \end{bmatrix} < 0,$$

then the inequality is preserved, implying

$$\dot{V}_i[z(t)] \leq -\alpha_i V_i[z(t)] + \beta_i, \quad t \in (t_{i-1}, t_i],$$

which completes the proof.   ∎

It is well known that the concept of an energetic function was rigorously formalized using Lyapunov stability theory as well as the notion of a positive invariant set. Here we just note that if there exists a set of solutions

$$(P_{A,t_i}, Q_{t_i}, K_{t_i}, L_{t_i}, \varepsilon_{1,i}, \varepsilon_{2,i}, \alpha_i)$$

within the time interval $(t_{i-1}, t_i]$ such that (11.15) holds, then the storage function (11.16) is not necessarily monotonically nonincreasing, that is, $V_i(z)$ is not a Lyapunov function for the considered system at least for this time interval. Below, we suggest the construction of a *Lyapunov-Like* function whose derivative on the trajectories of the considered controlled system is strictly negative outside of an ellipsoid. So below, we present a Lyapunov zone convergence analysis.

11.3.2   Zone-Convergence Analysis

Let us consider the function

$$G(t) := \sum_{i=1}^{\infty} \chi_i(t) \mathcal{G}_i(t),$$

$$\chi_i(t) := \begin{cases} 1, & t \in (t_{i-1}, t_i], \\ 0, & t \notin (t_{i-1}, t_i], \end{cases} \quad \sum_{i=1}^{\infty} \chi_i(t) = 1, \tag{11.19}$$

where

$$\mathcal{G}_i\left(t\right) = \left(\left[\sqrt{V_i[z(t)]} - \sqrt{\frac{\beta_i}{\alpha_i}}\,\right]_+\right)^2,\ t \in (t_{i-1}, t_i]\,,$$

$$[\gamma]_+ := \begin{cases} \gamma, & \gamma \geq 0, \\ 0, & \gamma < 0. \end{cases} \tag{11.20}$$

Observe that the function $[\gamma]_+$ is not differentiable at the point $\gamma = 0$, but the function $([\cdot]_+)^2$ is differential everywhere. In (11.20), the process $z(t)$ is defined by (11.13), and therefore, the function $G\left(t\right)$ is defined on all possible trajectories of (11.13).

Proposition 11.2 (On zone convergence). *If*

1. the collection $(P_{A,t_i}, Q_{t_i}, K_{t_i}, L_{t_i}, \varepsilon_{1,i}, \varepsilon_{2,i}, \alpha_i)$ satisfies the set of matrix inequalities in Proposition 11.1 within each time interval $(t_{i-1}, t_i]$, $i = 1, 2, \ldots$;
2. the following additional dynamic constraint is satisfied at each stage $i = 1, 2, \ldots$:

$$\mathcal{G}_{i-1}\left(t_i\right) = \left(\left[\sqrt{z^{\mathsf{T}}\left(t_i\right) P_{i-1}^{-1} z\left(t_i\right)} - \sqrt{\frac{\beta_{i-1}}{\alpha_{i-1}}}\,\right]_+\right)^2$$

$$\geq \left(\left[\sqrt{z^{\mathsf{T}}\left(t_i\right) P_i^{-1} z\left(t_i\right)} - \sqrt{\frac{\beta_i}{\alpha_i}}\,\right]_+\right)^2 = \mathcal{G}_i\left(t_i\right); \tag{11.21}$$

then the function $G\left(t\right)$ (11.19) is a Lyapunov function for the dynamical system (11.13), namely,

$$\frac{d}{dt} G\left(t\right) \leq 0 \tag{11.22}$$

for all $t \geq 0$, and

$$\frac{d}{dt} G\left(t\right) < 0 \text{ if } \sqrt{V_{i(t)}[z(t)]} > \mu_{i(t)}, \tag{11.23}$$

providing an "attractivity property" proof of Proposition 3.2:

$$\left[\sqrt{V_{i(t)}[z(t)]} - \mu_{i(t)}\right]_+ \to 0 \text{ as } t \to \infty,$$

$$\mu_{i(t)} := \sqrt{\beta_{i(t)}/\alpha_{i(t)}}, \ i\left(t\right) := \{i : t \in (t_{i-1}, t_i]\}\,.$$

Proof. Recall that the "generalized" derivative of the Heaviside function

$$\theta(t) := \begin{cases} 1, & \text{if } t \geq 0, \\ 0, & \text{if } t < 0, \end{cases}$$

is the Dirac delta function $\delta(t) = \theta'(t)$ with the property

$$\int_{t=-\infty}^{\infty} \delta(t) \, f(t) \, dt = f(0),$$

valid for every function $f(t)$ right-continuous at the origin. Since

$$\chi_i(t) = \theta(t - t_{i-1}) - \theta(t - t_i),$$

the differentiation of (11.19) leads to

$$\frac{d}{dt} G(t) = \sum_{i=1}^{\infty} \frac{d}{dt} [\chi_i(t) \, \mathcal{G}_i(t)] =$$
$$\sum_{i=1}^{\infty} \chi_i(t) \frac{d}{dt} \mathcal{G}_i(t) + \sum_{i=1}^{\infty} [\delta(t - t_{i-1}) - \delta(t - t_i)] \, \mathcal{G}_i(t).$$

This is a singularly perturbed differential equation, which in the equivalent integral form can be represented as follows:

$$G(t) - G(0) = \int_{s=0}^{t} \sum_{i=1}^{\infty} [\delta(s - t_{i-1}) - \delta(s - t_i)] \, \mathcal{G}_i(s) \, ds +$$

$$\int_{s=0}^{t} \left[\sum_{i=1}^{\infty} \chi_i(s) \frac{d}{ds} \mathcal{G}_i(s) \right] ds = \sum_{i=1}^{\infty} [\mathcal{G}_i(t_{i-1}) - \mathcal{G}_i(t_i)] +$$

$$\int_{s=0}^{t} \left[\sum_{i=1}^{\infty} \chi_i(s) \left[\sqrt{V_i(s)} - \sqrt{\frac{\beta_i}{\alpha_i}} \right]_+ \frac{\frac{d}{ds} V_i(s)}{\sqrt{V_i(s)}} \right] ds$$

$$\leq \mathcal{G}_0(t_0) + [-\mathcal{G}_0(t_1) + \mathcal{G}_1(t_1)] + [-\mathcal{G}_1(t_2) + \mathcal{G}_2(t_2)]$$

$$+ \cdots + \left[-\mathcal{G}_{i-1}\left(t_{i(t)}\right) + \mathcal{G}_i\left(t_{i(t)}\right) \right] + \cdots +$$

$$\int_{s=0}^{t} \left[\sum_{i=1}^{\infty} \chi_i(s) \left[\sqrt{V_i(s)} - \sqrt{\frac{\beta_i}{\alpha_i}} \right]_+ \frac{(-\alpha_i V_i(s) + \beta_i)}{\sqrt{V_i(s)}} \right] ds.$$

Taking into account the "monotonicity condition" (11.21), we get

$$G(t) - G(0) \leq \mathcal{G}_0(0) + I(t),$$

where

$$
\begin{aligned}
I(t) &:= \int_{s=0}^{t} \left[\sum_{i=1}^{\infty} \chi_i(s) \left[\sqrt{V_i(s)} - \sqrt{\tfrac{\beta_i}{\alpha_i}} \right]_+ \frac{(-\alpha_i V_i(s) + \beta_i)}{\sqrt{V_i(s)}} \right] ds = \\
&\quad - \int_{s=0}^{t} \sum_{i=1}^{\infty} \alpha_i \chi_i(s) \left[\sqrt{V_i(s)} - \sqrt{\tfrac{\beta_i}{\alpha_i}} \right]_+ \frac{\left(V_i(s) - \tfrac{\beta_i}{\alpha_i} \right)}{\sqrt{V_i(s)}} ds = \\
&\quad - \int_{s=0}^{t} \sum_{i=1}^{\infty} \alpha_i \chi_i(s) \left[\sqrt{V_i(s)} - \sqrt{\tfrac{\beta_i}{\alpha_i}} \right]_+ \frac{\left(\sqrt{V_i(s)} - \sqrt{\tfrac{\beta_i}{\alpha_i}} \right) \left(\sqrt{\tfrac{\beta_i}{\alpha_i}} + \sqrt{V_i(s)} \right)}{\sqrt{V_i(s)}} ds \\
&= - \int_{s=0}^{t} \sum_{i=1}^{\infty} \alpha_i \chi_i(s) \left[\sqrt{V_i(s)} - \sqrt{\tfrac{\beta_i}{\alpha_i}} \right]_+^2 \frac{\sqrt{\tfrac{\beta_i}{\alpha_i}} + \sqrt{V_i(s)}}{\sqrt{V_i(s)}} ds = \\
&\quad - \int_{s=0}^{t} \sum_{i=1}^{\infty} \alpha_i \chi_i(t) G_i(t) \frac{\left(\sqrt{V_i(t)} + \sqrt{\tfrac{\beta_i}{\alpha_i}} \right)}{\sqrt{V_i(t)}} ds \leq 0.
\end{aligned}
$$

We now introduce the so-called dominating process $\tilde{G}(t)$ satisfying

$$
\tilde{G}(t) - \tilde{G}(0) = \mathcal{G}_0(0) + I(t).
$$

It is clear that if

$$
G(0) = \tilde{G}(0),
$$

then

$$
G(t) \leq \tilde{G}(t).
$$

Differentiation of this last identity implies

$$
\frac{d}{dt} \tilde{G}(t) = - \sum_{i=1}^{\infty} \alpha_i \chi_i(t) G_i(t) \frac{\left(\sqrt{V_i(t)} + \sqrt{\tfrac{\beta_i}{\alpha_i}} \right)}{\sqrt{V_i(t)}} \leq 0.
$$

The right-hand side of the last expression is strictly negative if

$$
\sqrt{V_{i(t)}[z(t)]} > \mu_{i(t)}.
$$

Moreover, since $\tilde{G}(t)$ is a nonnegative monotonically nonincreasing function, it follows by Weierstrass's theorem that $\tilde{G}(t)$ has a limit:

$$
\lim_{t \to \infty} \tilde{G}(t) = \tilde{G}^*.
$$

From the identity above it follows that

$$
0 \leq \tilde{G}(t) + |I(t)| = \mathcal{G}_0(0) + \tilde{G}(0) = \text{const}.
$$

Taking $t \to \infty$, we obtain

$$0 \leq \tilde{G}^* + \limsup_{t \to \infty} |I(t)| < \infty,$$

implying

$$\limsup_{t \to \infty} |I(t)| < \infty.$$

This means that there exists a time subsequence $\{s_k\}_{k=1,2,\dots}$ such that

$$\sum_{i=1}^{\infty} \alpha_i \chi_i(s_k) \left[\sqrt{V_i(s_k)} - \sqrt{\frac{\beta_i}{\alpha_i}} \right]_+^2 \frac{\left(\sqrt{V_i(s_k)} + \sqrt{\frac{\beta_i}{\alpha_i}} \right)}{\sqrt{V_i(s_k)}} \underset{k \to \infty}{\to} 0,$$

and as a result,

$$\sum_{i=1}^{\infty} \alpha_i \chi_i(s_k) \left[\sqrt{V_i(s_k)} - \sqrt{\frac{\beta_i}{\alpha_i}} \right]_+^2 = G(s_k) \underset{k \to \infty}{\to} 0.$$

Hence from the continuity of $\tilde{G}(t)$, it follows that

$$\tilde{G}(s_k) \underset{k \to \infty}{\to} 0.$$

But the sequence $G(t)$ converges, and hence all its subsequences have the same limit point, providing

$$G^* = 0.$$

This completes the proof of the proposition.   ∎

Corollary 11.1. *If in Proposition 11.2 the numerical sequence* $\{\mu_{i(t)}\}$ *and the matrix sequence* $\{P_i\}$ *decrease monotonically, that is,*

$$\mu_{i(t')} \geq \mu_{i(t'')} \text{ for } t' > t'', \ P_{i-1} \geq P_i, \tag{11.24}$$

then, by Weierstrass's theorem, both have their limits

$$\lim_{t \to \infty} \mu_{i(t)} = \tilde{\mu}, \ \lim_{i \to \infty} P_i = \tilde{P},$$

which means that the ellipsoid $\mathcal{E}\left(0, \tilde{\mu}^2 \tilde{P}\right)$ *is* attractive *for all possible trajectories generated by (11.13) satisfying the inequality*

$$\limsup_{t \to \infty} z^{\mathsf{T}}(t) \left(\frac{1}{\tilde{\mu}^2} \tilde{P}^{-1} \right) z(t) \le 1. \tag{11.25}$$

Proof. Since $W_i < 0$ by the assumption of Proposition 11.1, we directly obtain (11.17). Note that the functions $\dot{V}_i$ satisfy (11.17) in the time intervals $(t_{i-1}, t_i]$ for all $i := 1, 2, \ldots$, and as a result,

$$V_i[z(t_i)] \le \beta_i / \alpha_i + \{V_i[z(t_{i-1})] - \beta_i / \alpha_i\} e^{-\alpha_i \tau_i}$$

$$= \frac{\beta_i}{\alpha_i} - \frac{\beta_i}{\alpha_i} e^{-\alpha_i \tau_i} + e^{-\alpha_i \tau_i} V_{i-1},$$

with $\tau_i := t_i - t_{i-1}$. Under the monotonicity condition (11.21) given in Corollary 11.1, we have

$$V_i[z(t_i)] \le V_{i-1}[z(t_{i-1})].$$

The use of Abel's identity (see, for example, Poznyak (2008), Sect. 12.2.2)

$$\prod_{s=i_0}^{i} \gamma_s + \sum_{s=i_0}^{i} (1 - \gamma_s) \prod_{l=s+1}^{i} \gamma_l = 1,$$

$$\prod_{s=i_0}^{i<i_0} (\cdot)_s \equiv 1, \quad \sum_{s=i_0}^{i<i_0} (\cdot)_s \equiv 0,$$

valid for every sequence $\{\gamma_s\}$ of real numbers, implies

$$V_i[z(t_i)] \le \frac{\beta_i}{\alpha_i} \left(1 - e^{-\alpha_i \tau_i} \right) + e^{-\alpha_i \tau_i} V_{i-1} \le$$

$$\frac{\beta_i}{\alpha_i} \left(1 - e^{-\alpha_i \tau_i} \right) + e^{-\alpha_i \tau_i} \frac{\beta_{i-1}}{\alpha_{i-1}} \left(1 - e^{-\alpha_{i-1} \tau_{i-1}} \right) + e^{-\alpha_i \tau_i - \alpha_{i-1} \tau_{i-1}} V_{i-2} \le$$

$$\cdots \le \sum_{s=i_0}^{i} \frac{\beta_s}{\alpha_s} \left(1 - e^{-\alpha_s \tau_s} \right) e^{-\sum_{l=s+1}^{i} \alpha_l \tau_l} + e^{-\sum_{l=i_0}^{i} \alpha_l \tau_l} V_{i_0} \le$$

$$\left(\max_{i_0 \le s \le i} \frac{\beta_s}{\alpha_s} \right) \sum_{s=i_0}^{i} \left(1 - e^{-\alpha_s \tau_s} \right) e^{-\sum_{l=s+1}^{i} \alpha_l \tau_l} + e^{-\sum_{l=i_0}^{i} \alpha_l \tau_l} V_{i_0} =$$

$$\left(\max_{i_0 \le s \le i} \frac{\beta_s}{\alpha_s} \right) \left[1 - e^{-\sum_{l=i_0}^{i} \alpha_l \tau_l} \right] + e^{-\sum_{l=i_0}^{i} \alpha_l \tau_l} V_{i_0} \le$$

$$\left(\max_{i_0 \le s \le i} \frac{\beta_s}{\alpha_s} \right) + e^{-\sum_{l=i_0}^{i} \alpha_l \tau_l} V_{i_0}$$

and

$$z^\mathsf{T}(t_i) P_{t_i}^{-1} z(t_i) \le \mu_{i(t)}^2,$$

or equivalently,

$$z^\mathsf{T}(t_i) \frac{P_{t_i}^{-1}}{\mu_{i(t)}^2} z(t_i) \le 1.$$

Taking the upper limit as $i \to \infty$ and using the monotonicity property (11.24) completes the proof ∎

11.3.3   The Attractive Ellipsoid of "Minimal Size"

As mentioned above, one can arrange the selection of the parameters

$$(P_{A,t_i}, Q_{t_i}, K_{t_i}, L_{t_i}, \varepsilon_{1,i}, \varepsilon_{2,i}, \alpha_i)$$

[which satisfy the conditions of Proposition 11.2 at each time-interval $(t_{i-1}, t_i]$ and additionally satisfy the monotonicity condition (11.24)] in such a way that the attractive ellipsoid $\mathcal{E}\left(0, \tilde{\mu}^2 \tilde{P}\right)$ will be of "minimal size." This process corresponds to the solution of the following optimization problem at each timeinterval $(t_{i-1}, t_i]$:

$$\mathrm{tr}\left\{\frac{\beta_{i(t)}}{\alpha_{i(t)}} P_i\right\} \to \inf_{P_{A,t_i}, Q_{t_i}, K_{t_i}, L_{t_i}, \varepsilon_{1,i}, \varepsilon_{2,i}, \alpha_i} \tag{11.26}$$
$$\text{subject to the constraints}$$
$$(11.15), (11.18), \text{ and } (11.24).$$

Denote the solution of this optimization problem (11.26) at each step $i = 1, 2, \ldots$ by

$$P_{A,t_i}^*, Q_{t_i}^*, K_{t_i}^*, L_{t_i}^*, \varepsilon_{1,i}^*, \varepsilon_{2,i}^*, \alpha_i^*.$$

From the monotonicity conditions (11.24), we may conclude that the limits

$$\lim_{i\to\infty} P_{A,t_i}^* := \tilde{P}_A^*, \quad \lim_{t\to\infty} \mu_{i(t)}^* := \tilde{\mu}^*$$

exist, so that we can consider the ellipsoid $\mathcal{E}\left(0, \left(\tilde{\mu}^*\right)^2 \tilde{P}^*\right)$, with $\tilde{P}^* = \begin{bmatrix} \tilde{P}_A^* & 0 \\ 0 & \tilde{P}_A^* \end{bmatrix}$
the asymptotically attractive ellipsoid of "minimal size" in the extended space of the variable

$$z_t^\mathsf{T} = \left(\hat{x}_t^\mathsf{T}, x_t - \hat{x}_t^\mathsf{T}\right).$$

To estimate the size of the "optimal attractive ellipsoid" in the state space of the system (11.1), observe that

$$x_t = H z_t, \quad H := \begin{bmatrix} I_{n \times n} & I_{n \times n} \end{bmatrix},$$

and therefore

$$x_t^\mathsf{T} P_x^{-1} x_t = z_t^\mathsf{T} H^\mathsf{T} P_x^{-1} H z_t.$$

On the other hand, it follows from (11.25) that the attractive ellipsoid in the extended z-space is $(\tilde{\mu}^*)^2 (\tilde{P}^*)^{-1}$. So the equality

$$\frac{1}{(\tilde{\mu}^*)^2} (\tilde{P}^*)^{-1} = H^\mathsf{T} P_x^{-1} H$$

should be satisfied. Unfortunately, this identity cannot be satisfied in general, since the size of the matrix P_x in which we are interested is less than the size of the matrix $\tilde{P}^*$. Therefore, we suggest estimating the "minimal ellipsoid" in the x-space as the solution P_x^* of the following optimization problem:

$$\left\| \frac{1}{(\tilde{\mu}^*)^2} (\tilde{P}^*)^{-1} - H^\mathsf{T} P_x^{-1} H \right\|^2 \rightarrow \min_{P_x > 0}, \tag{11.27}$$

where the norm in the Hilbert space of finite-dimensional matrices is defined as

$$\langle A, B \rangle := \operatorname{tr}\{A B^\mathsf{T}\}, \quad \|A\|^2 := \langle A, A \rangle = \operatorname{tr}\{A A^\mathsf{T}\}.$$

The solution P_x^* of the optimization problem (11.27) is as follows:

$$P_x^* = \frac{1}{(\tilde{\mu}^*)^2} (H^\mathsf{T})^+ (\tilde{P}^*)^{-1} H^+$$

(H^+ is the pseudoinverse matrix to H defined in the Moore–Penrose sense. Indeed, The solution X^* of the optimization problem

$$\|A - XB\|^2 \rightarrow \min_X \tag{11.28}$$

satisfies the identity

$$\frac{\partial}{\partial X} \|A - XB\|^2 = \frac{\partial}{\partial X} \operatorname{tr}\{(A - XB)(A^\mathsf{T} - B^\mathsf{T} X^\mathsf{T})\}$$

$$= -2\left(A B^\mathsf{T} - X^+ B B^\mathsf{T}\right) = 0,$$

or equivalently,

$$X^* B B^\mathsf{T} = A B^\mathsf{T},$$

whose solution (in the case $B B^\mathsf{T} > 0$) is

$$X^* = A B^+ + Y \left(I - B B^+ \right),$$

$$B^+ := B^\mathsf{T} \left(B B^\mathsf{T} \right)^{-1},$$

where Y is any matrix of the corresponding size. So one has

$$\left\| X^* \right\|^2 = \left\| A B^+ \right\|^2 + \left\| Y \left(I - B B^+ \right) \right\|^2 +$$

$$2\operatorname{tr} \left\{ A B^+ \left(I - \left(\left(B B^\mathsf{T} \right)^{-1} B \right) B^\mathsf{T} \right) Y^\mathsf{T} \right\} = \left\| A B^+ \right\|^2 + \left\| Y \left(I - B B^+ \right) \right\|^2 \geq \left\| A B^+ \right\|^2.$$

This means that the solution of the optimization problem (11.28) of minimal norm is

$$X^* = A B^+.$$

Since in our case,

$$H^+ = \left[\, I_{n \times n} \ \ I_{n \times n} \, \right]^+ = \frac{1}{2} \begin{bmatrix} I_{n \times n} \\ I_{n \times n} \end{bmatrix},$$

we obtain

$$\left(P_x^* \right)^{-1} = \frac{1}{4 \left(\tilde{\mu}^* \right)^2} \left[\, I_{n \times n} \ \ I_{n \times n} \, \right] \begin{bmatrix} \left(\tilde{P}_A^* \right)^{-1} & 0 \\ 0 & \left(\tilde{P}_A^* \right)^{-1} \end{bmatrix} \begin{bmatrix} I_{n \times n} \\ I_{n \times n} \end{bmatrix}$$

$$= \frac{1}{2 \left(\tilde{\mu}^* \right)^2} \left(\tilde{P}_A^* \right)^{-1}.$$

Hence the attractive ellipsoid of "minimal size" in the state space that guarantees the property $\limsup_{t \to \infty} x_t^\mathsf{T} \left(P_x^* \right)^{-1} x_t \leq 1$ is defined by the ellipsoid matrix

$$\left(P_x^* \right)^{-1} = \frac{1}{2 \left(\tilde{\mu}^* \right)^{-2}} \left(\tilde{P}_A^* \right)^{-1}. \tag{11.29}$$

So the attractive ellipsoid in x-space is twice the size of the corresponding minimal ellipsoid in z-space.

11.4 Numerical Aspects

11.4.1 Transformation of BMI Constraints into LMI Constraints

The optimization problem (11.26) is a nonlinear optimization problem, subject (with fixed $\alpha_i, \varepsilon_{1,i}, \varepsilon_{2,i}, \lambda_i$) to the bilinear matrix inequality (BMI) (11.15). This bilinear (under fixed scalars) optimization problem can be transformed to a linear one (containing only LMI constraints) using the transformation of variables given in the following proposition.

Proposition 11.3. *The solution* $(P_{A,t_i}, Q_{t_i}, K_{t_i}, L_{t_i})$ *of the optimization problem (11.26) under fixed scalar parameters* α_i, $\varepsilon_{1,i}$, $\varepsilon_{2,i}$, *is "isomorphic" to the set of variables*

$$X_{t_i} := \mathrm{diag}(X_{11}^{t_i}, X_{22}^{t_i}),$$

$$Y_1^{t_i} := P_{A,t_i}^{-1} B K_{t_i}, \ \ Y_2^{t_i} := P_{A,t_i}^{-1} L_{t_i},$$

and uniquely related to the previous one by

$$X_{11}^{t_i} := P_{t_i}^{B,11} > 0, \ \ X_{22}^{t_i} := P_{t_i}^{B,22} > 0,$$

$$Y_1^{t_i} := \begin{bmatrix} 0_{(n-m)\times n} \\ Y_i^{1,2} \end{bmatrix}, \ \ Y_i^{1,2} := P_{t_i}^{B,22} K_{t_i}, \ \ Y_{i,2} := P_{A,t_i}^{-1} L_{t_i},$$

satisfying the LMIs

$$\begin{bmatrix} \bar{W}_{11} & \bar{W}_{12} & \bar{W}_{13} \\ \bar{W}_{12}^{\mathsf{T}} & -\varepsilon_1 I_{(n+p)\times(n+p)} & 0_{(n+p)\times 2m} \\ \bar{W}_{13}^{\mathsf{T}} & 0_{2m\times(n+p)} & -\varepsilon_2 I_{2m\times 2m} \end{bmatrix} < 0 \tag{11.30}$$

$$\begin{bmatrix} (u^+)^2 & x_{t_{i-1}}^{\mathsf{T}} Q_{t_i} \\ Q_{t_i} x_{t_{i-1}} & Q_{t_i} \end{bmatrix} > 0, \ t \in [t_{i-1}, t_i), \tag{11.31}$$

with the following elements:

$$\bar{W}_{11} = \begin{bmatrix} X_{t_i} A_\alpha + A_\alpha^\mathsf{T} X_{t_i} + Y_1^{t_i} + \left(Y_1^{t_i}\right)^\mathsf{T} & Y_2^{t_i} C \\ \quad + d_1 \varepsilon_{1,i} I_{n\times n} + \varepsilon_{2,i} Q_{t_i} & \\ & \\ C^\mathsf{T}\left(Y_2^{t_i}\right)^\mathsf{T} & X_{t_i} A_\alpha + A_\alpha^\mathsf{T} X_{t_i} - Y_2^{t_i} C - C^\mathsf{T}\left(Y_2^{t_i}\right)^\mathsf{T} \\ & \quad + d_1 \varepsilon_{1,i} I_{n\times n} \end{bmatrix},$$

$$\bar{W}_{12} = \begin{bmatrix} 0_{n\times n} & Y_2^{t_i} \\ X_{t_i} & -Y_2^{t_i} \end{bmatrix}, \quad \bar{W}_{13} = \begin{bmatrix} X_{t_i} B & 0_{n\times m} \\ 0_{n\times m} & 0_{n\times m} \end{bmatrix}.$$

The solution $P_{t_i}^*$, $K_{t_i}^*$, $L_{t_i}^*$ is obtained using the so-called regular form (Utkin 1993). The matrix $P_{A,t_i}^{-1} := X_{11}^{t_i}$ can be found as follows:

$$P_{A,t_i}^{-1} := G^\mathsf{T} P_{B,t_i} G, \quad B := \begin{bmatrix} B_1 \\ B_2 \end{bmatrix},$$

$$G := \begin{bmatrix} I_{(n-m)\times(n-m)} & -B_1 B_2^{-1} \\ 0_{m\times(n-m)} & B_2^{-1} \end{bmatrix}, \quad P_{B,t_i} := \begin{bmatrix} P_{B,1}^{t_i} & 0_{(n-m)\times m} \\ 0_{m\times(n-m)} & P_{B,2}^{t_i} \end{bmatrix},$$

where

$$B_1 \in \mathbb{R}^{(n-m)\times m}, \, B_2 \in \mathbb{R}^{m\times m}$$

and

$$\det(B_2) \neq 0.$$

Finally, the solution of the problem (11.26) becomes

$$K_{t_i}^* := \frac{\mathcal{X}_i^{B,22*}}{\det\left(B_2^{-1}\right)} \left[B_1 B_2^{-1}, \ B_2^{-1}\right] Y_i^{1,2*},$$

$$L_{t_i}^* := \left(\mathcal{X}_i^*\right)^{-1} Y_{i,2}^*.$$

Remark 11.1. Observe that the optimization problem (11.26) as is formulated in the proposition above contains the additional constraint (11.31), which is introduced here to update the initial value of the matrix Q_{t_i}, which restricts the admissible set of gain matrices K_{t_i} by

$$K_{t_i}^\mathsf{T} K_{t_i} \leq Q_{t_i}$$

at each time interval $(t_{i-1}, t_i]$.

11.4.2   Computational Aspects

The problem (11.26) can be solved numerically with the MATLAB toolboxes
SeDuMi and Yalmip. The calculation of K_{t_i}, P_{t_i}, $i = 1, 2, \ldots$, can be obtained
recursively using the following procedure:

1. First, fixing some initial values of scalar parameters

$$\alpha_i = \alpha_i^0, \ \varepsilon_{1,i} = \varepsilon_{1,i}^0, \ \varepsilon_{2,i} = \varepsilon_{2,i}^0,$$

   we apply the MATLAB toolbox SeDuMi to solve the corresponding constraint
   optimization problem (11.26). As a result, we obtain the matrices P_{t_0}, Q_{t_0}, K_{t_0},
   and L_{t_0}.
2. Fixing matrices P_{t_0}, Q_{t_0}, K_{t_0}, and L_{t_0}, we suggest increasing the parameter α_i,
   taking

$$\alpha_i^1 = \alpha_i^0 + \Delta\alpha_i, \ 0 < \Delta\alpha_i \ll 1,$$

   and decreasing $\varepsilon_{1,i}$ and $\varepsilon_{2,i}$, making

$$\varepsilon_{1,i}^1 = \varepsilon_{1,i}^0 - \Delta\varepsilon_{1,i} > 0, \ 0 < \Delta\varepsilon_{1,i} \ll 1,$$

$$\varepsilon_{2,i}^1 = \varepsilon_{2,i}^0 - \Delta\varepsilon_{2,i} > 0, \ 0 < \Delta\varepsilon_{2,i} \ll 1.$$

3. When the SeduMi toolbox "informs" us that the current LMIs (11.30) have no
   solution, we stop the procedure. The last admissible parameters are declared to
   be optimal:

$$(P_{A,t_i}^*, Q_{t_i}^*, K_{t_i}^*, L_{t_i}^*, \alpha_i^* \varepsilon_{1,i}^*, \varepsilon_{2,i}^*)$$

   for the time interval $t \in (t_{i-1}, t_i]$.
4. Apply the switched controller (11.8) in (11.9) for the current time interval
   $(t_{i-1}, t_i]$.
5. Increase $i = i + 1$, and return to step 1, verifying the following conditions:

   (a) If the condition (11.24) holds, we may conclude that the set of solutions
       P_{A,t_i}, Q_{t_i}, K_{t_i}, L_{t_i}, α_i, $\varepsilon_{1,i}, \varepsilon_{2,i}$ is the final solution and is declared the
       suboptimal solution set $(P_{A,t_i}^*, Q_{t_i}^*, K_{t_i}^*, L_{t_i}^*, \varepsilon_{1,i}^*, \varepsilon_{2,i}^*)$ for each time interval
       $t \in (t_{i-1}, t_i]$.
   (b) If (11.24) does not hold, return to step 1 with

$$\alpha_i^0 = \alpha_i^*, \varepsilon_{1,i}^0 = \varepsilon_{1,i}^*, \varepsilon_{2,i}^0 = \varepsilon_{2,i}^*.$$

6. Since $\bar{P}_{A,t_i} \rightarrow \bar{P}_A$, the minimal size of the ellipsoidal matrix $\bar{P}_A$ is declared as
   $\bar{P}_{A,t_i}$ for large enough i.

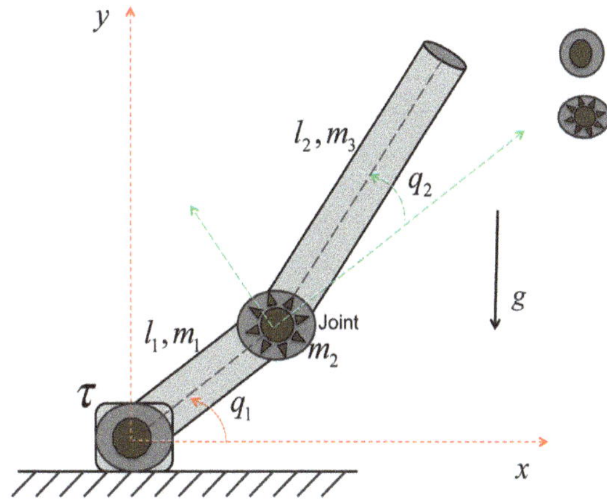

Fig. 11.2 Two-degree-of-freedom flexible pendulum

11.5   Illustrative Example

In this section, we consider the flexible arm robot (two-degree-of-freedom flexible pendulum) depicted in Fig. 11.2. The flexibility of the link is a result of lightening the robot arm (for example,in space applications). The study of link flexibility is also applicable to certain types of heavy manipulators, such as large-scale systems. If the spring constant of the flexible arm is 0, the system is an extreme case of the underectuated two-link robot (pendubot system). The control to be designed is intended to stabilize the pendulum in the vertical position using shoulder torque in the first (lowest) joint.

11.5.1   Dynamic Model

The mathematical model of the considered systems can be presented as

$$D(q)\ddot{q} + C(q,\dot{q})\dot{q} + G(q) = Q, \tag{11.32}$$

where the position coordinates $q \in \mathbb{R}^M$ with associated velocities $\dot{q}$ and accelerations $\ddot{q}$ are controlled by the vector $Q \in \mathbb{R}^M$ of driving forces. The generalized moment of inertia $D(q) \in \mathbb{R}^{M \times M}$ is a symmetric and positive definite matrix, the Coriolis (centripetal) forces are $C(q,\dot{q})\dot{q} \in \mathbb{R}^M$, and the gravitational forces are denoted by $G(q) \in \mathbb{R}^M$. All vary along the trajectories; $M = 2$ is the degree of freedom. We can represent (11.32) in the standard Cauchy affine (with respect to the control) form

$$\dot{\bar{x}} = f(\bar{x}) + g(\bar{x})u + \zeta(t),$$

$$f(x) := \begin{bmatrix} f_1(\bar{x}) & f_2(\bar{x}) & f_3(\bar{x}) & f_4(\bar{x}) \end{bmatrix}^{\mathsf{T}},$$

$$g(\bar{x}) := \begin{bmatrix} 0 & 0 & g_1(\bar{x}) & g_2(\bar{x}) \end{bmatrix}^{\mathsf{T}}.$$

The problem to be solved is to stabilize this system at the upper equilibrium point of the desired unstable position. For flexible pendulum systems, we use the following change in variable coordinates:

$$\bar{x}_{1:} = q_1, \ \bar{x}_2 := q_2, \ \bar{x}_3 := \dot{q}_1, \ \bar{x}_4 := \dot{q}_2.$$

The top position is

$$\bar{x}_{\mathrm{eq}} = (\bar{x}_1, \bar{x}_2, \bar{x}_3, \bar{x}_4) = (\pi/2, 0, 0, 0).$$

The functions $f(x)$ and $g(x)$ have the following structure:

$$\begin{bmatrix} f_1(\bar{x}) \\ f_2(\bar{x}) \end{bmatrix} = \begin{bmatrix} \bar{x}_3 \\ \bar{x}_4 \end{bmatrix},$$

$$\begin{bmatrix} f_3(\bar{x}) \\ f_4(\bar{x}) \end{bmatrix} = D^{-1}(\bar{x}) \left[-C(\bar{x}) \begin{bmatrix} \bar{x}_3 \\ \bar{x}_4 \end{bmatrix} - G(\bar{x}) - F(x) \right],$$

$$g_1(\bar{x}) = \frac{\theta_2}{\det(D(\bar{x}))},$$

$$g_2(\bar{x}) = -\frac{\theta_2 + \theta_3 \cos \bar{x}_2}{\det(D(\bar{x}))},$$

$$\det(D(\bar{x})) = \theta_1 \theta_2 + \theta_3^2 \cos^2 \bar{x}_2,$$

$$\begin{aligned}
d_{11} &= \theta_1 + \theta_2 + \theta_3 \cos(\bar{x}_2), \ d_{12} = d_{21} = \theta_2 + \theta_3 \cos(\bar{x}_2) \\
d_{22} &= \theta_2, &\quad c_{11} &= -\theta_3 \sin(\bar{x}_2)\bar{x}_3, \\
c_{12} &= -\theta_3 \sin(\bar{x}_2)(\bar{x}_3 + \bar{x}_4), &\quad c_{21} &= -\theta_3 \sin(\bar{x}_2)\bar{x}_3, \\
c_{22} &= 0, &\quad G_1(x) &= \theta_4 \cos(\bar{x}_1) + G_2(x), \\
G_2(x) &= \theta_5 \cos(\bar{x}_1 + \bar{x}_2) &\quad F_1(x) &= 0, \\
F_2(x) &= F_r \bar{x}_2.
\end{aligned}$$

Note that this example has the underactuated property, i.e.,

$$Q = \begin{bmatrix} \tau \\ 0 \end{bmatrix} = \begin{bmatrix} 1 \\ 0 \end{bmatrix} u.$$

The parameters of this system are given by

$$\theta_1 = \mathbb{I}_{zz,1} + l_1^2 \left[M_2 + M_3 \right], \; \theta_2 = \mathbb{I}_{zz,2}, \quad \theta_3 = M_2 l_1 \tfrac{l_2}{2},$$
$$\theta_4 = \tfrac{3}{2} g M_1 l_1 + g M_3, \qquad \theta_5 = g M_2 \tfrac{l_2}{2}.$$

Here $M_1 = 0.0832\,\mathrm{kg}$, $M_3 = 0.12899\,\mathrm{kg}$ are the mass of ith tube, $M_2 = 0.1659\,\mathrm{kg}$ is the mass of joint, $l_1 = 0.275\,\mathrm{m}$, $l_2 = 0.467\,\mathrm{m}$ is the length of the ith barr, $\mathbb{I}_{zz,1} = 0.0005\,\mathrm{kg \cdot m^2}$, $\mathbb{I}_{zz,2} = 0.00045\,\mathrm{kg \cdot m^2}$ is the inertia of one of the central moments of the ith barr, $i = 1, 2$, the acceleration of the gravity constant $g = 9.81\,\mathrm{m \cdot s^{-2}}$, and the spring constant of the flexible arm $F_r := 3.56$.

The external perturbations $\zeta_x(t)$ are generated by sensor noise, and $\zeta_y(t)$ by communication noise. Defining the deviation vector as $x(t) = \bar{x}(t) - x_{\mathrm{eq}}$ and introducing artificial perturbations

$$\zeta_x(t) = [0.592 \sin(\omega t), \; 0.52 \sin(\omega t), \; 0.252 \cos(\omega t), \; 0.195 \cos(\omega t)]^\mathsf{T},$$

$$\zeta_y(t) = [0.02 \sin(\omega t) . \, 0.004 \cos(\omega t)]^\mathsf{T},$$

with $\omega = 60\,\mathrm{rad}$, and taking into account that in a neighborhood of the equilibrium point, $x_1 \simeq 0$ and $x_2 \simeq 0$, we may conclude that

$$g(x) \simeq B = [\,0 \; 0 \; B_{31} \; B_{41}\,]^\mathsf{T},$$

$$B_{31} = \frac{\theta_2}{\theta_1 \theta_2 - 2\theta_3^2}, \; B_{41} = -\frac{\theta_2 + \theta_3}{\theta_1 \theta_2 - 2\theta_3^2}. \tag{11.33}$$

By the properties of the inertia matrix $D(\bar{x})$ and the physical construction, all denominators in (11.33) are nonsingular.

11.5.2   Numerical Results

We Apply the suggested technique for the flexible link system and use the initial conditions

$$A = \begin{bmatrix} 0 & 1 & 0 & 0 \\ 0 & 0 & 1 & 0 \\ 51 & -52 & 0 & 0 \\ -81 & -145 & 0 & 0 \end{bmatrix}.$$

First, we fix the positive scalar parameters and solve our problem with respect to the matrix variables that satisfy LMI constraints. If the toolbox says that the LMI constraint is not feasible, it is suggested that one select a 10 %-smaller parameter α_0 and a 10 %-bigger parameter $\varepsilon_{1,0}$, $\varepsilon_{2,0}$, etc. Such parameters, which provide the feasibility of the considered LMI, must exist, since by our assumptions, the pair

(A, B) is controllable and the pair (C, A) is observable. After 25 recurrent steps of the numerical procedure, we obtained

$$\alpha_1^* = 0.8, \ \varepsilon_{1,1}^* = 0.23, \ \varepsilon_{2,1}^* = 0.23,$$

$$K_{t_1}^* = 10^3 \begin{bmatrix} 14.8717 \\ 24.4791 \\ 3.6508 \\ 3.1516 \end{bmatrix}^\mathsf{T}, \ L_{t_0}^* = \begin{bmatrix} 82.5301 & -27.1871 \\ -27.1869 & 44.9827 \\ -147.1981 & -138.3028 \\ 0.0542 & 0.3672 \end{bmatrix},$$

$$\left(P_{A,t_1}^* \right)^{-1} = 10^2 \begin{bmatrix} 65.9240 & 91.7881 & 16.8821 & 8.0148 \\ 91.7881 & 142.8598 & 25.0746 & 11.9062 \\ 16.8821 & 25.0746 & 4.8337 & 2.2943 \\ 8.0148 & 11.9062 & 2.2943 & 1.0902 \end{bmatrix}.$$

In order to illustrate the numerical results, we chose the initial conditions x_0 as follows:

$$\hat{x}_0 = \begin{bmatrix} 0.52 \\ 0.5 \\ 0 \\ 0 \end{bmatrix}, \ x_0 = \begin{bmatrix} 0.7 \\ 0.2 \\ 0.3 \\ 0.6 \end{bmatrix}.$$

Observe that such initial conditions correspond to the internal part of the constructed attractive ellipsoid. The condition (11.24) holds during the fourth iteration ($i = 4$), and the current algorithm stopped at iteration $i = 8$, yielding the following result:

$$\alpha_8^* = 0.1, \ \varepsilon_{1,8}^* = 0.23, \ \varepsilon_{2,8}^* = 0.23,$$

$$K_{t_8}^* = 10^3 \begin{bmatrix} 66.0606 \\ 108.5089 \\ 16.2737 \\ 14.0701 \end{bmatrix}^\mathsf{T}, \ L_{t_8}^* = \begin{bmatrix} 57.5370 & -14.5432 \\ -14.5426 & 36.1554 \\ -139.5972 & -128.8544 \\ 0.1045 & 0.3347 \end{bmatrix},$$

$$\left(P_{A,t_1}^* \right)^{-1} = 10^2 \begin{bmatrix} 92.8922 & 118.2234 & 22.6702 & 10.7640 \\ 118.2234 & 170.6464 & 30.9422 & 14.6938 \\ 22.6702 & 30.9422 & 6.1191 & 2.9048 \\ 10.7640 & 14.6938 & 2.9048 & 1.3802 \end{bmatrix}.$$

11.5.3 Simulation Results

The illustrative plots are divided into two figures: Fig. 11.3 presents the trajectories $x_1(t)$ and $x_3(t)$ corresponding to the position and velocity of the first link, and Fig. 11.4 represents the position and velocity corresponding to the second link of the system.

Figures 11.5 and 11.6 illustrate how the ellipsoid changes in the time intervals, and how they converge to an ellipsoid of minimal size.

Finally, Fig. 11.7 represents the evolution of the bounded control law (11.8) over all time intervals. In this figures one can see how the control action is saturated by the upper control estimate u^+.

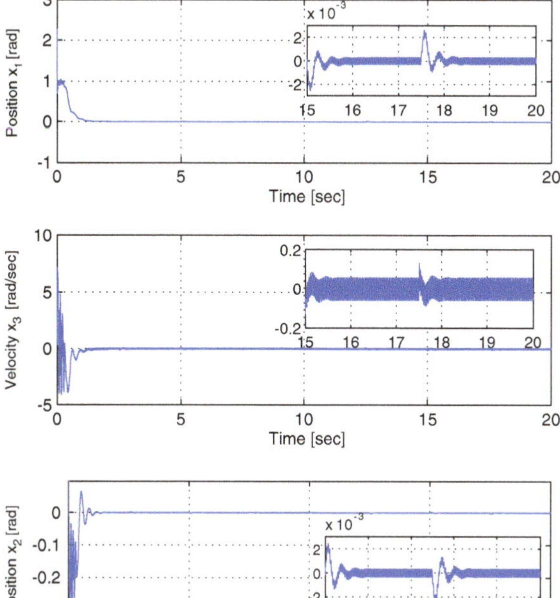

Fig. 11.3 Trajectories of the first link coordinates

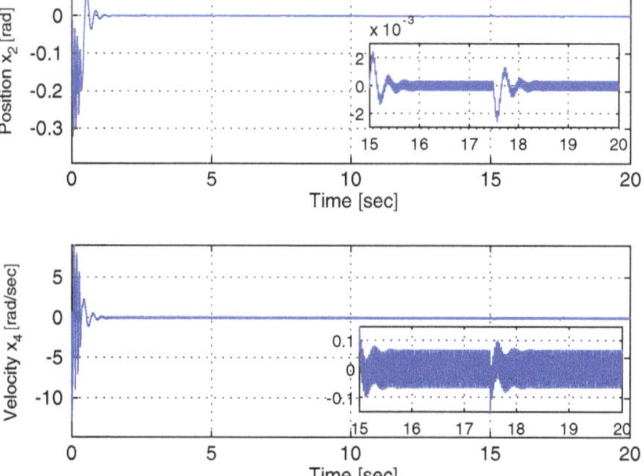

Fig. 11.4 Trajectories of the second link coordinates

11.6 Conclusion

- Here we have suggested a methodology (numerical procedure) that permits us to apply AEM for designing a nonlinear saturated output control based on only sample data outputs within a class of nonlinear systems without exact knowledge of the dynamic mathematical model and in the presence of external bounded perturbations.
- The notion "optimal robust output feedback" is associated with a set of feedback parameters that guarantee the "minimal size" of the attractive ellipsoid among all ellipsoids containing all possible trajectories of a controlled nonlinear system.
- The suggested methodology was effectively applied to a vertical underactuated pendulum (commonly known as a flexible arm) of two degrees of freedom.

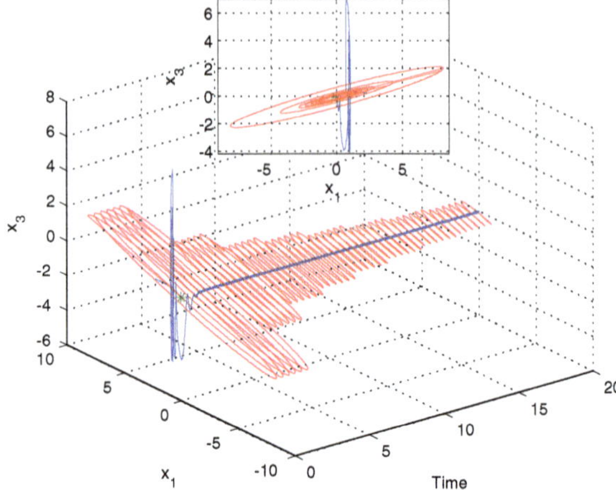

Fig. 11.5 The ellipsoids in the $x_1 - x_3$ plane

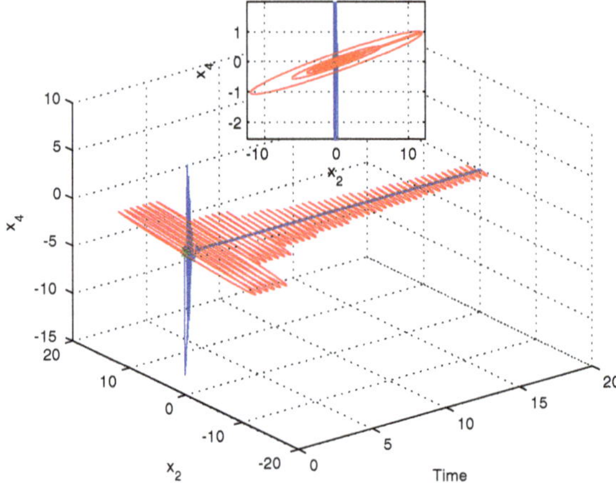

Fig. 11.6 The ellipsoids in the $x_2 - x_4$ plane

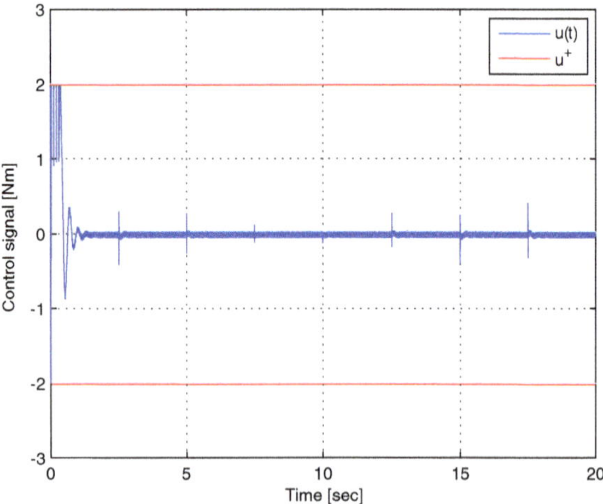

Fig. 11.7 The control signal

Chapter 12
Attractive Ellipsoid Method with Adaptation

Abstract This chapter deals with the development of a state estimator and adaptive controller based on the attractive ellipsoid method (AEM) for a class of uncertain nonlinear systems having "quasi-Lipschitz" nonlinearities as well as external perturbations. The set of stabilizing feedback matrices is given by a specific matrix inequality including the characteristic matrix of the attractive ellipsoid that contains all possible bounded trajectories around the origin. Here we present two modifications of the AEM that allow us to use online information obtained during the process and to adjust matrix parameters participating in constraints that characterize the class of adaptive stabilizing feedbacks. The proposed method guarantees that under a specific persistent excitation condition, the controlled system trajectories converge to an ellipsoid of "minimal size" having a minimal trace of the corresponding inverse ellipsoidal matrix.

Keywords Robust adaptive control • Attractive ellipsoids • Quasi-Lipschitz systems

This chapter deals with the development of a state estimator and adaptive controller based on the attractive ellipsoid method (AEM) for a class of uncertain nonlinear systems having "quasi-Lipschitz" nonlinearities as well as external perturbations. The set of stabilizing feedback matrices is given by a specific matrix inequality including the characteristic matrix of the attractive ellipsoid that contains all possible bounded trajectories around the origin. Here we present two modifications of the AEM that allow us to use online information obtained during the process and to adjust matrix parameters participating in constraints that characterize the class of adaptive stabilizing feedbacks. The proposed method guarantees that under a specific persistent excitation condition, the controlled system trajectories converge to an ellipsoid of "minimal size" having a minimal trace of the corresponding inverse ellipsoidal matrix.

© Springer International Publishing Switzerland 2014

A. Poznyak et al., *Attractive Ellipsoids in Robust Control*, Systems & Control: Foundations & Applications, DOI 10.1007/978-3-319-09210-2_12

12.1 Introduction

The study of *robust controls* using input–output measurements and governed by an uncertain dynamic model with arbitrary bounded disturbances still attracts considerable attention in the literature because of its importance in many theoretical aspects and practical applications. Many publications deal with robust control problems in which not all measurement data of all state variables are available (Gonzalez-Garcia, Polyakov, & Poznyak, 2011; Nazin, Polyak, & Topunov, 2007; Kurzhanski & Veliov 1994). Basically, *robust control* is one of the principal problems in control theory because it provides a workable instrument for designing controllers that are able to operate successfully without complete information (see, e.g., Duncan & Schweppe 1971; Kurzhanski & Veliov 1994).

Several approaches that work successfully under incomplete model information were developed in the last decades of the previous century. One dealing with H^∞-theory using the explicit form of the optimal controller for the L^2-class of disturbances was presented in Francis, Helton, and Zames (1984), but that technique involved knowledge of the linear model approximation. The requirement of the boundedness of the disturbance energy was relaxed for discrete systems by Francis et al. (1984) and Barabanov and Granichin (1984), where only the maximal disturbance amplitude, that is, the estimate in the norm l_1 (in the discrete-time case), was allowed. Unfortunately, the so-designed optimal controller often is of high order. Recently, the *attractive ellipsoid* concept was used by Nazin et al. (2007) and Polyakov and Poznyak (2011), who present an innovative approach that minimizes the effect of unmatched perturbations (which act in a subspace other than that of the control) in linear systems. But the results mentioned above ignore later information obtained online during the control process.

On the other hand, there exists the "classical" *adaptive control* approach (based on the time-varying linear approximation of black-box models). It has been recently discussed in terms of theoretical design, experimental control, and analytic tools (such as the model reference method and pole placement, among others). Every adaptive controller contains as part of applicable feedback the parameter estimates obtained online with the controlled plant evaluation. Different methods have been proposed for the identification of online parameters: maximum likelihood, extended Kalman filter, least squares method (LSM), instrumental variables, LSM with forgetting, among others (Ljung 1999; Poznyak 2008). Such adaptive controllers permit us to control real systems without complete a priori knowledge of all parameters participating in the dynamic description of the considered model. Traditionally, these controllers are analyzed for models that do not contain any deterministic external disturbance (see, e.g., Ioannou & Sun 1996; Sastry 1999). Usually, adaptive feedbacks involve current parameter estimates obtained by an appropriate identification scheme. The identification procedure works under the assumption that all unknown parameters are given as a regressor vector to be estimated (see Sastry & Bodson 1994; Narendra & Annaswamy 2005). The key point of adaptive techniques for estimating the regressor vector for parameter identification is the so-called persistent excitation condition. This condition guarantees that the parameter estimates converge to the real parameters (Anderson 1985; Ljung 1999).

Our objective in this section is to present a theoretical adaptive control-based AEM concept and state estimation. Here we deal with the appropriate design of a linear feedback for a class of quasi-Lipschitz uncertain nonlinear systems containing external perturbations as well. This class C of nonlinear models is assumed to be given a priori and to be characterized by the matrix A corresponding to a linear nominal plant contained in this class and two scalar parameters. These parameters define a permitted deviation of nonlinearities in model dynamics with respect to this nominal linear model (Polyakov & Poznyak 2011; Ordaz & Poznyak 2012; Davila & Poznyak 2011). Ordaz and Poznyak (2012) present a simple idea of adaptive controller: an intelligent adjustment (or learning) of the gain matrices K in the feedback and L in the Luenberger-type observer are introduced to reduce the volume of the attractive ellipsoid (having a minimal trace of the corresponding ellipsoidal matrix) characterizing a "convergence zone." This adaptation process is referred to as KL-adaptation. Additionally, here we study the effect of "A-adaptation," which describes a procedure for adjusting the matrix parameters (in this case, the matrix A) participating in the matrix constraints characterizing the class of stabilizing feedbacks. This adjustment (or adaptation) is made only in some fixed sampletimes, so that the corresponding gain matrix of the controller remains constant within any interval between two neighboring sample times. This adaptive scheme applied to traditional AEM results in a new *adaptive robust control method.*

We discuss here in detail two *adaptive versions* of the AEM for a class of nonlinear systems in which only outputs (but not states) are available online. Moreover, these measurements may be disturbed by bounded perturbations.

The first version of the suggested robust control law contains the adaptation (learning) laws for the feedback matrix gain and for the observer gain matrix as well, referred to below as KL-adaptation.

The second suggested robust adaptive controller, called A-adaptation, contains the adjustment (or adaptation) of a specific matrix parameter only at some fixed sample times, so that the corresponding gain matrix of the controller remains constant within every interval between two neighboring sample times.

12.2 Attractive Ellipsoid Method with KL-Adaptation

In this section, we propose an extension of the AEM with time-varying gain matrices for the observer and a controller. Moreover, we present the combined "observer–controller" as an extension of the adaptive feedback using state estimates. The suggested adaptive AEM provides a significantly smaller attractive ellipsoid compared with the nonadaptive technique (see Gonzalez-Garcia et al. 2011).

12.2.1 Basic Assumptions and Constraints

Here we assume that:

- The states of the process cannot be measured directly, requiring the use of a state estimator generating online the state estimates $\hat{x}(t)$.
- The observer is driven by the output $y(t)$ and the input $u(t)$, and its corresponding gain matrix $L(t)$ is a time-varying matrix adjusted online.
- The feedback control $u(t)$ to be designed is driven by a state estimate as

$$u(t) = K(t)\hat{x}(t),$$

where $K(t)$ is also a time-varying matrix adjusted online according to the adaptive procedure suggested in this paper.

With respect to the constraints, we assume that:

- The dynamic and the output system model are assumed to have unknown nonlinearities from a given classes of "quasi-Lipschitz functions."
- The external bounded perturbations of the considered dynamical system are applied to the right-hand side of the differential model and to its output as well.

12.2.2 System Description and Problem Formulation

Consider the nonlinear dynamical system given by

$$\dot{x}(t) = f(x(t)) + Bu(t) + \zeta_x(t), \ x(0) = x_0,$$

$$y(t) = h(x(t)) + \zeta_y(t),$$

(12.1)

where

$x(t) \in \mathbb{R}^n$ is the state vector at time $t \geq 0$,
$y(t) \in \mathbb{R}^p$ is the output system at time t,

$f : \mathbb{R}^n \rightarrow \mathbb{R}^n$ and $h : \mathbb{R}^n \rightarrow \mathbb{R}^p$ are unknown nonlinear vector functions participating in the right-hand side of the dynamics and output system (12.1), respectively,

$B \in \mathbb{R}^{n \times m}$ is the matrix realizing the actuator mapping,
$u \in \mathbb{R}^m$ is the control input at time t, and
$\zeta_x(t) \in \mathbb{R}^n$ and $\zeta_y \in \mathbb{R}^p$ are external perturbations.

12.2.3 Main Assumptions

Hereinafter, we suppose that

A1. The unknown nonlinearities $f : \mathbb{R}^n \to \mathbb{R}^n$ and $h : \mathbb{R}^n \to \mathbb{R}^p$ belong to the wide class $\mathcal{C}$ of the so-called quasi-Lipschitz functions, which may include discontinuous and hysteresis functions as well. The formal definition of this class can be formulated as follows: for every vector function $g : \mathbb{R}^n \to \mathbb{R}^k$ of this class, there exist a matrix $G \in \mathbb{R}^{k \times n}$ and nonnegative constants δ_1 and δ_1 such that for every $x \in \mathbb{R}^n$, the following inequality holds:

$$\|g(x) - Gx\|^2 \leq \delta_1 + \delta_2 \|x\|^2,$$

which implies that the growth rate of $g(x)$ as $\|x\| \to \infty$ is not faster than linear. We will denote this class by $\mathcal{C}(G, \delta_1, \delta_2)$. So we suppose that

$$f \in \mathcal{C}(A, c_1, c_2), \; h \in \mathcal{C}(C, c_3, c_4)$$

(see Fig. 2.1 illustrating the single-dimensional case $n = k = 1, a > c_1 > 0$).

A2. The external perturbations $\zeta_x(t)$ and $\zeta_y(t)$ are assumed to be bounded:

$$\|\zeta_x(t)\|^2 \leq c_5 < \infty, \; \left\|\zeta_y(t)\right\|^2 \leq c_6 < \infty.$$

A3. The state estimate $\hat{x}(t)$ is generated by (for a fixed $\hat{x}(0) = \hat{x}_0$)

$$\dot{\hat{x}}(t) = A\hat{x}(t) + Bu(t) + L(t)(y(t) - C\hat{x}(t)), \tag{12.2}$$

where $L(t) \in \mathbb{R}^{n \times p}$ is the time-varying observer gain matrix to be designated.

A4. The control action $u(t) \in \mathbb{R}^m$ is the linear nonstationary feedback of the current state estimates, i.e.,

$$u(\hat{x}(t), t) = K(t)\hat{x}(t), \tag{12.3}$$

where $K(t) \in \mathbb{R}^{m \times n}$ is a time-varying gain matrix that also should be chosen to make the attractive ellipsoid of minimal possible "size" (in fact, the minimization of the trace of the inverse ellipsoid matrix).

A5. The original nonlinear system (12.1) is assumed to be controllable and observable; it is also assumed that the pair (A, B) is *controllable*, and the pair (C, A) is *observable*.

Remark 12.1. Certainly, knowledge of the matrix G (characterizing the "nominal linear plant" within $\mathcal{C}(G, \delta_1, \delta_2)$) as well as two scalar parameters gives very "approximate" information about the nonlinear function $g \in \mathcal{C}(G, \delta_1, \delta_2)$. Nevertheless, the approximate values of these class parameters can be estimated a priori based on some preliminary information concerning the "behavior" of the considered plant in a neighborhood of the origin $x = 0$.

12.2.4   Extended Quasilinear Format

The system (12.1) can be represented in "quasilinear format" as

$$\dot{x}(t) = Ax(t) + Bu(t) + \xi_x(x(t), t), \, x(0) = x_0,$$

$$y(t) = Cx(t) + \xi_y(x(t), t),$$

$$\xi_x(x(t), t) := \Delta f(x(t)) + \zeta_x(t), \quad \Delta f(x) := f(x) - Ax,$$

$$\xi_y(x(t), t) := \Delta h(x(t)) + \zeta_y(t), \quad \Delta h(x) := h(x) - Cx,$$

(12.4)

where the extended uncertain term $\xi(x, t) := [\xi_x^\mathsf{T}(x, t) \, \xi_y^\mathsf{T}(x, t)]^\mathsf{T}$ satisfies for every $t \geq 0$ and $x \in \mathbb{R}^{m \times n}$ the following inequality:

$$\|\xi(x, t)\|^2 \leq C_1 + C_2 \|x\|^2,$$

$$C_1 = d_1 + d_3 = 2(c_1 + c_3 + c_5 + c_6), \, C_2 = 2(d_2 + d_4),$$

(12.5)

$$d_1 = 2(c_1 + c_5), \, d_2 = 2c_2, \, d_3 = 2(c_3 + c_6), d_4 = 2c_4.$$

From (12.2), using the output feedback (12.3), we obtain

$$\dot{\hat{x}}(t) = [A + BK(t)]\hat{x}(t) + L(t)[y(t) - C\hat{x}(t)],$$

so that for the estimation error

$$e(t) = x(t) - \hat{x}(t),$$

(12.6)

we have

$$\dot{e}(t) = [A - L(t)C]e(t) + \xi_x(x(t), t) - L(t)\xi_y(x(t), t).$$

(12.7)

Using an extended vector

$$z^\mathsf{T}(t) = [\hat{x}^\mathsf{T}(t), \, e^\mathsf{T}(t)],$$

(12.8)

we get

$$\dot{z}(t) = \bar{A}(t)z(t) + \bar{L}(t)\xi(x(t), t), \, z(0) = z_0,$$

$$\bar{A}(t) = \begin{bmatrix} A + BK(t) & L(t)C \\ 0 & A - L(t)C \end{bmatrix}, \, \bar{L}(t) = \begin{bmatrix} 0 & L(t) \\ I & -L(t) \end{bmatrix}.$$

(12.9)

12.2.5 Problem Formulation

To formulate the problem that we intend to solve, we need the following definition.

Definition 1. We say that a trajectory $\{x(t)\}_{t\geq 0}$ belongs asymptotically to the attractive ellipsoid

$$\mathcal{E}\,(0,P) = \left\{x \in \mathbb{R}^n : x^\mathsf{T} P^{-1} x \leq 1,\, P = P^\mathsf{T} > 0\right\}$$

with center at the origin and with corresponding matrix P if

$$\limsup_{t\to\infty} x^\mathsf{T}(t)\, P^{-1} x\,(t) \leq 1.$$

In view of this definition, our problem is to designate the gain matrices $K(t) \in \mathbb{R}^{m\times n}$ and $L(t) \in \mathbb{R}^{n\times p}$ providing the *zone-stability* (zone-convergence) property (Lakshmikantham, Leela, & Martynyuk, 1990) for the class of uncertain systems (12.9) containing a bounded uncertain term (12.5), which guarantees the approach of each possible trajectory of the closed-loop system (12.9) to an attractive ellipsoid (AE) of minimal possible "size" (corresponding to the maximization of $\mathrm{tr}\left\{P^{-1}\right\}$ or minimization of $\mathrm{tr}\left\{P\right\}$).

12.2.6 Learning Laws, Storage Function Properties, and the "Minimal Size" Ellipsoid

Let $\left(P_A^*, \Lambda^*, K^*, L^*, \alpha^*, \varepsilon^*, \rho^*\right)$ be an approximate solution (if it exists) of the following constraint optimization problem:

Problem 12.1.

$$\frac{\varepsilon}{\alpha}\mathrm{tr}\,P \;\to\; \inf_{0<\alpha,\varepsilon,\rho;\; K,L;\; 0<P=P^\mathsf{T},0<\Lambda=\Lambda^\mathsf{T}} \tag{12.10}$$

subject to the matrix constraint

$$W\,(P_A, K, L, \Lambda, \alpha, \varepsilon, \rho) = \overset{\circ}{W}\,(P_A, K, L, \alpha, \varepsilon, \rho)$$

$$+ \,\mathrm{diag}(0_{n\times n}, \Lambda, 0_{2n\times 2n}) < 0 \tag{12.11}$$

with

$$\overset{\circ}{W}\left(P_A, K, L, \alpha, \varepsilon, \rho\right) :=$$

$$\begin{bmatrix} P^{-1}\bar{A}_\alpha + \bar{A}_\alpha^\mathsf{T} P^{-1} + R & P^{-1}\bar{L} \\ \bar{L}^\mathsf{T} P^{-1} & -\varepsilon I_{2n} \end{bmatrix},$$

$$\bar{A}_\alpha := \bar{A} + \tfrac{\alpha}{2} I_{2n},$$ (12.12)

$$P = \begin{bmatrix} P_A & 0_{n\times n} \\ 0_{n\times n} & P_A \end{bmatrix} \in \mathbb{R}^{2n\times 2n},$$

$$R := \varepsilon C_2 \mathcal{R}, \quad \mathcal{R} = \begin{bmatrix} (1+\rho)\, I_n & 0_{n\times n} \\ 0_{n\times n} & \left(1+\rho^{-1}\right) I_n \end{bmatrix},$$

where the matrix $\bar{A}$ is given by (12.9) with the constant matrices $K(t) = K$ and $L(t) = L$.

We emphasize that the optimization problem (12.10)–(12.12) has a constraint that is bilinear (under fixed scalars) with respect to the matrix variables. Fortunately, there exist two ways to resolve this problem numerically:

1. We can use the MATLAB toolbox Tomlab (Holmstrom, Goran, & Edvall 2009) to solve optimization problems with bilinear matrix constraints.
2. We can transform this bilinear matrix inequality (BMI) into a linear one (LMI) keeping the isomorphism of this transformation: the solution of the original optimization problem with BMI constraint is uniquely defined by the solution of the optimization problem with LMI constraint, and conversely. The solution of this equivalent optimization problem with linear matrix constraints (actually, an acceptable approximation) can be found by implementing the standard MATLAB toolbox SeDuMi (Peaucelle, Henrion, Labit, & Taitz 2002). This transformation is discussed below in detail.

Now let us formulate the main theorem on the so-called storage function that is used extensively during analysis of the designed adaptive controller.

Theorem 12.1 (On the "storage function" properties). *If on the trajectories of (12.9), the matrices $K(t) \in \mathbb{R}^{m\times n}$ and $L(t) \in \mathbb{R}^{n\times p}$ are adapted by the **learning laws***

$$-\dot{K}(t) = \frac{\alpha^*}{2}\hat{K}(t) + \frac{2}{k_1}B^{\mathsf{T}}\left(P_A^*\right)^{-1}\hat{x}(t)\hat{x}^{\mathsf{T}}(t), \ \hat{K}(t) = K(t) - K^*, \ K(0) = K^*,$$

$$-\dot{L}(t) = \frac{1}{k_2}\left[y(t) - Cx(t)\right]\left\{[y(t) - C\hat{x}(t)]^{\mathsf{T}}\hat{L}^{\mathsf{T}}(t)\left(P_A^*\right)^{-1}\Lambda^{-1} + 2\hat{x}(t)^{\mathsf{T}}P_A^*\right\} + \frac{\alpha^*}{2}\hat{L}(t),$$

$$\hat{L}(t) = L(t) - L^*, \ L(0) = L^*, \ k_1 > 0, \ k_2 > 0,$$

$$\tag{12.13}$$

then the "storage" (energetic) function

$$V(z(t), t) := z^{\mathsf{T}}(t)\left(P^*\right)^{-1}z(t) + \frac{k_1}{2}\mathrm{tr}\left\{\hat{K}_t^{\mathsf{T}}\hat{K}_t\right\} + \frac{k_2}{2}\mathrm{tr}\left\{\hat{L}_t^{\mathsf{T}}\hat{L}_t\right\}$$

has the following properties:

$$\dot{V}(z(t), t) \leq -\alpha^* V(z(t), t) + \varepsilon^* C_1,$$

$$V(z(t), t) \leq \frac{\varepsilon^*}{\alpha^*}C_1 + \left[V(z(0), 0) - \frac{\varepsilon^*}{\alpha^*}C_1\right]e^{-\alpha^* t}, \tag{12.14}$$

$$\limsup_{t \to \infty} V(z(t), t) \leq \frac{\varepsilon^*}{\alpha^*}C_1.$$

Proof. Using the estimate $\|x\|^2 \leq z^{\mathsf{T}}Rz$, we have

$$\dot{V}(z(t), t) \leq v^{\mathsf{T}}(t)\mathring{W}^* v(t) + v^{\mathsf{T}}(t)\tilde{W}(t)v(t) - \alpha^* V(z(t), t) +$$

$$k_1 \mathrm{tr}\left\{\hat{K}_t^{\mathsf{T}}\dot{\hat{K}}_t\right\} + \alpha^* \frac{k_1}{2}\mathrm{tr}\left\{\hat{K}_t^{\mathsf{T}}\hat{K}_t\right\} +$$

$$k_2 \mathrm{tr}\left\{\hat{L}^{\mathsf{T}}_t\dot{\hat{L}}_t\right\} + \alpha^* \frac{k_2}{2}\mathrm{tr}\left\{\hat{L}_t^{\mathsf{T}}\hat{L}_t\right\} + \varepsilon^* C_1$$

with the extended vector

$$v^{\mathsf{T}}(t) := \left[z^{\mathsf{T}}(t) \ \xi^{\mathsf{T}}(x(t), t)\right]$$

and the matrices

$$\mathring{W}^* := \mathring{W}\left(P_A^*, K^*, L^*, \alpha^*, \varepsilon^*, \rho^*\right),$$

$$\tilde{W}(t) := \begin{bmatrix} \left(P^*\right)^{-1}\mathcal{A}(t) + \mathcal{A}(t)^{\mathsf{T}}\left(P^*\right)^{-1} & \left(P^*\right)^{-1}\mathcal{L}(t) \\ \left(P^*\right)^{-1}\mathcal{L}(t)^{\mathsf{T}} & 0_{2n \times 2n} \end{bmatrix}, \ P^* = \mathrm{diag}\left[P_A^*, P_A^*\right],$$

$$\mathcal{A}(t) := \begin{bmatrix} B\hat{K}(t) & \hat{L}(t)C \\ 0 & -L(t)C \end{bmatrix}, \ \mathcal{L}(t) := \begin{bmatrix} 0 & \hat{L}(t) \\ 0 & -\hat{L}(t) \end{bmatrix}.$$

The quadratic form $v^{\mathsf{T}}(t)\tilde{W}(t)v(t)$ may be represented as

$$v^{\mathsf{T}}(t)\tilde{W}(t)v(t) = 2z^{\mathsf{T}}(t)\,(P^*)^{-1}\,A(t)z(t) + 2\xi^{\mathsf{T}}(x(t),t)\,(P^*)^{-1}\,L(t)z(t) =$$

$$2\hat{x}^{\mathsf{T}}(t)\left(P_A^*\right)^{-1} B\hat{K}(t)\hat{x}(t) + 2\hat{x}^{\mathsf{T}}(t)\left(P_A^*\right)^{-1}\hat{L}_t Ce(t) - 2e(t)^{\mathsf{T}}\left(P_A^*\right)^{-1}\hat{L}_t C(t)e$$

$$+ 2\hat{x}^{\mathsf{T}}(t)\left(P_A^*\right)^{-1}\hat{L}_t\xi_y(x(t),t) - 2e^{\mathsf{T}}(t)\left(P_A^*\right)^{-1}\hat{L}_t\xi_y(x(t),t).$$

Since

$$\hat{x}^{\mathsf{T}}(t)\left(P_A^*\right)^{-1} B\hat{K}(t)\hat{x}(t) = \operatorname{tr}\left\{\hat{K}^{\mathsf{T}}(t)\left[B^{\mathsf{T}}\left(P_A^*\right)^{-1}\hat{x}(t)\hat{x}^{\mathsf{T}}(t)\right]\right\}$$

and

$$Ce(t) + \xi_y(x(t),t) = Cx(t) - C\hat{x}(t) + \xi_y(x(t),t) = y(t) - C\hat{x}(t),$$

one has

$$\dot{V}(z(t),t) \le v^{\mathsf{T}}(t)\mathring{W}^* v(t) - \alpha^* V(z(t),t) - 2e^{\mathsf{T}}(t)\left(P_A^*\right)^{-1}\hat{L}_t\,(y(t) - C\hat{x}(t))$$

$$+ \varepsilon^* C_1 + \operatorname{tr}\left\{\hat{K}_t^{\mathsf{T}}\left[k_1\dot{\hat{K}}_t + \alpha^*\frac{k_1}{2}\hat{K}_t + 2B^{\mathsf{T}}\left(P_A^*\right)^{-1}\hat{x}(t)\hat{x}(t)^{\mathsf{T}}\right]\right\} +$$

$$\operatorname{tr}\left\{\hat{L}_t^{\mathsf{T}}\left[k_2\dot{\hat{L}}_t + \alpha^*\frac{k_2}{2}\hat{L}_t + 2\,(y(t) - C\hat{x}(t))\,\hat{x}^{\mathsf{T}}(t)P_A^*\right]\right\}.$$

Using the inequality

$$2Z^T Y \le Z^T\Lambda Z + Y^T\Lambda^{-1}Y$$

for

$$Z := -e(t),\ Y := P_A^*\hat{L}_t\,(y(t) - C\hat{x}(t))$$

and some $\Lambda = \Lambda^{\mathsf{T}} > 0$ leads to the estimate

$$-2e^{\mathsf{T}}(t)\left(P_A^*\right)^{-1}\hat{L}_t\,(y(t) - C\hat{x}(t)) \le$$

$$e^{\mathsf{T}}(t)\Lambda e(t) + (y(t) - C\hat{x}(t))^{\mathsf{T}}\hat{L}_t^{\mathsf{T}}\left(P_A^*\right)^{-1}\Lambda^{-1}\left(P_A^*\right)^{-1}\hat{L}_t\,(y(t) - C\hat{x}(t)),$$

and as a result, we get

$$\dot{V}(z(t),t) \le v^{\mathsf{T}}(t)W^{*}v(t) - \alpha^{*}V(z(t),t) + \varepsilon^{*}C_{1} +$$

$$\mathrm{tr}\left\{\hat{K}^{\mathsf{T}}(t)\left[k_{1}\dot{\hat{K}}_{t} + \alpha^{*}\frac{k_{1}}{2}\hat{K}_{t} + 2B^{\mathsf{T}}\left(P_{A}^{*}\right)^{-1}\hat{x}(t)\hat{x}(t)^{\mathsf{T}}\right]\right\} +$$

$$\mathrm{tr}\left\{\hat{L}^{\mathsf{T}}(t)\left[k_{2}\dot{\hat{L}}_{t} + \alpha^{*}\frac{k_{2}}{2}\hat{L}_{t} + (y - C\hat{x})\left(2\hat{x}^{\mathsf{T}} + (y - C\hat{x})^{\mathsf{T}}\hat{L}^{\mathsf{T}}_{t}\left(P_{A}^{*}\right)^{-1}\Lambda^{-1}\right)\left(P_{A}^{*}\right)^{-1}\right]\right\},$$

where

$$W^{*} = \mathring{W}^{*} + \mathrm{diag}(0_{n\times n}, \Lambda, 0_{2n\times 2n}).$$

In view of (12.12) and (12.13), the last inequality becomes

$$\dot{V}(z(t),t) \le v^{\mathsf{T}}(t)W^{*}v(t) - \alpha^{*}V(z(t),t) + \varepsilon^{*}C_{1},$$

and finally,

$$\dot{V}(z(t),t) \le -\alpha^{*}V(z(t),t) + \varepsilon^{*}C_{1},$$

which completes the proof of the theorem.   ∎

12.2.7   Attractive Ellipsoid for Robust Control with KL-Adaptation

Now we are ready to present the main result (see Ordaz & Poznyak 2014) concerning the workability of a designed robust controller with adaptation.

Theorem 12.2. *Under the assumptions of Theorem 12.1, the corresponding attractive ellipsoid $\mathcal{E}\left(0, \bar{P}_{ad}\right)$ has the following ellipsoidal matrix:*

$$\left(\bar{P}_{ad}\right)^{-1} = \frac{\alpha^{*}}{\varepsilon^{*}C_{1} - \alpha^{*}\varkappa}\left(P^{*}\right)^{-1}, \quad \varkappa = \varkappa_{1} + \varkappa_{2} < \frac{\varepsilon^{*}}{\alpha^{*}}C_{1},$$

$$\varkappa_{1} = \frac{k_{1}}{2}\liminf_{t\to\infty}\mathrm{tr}\left\{\hat{K}^{\mathsf{T}}(t)\hat{K}(t)\right\}, \tag{12.15}$$

$$\varkappa_{2} = \frac{k_{2}}{2}\liminf_{t\to\infty}\mathrm{tr}\left\{\hat{L}^{\mathsf{T}}(t)\hat{L}(t)\right\},$$

satisfying

$$\limsup_{t \to \infty} z^{\mathsf{T}}(t) \left(\bar{P}_{ad}\right)^{-1} z(t) \leq 1,$$

or in other words,

$$z(t) \to \mathcal{E}\left(0, \bar{P}_{ad}\right)$$

when $t \to \infty$.

Proof. Theorem 12.1 implies that

$$\limsup_{t \to \infty} \left\{ z^{\mathsf{T}}(t) \left(P^*\right)^{-1} z(t) + \frac{k_1}{2}\mathrm{tr}\left\{\hat{K}_t^{\mathsf{T}}\hat{K}_t\right\} + \frac{k_2}{2}\mathrm{tr}\left\{\hat{L}_t^{\mathsf{T}}\hat{L}_t\right\}\right\} \leq \frac{\varepsilon^*}{\alpha^*}C_1,$$

resulting in

$$\limsup_{t \to \infty} \left\{ z^{\mathsf{T}}(t) \left(P^*\right)^{-1} z(t)\right\} \leq \frac{\varepsilon^*}{\alpha^*}C_1 - \varkappa_1 - \varkappa_2,$$

or equivalently,

$$\limsup_{t \to \infty} \left\{ z^{\mathsf{T}}(t) \left(\frac{\alpha^*\left(P^*\right)^{-1}}{\varepsilon^*C_1 - \alpha^*\varkappa}\right) z(t)\right\} \leq 1. \qquad \blacksquare$$

The following corollary states that this attractive ellipsoid is also an invariant one, that is, *starting within it, one never leaves it.*

Corollary 12.1 (On the attractive-invariant ellipsoid). *Under the hypotheses of Theorem 12.1, the attractive ellipsoid* $\mathcal{E}\left(0, \bar{P}_{ad}\right)$ *is at same time an* **invariant** *one such that the function* $G\left(V\left(z\right)\right)$ *defined by*

$$G\left(V\left(z\right)\right) := \left(\left[\sqrt{V(z)} - \sqrt{\frac{\varepsilon^*C_1}{\alpha^*}}\,\right]_+\right)^2, \quad [\gamma]_+ := \begin{cases} \gamma & \text{if} \quad \gamma \geq 0, \\ 0 & \text{if} \quad \gamma < 0, \end{cases}$$

is a **Lyapunov function** *for the dynamical system (12.9) having invariant set*

$$\mathcal{D} := \{z \in \mathbb{R}^n : G\left(V(z)\right) = 0\}$$

and satisfying on the trajectories of (12.9) the following differential inequality:

$$\frac{d}{dt}G\left(V\left(z\left(t\right)\right)\right) < 0 \text{ if } V(z(t)) > \frac{\varepsilon^*}{\alpha^*}C_1. \qquad (12.16)$$

Proof. Observe that the function $[\gamma]_+$ is not differentiable at the point $\gamma = 0$, but the function $\left([\gamma]_+\right)^2$ is twice differentiable everywhere. Based on this fact, and following (Poznyak 2004), we may conclude that the function

$$G\left(V(z)\right) := \left(\left[\sqrt{V(z)} - \sqrt{\beta/\alpha}\right]_+\right)^2 \tag{12.17}$$

is also twice differentiable everywhere. Therefore,

$$\frac{d}{dt}G\left(V\left(z(t)\right)\right) = \frac{d}{dV}G\left(V\left(z(t)\right)\right)\dot{V}\left(z(t)\right) = \left[\sqrt{V\left(z(t)\right)} - \sqrt{\beta/\alpha}\right]_+ \frac{\dot{V}\left(z(t)\right)}{\sqrt{V\left(z(t)\right)}}.$$

Using the inequality (12.14) with $\varepsilon^* C_1 = \beta$ and $\alpha^* = \alpha$, we obtain

$$\frac{d}{dt}G\left(V\left(z(t)\right)\right) \le \left[\sqrt{V\left(z(t)\right)} - \sqrt{\beta/\alpha}\right]_+ \frac{-\alpha V\left(z(t)\right) + \beta}{\sqrt{V\left(z(t)\right)}} =$$

$$-\alpha\left[\sqrt{V\left(z(t)\right)} - \sqrt{\beta/\alpha}\right]_+ \frac{V\left(z(t)\right) - \beta/\alpha}{\sqrt{V\left(z(t)\right)}} = \tag{12.18}$$

$$-\alpha\left(\left[\sqrt{V\left(z(t)\right)} - \sqrt{\beta/\alpha}\right]_+\right)^2 \frac{\sqrt{V\left(z(t)\right)} + \sqrt{\beta/\alpha}}{\sqrt{V\left(z(t)\right)}} \le 0,$$

which proves (12.16). Since the nonnegative function $G\left(V\left(z(t)\right)\right)$ is monotonically nonincreasing, by Weierstrass's theorem, it has a limit, that is, there exists

$$0 \le G^* = \lim_{t\to\infty} G\left(V\left(z(t)\right)\right). \tag{12.19}$$

Integration of (12.18) implies

$$G\left(V\left(z(t)\right)\right) - G\left(V\left(z(0)\right)\right) \le$$

$$-\alpha \int_{\tau=0}^{t} \left(\left[\sqrt{V\left(z(\tau)\right)} - \sqrt{\beta/\alpha}\right]_+\right)^2 \frac{\sqrt{V\left(z(\tau)\right)} + \sqrt{\beta/\alpha}}{\sqrt{V\left(z(\tau)\right)}} d\tau,$$

or equivalently,

$$\alpha \int_{\tau=0}^{t} \left(\left[\sqrt{V\left(z(\tau)\right)} - \sqrt{\beta/\alpha}\right]_+\right)^2 \left(1 + \frac{\sqrt{\beta/\alpha}}{\sqrt{V\left(z(\tau)\right)}}\right) d\tau$$

$$\le G\left(V\left(z(0)\right)\right) - G\left(V\left(z(t)\right)\right) \le G\left(V\left(z(0)\right)\right) = \text{const},$$

which for $t \to \infty$ and in view of (12.17), leads to the following conclusion:

$$\int\limits_{t=0}^{\infty} G\left(V\left(z\left(t\right)\right)\right)\left(1 + \frac{\sqrt{\beta/\alpha}}{\sqrt{V\left(z\left(\tau\right)\right)}}\right) d\tau < \infty.$$

The convergence of the last integral permits us to state that there exists a subsequence $\{t_k\}_{k=1,2,\ldots}$ such that $G\left(V\left(z\left(t_k\right)\right)\right) \underset{k\to\infty}{\to} 0$. But the sequence $G\left(V\left(z\left(t\right)\right)\right)$ converges [see (12.19)], and hence all its subsequences have the same limit point, which proves that $G^* = 0$, and as a result, we have that

$$\{z \in \mathbb{R}^{2n} : G\left(V(z)\right) = 0\}$$

is a positive invariant ellipsoid.   ∎

12.2.8   *On the Attractive Ellipsoid in the State Space*

To estimate the size of the "optimal attractive ellipsoid" in the state space of the system (12.1), note that

$$x_t = Hz_t, \quad H := \left[\, I_{n\times n} \ \ I_{n\times n} \,\right],$$

and therefore

$$x_t^{\mathsf{T}} \left(P_x\right)^{-1} x_t = z_t^{\mathsf{T}} H^{\mathsf{T}} \left(P_x\right)^{-1} H z_t.$$

On the other hand, as follows from Theorem 12.2 and (12.12), the attractive ellipsoid in the extended z-space is defined by

$$\left(\bar{P}_{ad}\right)^{-1} = \frac{1}{\mu^*}\left(P^*\right)^{-1},$$

$$\mu^* := \frac{\varepsilon^* C_1 - \alpha^* \varkappa}{\alpha^*}, \quad P^* = \begin{bmatrix} P_A^* & 0 \\ 0 & P_A^* \end{bmatrix},$$

so the equality

$$\left(\bar{P}_{ad}\right)^{-1} = H^{\mathsf{T}} \left(P_x\right)^{-1} H$$

must be satisfied. Unfortunately, in general, this identity cannot be satisfied, since the size of the matrix P_x in which we are interested is less than that of the matrix

$\bar{P}_{ad}$. Therefore, we suggest estimating the "minimal ellipsoid" in the x-space as the solution P_x^* of the following optimization problem:

$$\left\| \left(\bar{P}_{ad} \right)^{-1} - H^{\mathsf{T}} \left(P_x \right)^{-1} H \right\|^2 \to \min_{P_x \geq 0}, \tag{12.20}$$

where the norm in the Hilbert space of finite-dimensional matrices is defined as

$$\|A\|^2 := \langle A, A \rangle = \mathrm{tr}\{AA^{\mathsf{T}}\}, \quad \langle A, B \rangle := \mathrm{tr}\{AB^{\mathsf{T}}\}.$$

The solution P_x^* of the optimization problem (12.20) is as follows:[1]

$$\left(P_x^* \right)^{-1} = \left(H^{\mathsf{T}} \right)^{+} \left(\bar{P}_{ad} \right)^{-1} H^{+}$$

[H^{+} is the pseudoinverse matrix to H defined in the Moore–Penrose sense (Poznyak 2008)]. Since in our case,

$$H^{+} = \left[I_{n \times n} \; I_{n \times n} \right]^{+} = \frac{1}{2} \left[\begin{array}{c} I_{n \times n} \\ I_{n \times n} \end{array} \right],$$

we obtain

$$\left(P_x^* \right)^{-1} = \frac{1}{4\mu^*} \left[I_{n \times n} \; I_{n \times n} \right] \left[\begin{array}{cc} \left(P_A^* \right)^{-1} & 0 \\ 0 & \left(P_A^* \right)^{-1} \end{array} \right] \left[\begin{array}{c} I_{n \times n} \\ I_{n \times n} \end{array} \right] =$$

$$\left(P_x^* \right)^{-1} = \frac{1}{4\mu^*} \left[I_{n \times n} \; I_{n \times n} \right] \left[\begin{array}{c} \left(P_A^* \right)^{-1} \\ \left(P_A^* \right)^{-1} \end{array} \right] = \frac{1}{2} \left(P_A^* \right)^{-1}.$$

---

[1] The solution X^* of the optimization problem

$$\|A - XB\|^2 \to \min_{X} \tag{12.21}$$

satisfies the identity

$$\frac{\partial}{\partial X} \|A - XB\|^2 = \frac{\partial}{\partial X} \mathrm{tr}\{(A - XB)(A^{\mathsf{T}} - B^{\mathsf{T}} X^{\mathsf{T}})\}$$
$$= -2 \left(AB^{\mathsf{T}} - X^{+} BB^{\mathsf{T}} \right) = 0,$$

or equivalently, $X^* BB^{\mathsf{T}} = AB^{\mathsf{T}}$, whose solution (in the case $BB^{\mathsf{T}} > 0$) is

$$X^* = AB^{+} + Y \left(I - BB^{+} \right), \quad B^{+} := B^{\mathsf{T}} \left(BB^{\mathsf{T}} \right)^{-1},$$

where Y is any matrix of the corresponding size. So one has

$$\|X^*\|^2 = \|AB^{+}\|^2 + \|Y(I - BB^{+})\|^2$$
$$+ 2\mathrm{tr}\left\{ AB^{+} \left(I - \left((BB^{\mathsf{T}})^{-1} B \right) B^{\mathsf{T}} \right) Y^{\mathsf{T}} \right\} =$$
$$\|AB^{+}\|^2 + \|Y(I - BB^{+})\|^2 \geq \|AB^{+}\|^2.$$

This means that the solution of the optimization problem (12.21) of minimal norm is $X^* = AB^{+}$.

Hence the attractive ellipsoid of "minimal size" in the state space that guarantees the property

$$\limsup_{t\to\infty} x_t^\mathsf{T} \left(P_x^*\right)^{-1} x_t \le 1$$

has the ellipsoid matrix

$$P_x^* = \frac{1}{2\mu^*} P_A^*. \tag{12.22}$$

So the attractive ellipsoid in x-space is twice the size of the corresponding minimal ellipsoid in $\hat{x}$-space (or equivalently, in e-space) and has the same orientation.

12.2.9   On the Effectiveness of the Adaptation Process

As follows from (12.15), the condition $\varkappa > 0$ guarantees that the attractive ellipsoid $\mathcal{E}\left(0, \bar{P}_{ad}\right)$ corresponding the AEM with adaptation is always "less" than one without adaptation, that is,

$$\mathcal{E}\left(0, \bar{P}_{ad}\right) \subset \mathcal{E}\left(0, \bar{P}\right),$$

so that each semiaxis $r_i^{ad} := \lambda_i \left(\bar{P}_{ad}\right)$ of $\mathcal{E}\left(0, \bar{P}_{ad}\right)$ is strictly less than the semiaxis $r_i := \lambda_i \left(\bar{P}\right)$ of $\mathcal{E}\left(0, \bar{P}\right)$ satisfying

$$r_i^{ad} / r_i = \sqrt{\frac{\lambda_i \left(\bar{P}_{ad}\right)}{\lambda_i \left(\bar{P}\right)}} = \sqrt{1 - \frac{\alpha^* \varkappa}{\varepsilon^* C_1}} < 1.$$

As follows from the consideration above, the gain adaptation process is *effective* (leading to a "smaller" ellipsoid) if

$$\varkappa = \varkappa_1 + \varkappa_2 > 0.$$

To guarantee the effectiveness of the considered robust adaptive controller, we need the additional definitions and resulting statements presented in the following section.

Specific Persistent Excitation Condition

Definition 12.1. We say that the "specific persistent excitation condition" occurs in the considered controlled dynamical system (12.9) if there exists a positive constant h_1 such that

$$\liminf_{t \to \infty} \left(\int_{s=t-h_1}^{t} \|\hat{x}(s)\|^2 \, ds \right) > 0. \tag{12.23}$$

Obviously, the property is satisfied if

$$\int_{s=t-h_1}^{t} \|\hat{x}(s)\|^2 \, ds \geq \epsilon > 0$$

for all $t > h_1$.

Definition 12.2. The external output perturbation $\zeta_y(t)$ [participating in (12.1)] is said to be "independent of the current state $x(t)$" if there exists a positive constant h_2 such that

$$\liminf_{t \to \infty} \int_{s=t-h_2}^{t} \|L(s)[h(x(s)) + \zeta_y(s)]\|^2 \, ds > 0. \tag{12.24}$$

This property holds, for example, if for all sufficiently large t within some small time interval satisfy

$$\zeta_y(t) \neq -h(x(t)).$$

Now we are ready to formulate our main result on the effectiveness of the KL-gains adaptation process.

Theorem 12.3. *If the external output perturbation $\zeta_y(t)$ is "independent" of the current state $x(t)$, that is, the property (12.24) is satisfied, then the* **specific persistent excitation condition** *(12.23) holds, and the KL-gain adaptation process is effective, i.e.,*

$$\varkappa \geq \varkappa_1 > 0.$$

Proof. 1. First, note that

$$\varkappa_1 := \frac{k_1}{2} \lim_{t \to \infty} \operatorname{tr}\left\{\hat{K}_t \hat{K}_t^{\mathsf{T}}\right\} = \frac{2}{k_1} \lim_{t \to \infty} \operatorname{tr}\left\{\Theta(t)\left(P_A^*\right)^{-1} BB^{\mathsf{T}}\left(P_A^*\right)^{-1}\Theta(t)\right\}$$

$$\geq \frac{2}{k_1} \lambda_{\min}\left\{B^{\mathsf{T}}\left(P_A^*\right)^{-2} B\right\} \lim_{t \to \infty} \operatorname{tr}\left(\Theta^2(t)\right) \geq$$

$$\frac{2}{k_1} \lambda_{\min}\left\{B^{\mathsf{T}}\left(P_A^*\right)^{-2} B\right\} \lim_{t \to \infty} \lambda_{\max}^2\left(\Theta(t)\right),$$

where

$$\Theta(t) := \int_{s=0}^{t} e^{-\alpha(t-s)} \hat{x}(s)\hat{x}^{\mathsf{T}}(s)ds.$$

To demonstrate that $\varkappa_1 > 0$, it is sufficient to show that

$$\liminf_{t\to\infty} \lambda_{\max}^2 (\Theta(t)) > 0,$$

which holds if the "specific persistent excitation condition" (12.23) is satisfied. This follows from the following relations:

$$\lim_{t\to\infty} \lambda_{\max}^2 (\Theta(t)) \geq \lim_{t\to\infty} \lambda_{\max}^2 \left(\int_{s=t-h}^{t} e^{-\alpha(t-s)} \hat{x}(s)\hat{x}^{\mathsf{T}}(s)\, ds \right) \geq$$

$$e^{-\alpha h} \lim_{t\to\infty} \left(\int_{s=t-h}^{t} \|\hat{x}(s)\|^2 ds \right)^2 > 0.$$

2. To show that (12.24) implies (12.23), suppose that (12.24) holds, but (12.23) does not. This means that there exists a time sequence t_k such that

$$\int_{s=t_k-h}^{t_k} \|\hat{x}(s)\|^2 ds = 0,$$

or equivalently, for every $s \in [t_k - h, t_k]$, we have

$$\hat{x}(s) = \frac{d}{ds}\hat{x}(s) = 0,$$

which, in view of the relation

$$\dot{\hat{x}}(s) = [A + BK(s) - L(s)C]\hat{x}(s) + L(s)\left[h(x(s)) + \zeta_y(s)\right],$$

implies

$$0 = L(s)\left[h(x(s)) + \zeta_y(s)\right].$$

Integration of last identity leads to

$$\int_{s=t_k-\bar{h}}^{t_k} \|L(s)\left[h(x(s)) + \zeta_y(s)\right]\|^2 ds = 0,$$

which contradicts (12.24). This completes the proof of the theorem.

∎

12.2.10   On Transformation BMI Constraints into LMI Constraints

In this subsection, we present a transformation that allows us to convert the nonlinear matrix inequality (12.12) into a linear one. The solution of the optimization problem given by Theorem 12.1 satisfies a set of BMIs (under fixed scalar parameters) having the following structure:

$$W = \begin{bmatrix} \Omega_1 & P_A L C & 0_{n\times n} & (P_A)^{-1} L \\ C^\mathsf{T} L^\mathsf{T} (P_A)^{-1} & \Omega_2 & (P_A)^{-1} & -(P_A)^{-1} L \\ 0_{n\times n} & (P_A)^{-1} & -\varepsilon I_{n\times n} & 0_{n\times p} \\ L^\mathsf{T} (P_A)^{-1} & -L^\mathsf{T} (P_A)^{-1} & 0_{p\times n} & -\varepsilon I_{p\times p} \end{bmatrix} < 0, \qquad (12.25)$$

where

$$\Omega_1 = (P_A)^{-1} A + (P_A)^{-1} BK + A^\mathsf{T} (P_A)^{-1} +$$

$$K^\mathsf{T} B^\mathsf{T} (P_A)^{-1} + \alpha (P_A)^{-1} + \varepsilon C_2 (1 + \rho) I_{n\times n}$$

$$\Omega_2 = (P_A)^{-1} A - (P_A)^{-1} LC + A^\mathsf{T} (P_A)^{-1} -$$

$$C^\mathsf{T} L^\mathsf{T} (P_A)^{-1} + \alpha (P_A)^{-1} + \varepsilon C_2 \left(1 + \rho^{-1}\right) I_{n\times n}.$$

$$(12.26)$$

Using the so-called *regular form representation* for the quasilinear model (12.9) (see, for example, Poznyak (2008), Sect. 19.4.3.2), one may represent the block matrix as

$$B := \begin{bmatrix} B_1 \in R^{(n-m)\times m} \\ B_2 \in R^{m\times m} \end{bmatrix},$$
$$\det(B_2) \neq 0.$$

Defining then the nonsingular matrix G as

$$G := \begin{bmatrix} I_{(n-m)\times(n-m)} & -B_1 B_2^{-1} \\ 0_{m\times(n-m)} & B_2^{-1} \end{bmatrix} \in \mathbb{R}^{n\times n},$$

let us try to find the matrix $(P_A)^{-1}$ in the form

$$(P_A)^{-1} = G^\mathsf{T} P_B G,$$

where P_B is the block-diagonal matrix

$$P_B := \begin{bmatrix} P_{B,11} & 0_{(n-m)\times m} \\ 0_{m\times(n-m)} & P_{B,22} \end{bmatrix}, \; 0 < P_{B,11}, \; 0 < P_{B,22}.$$

Lemma 12.1. *Under fixed scalar parameters $(\alpha, \varepsilon, \rho)$, the set of matrix variables P_B, K, and L satisfying (12.25) and (12.26) is isomorphic to the set of variables $X := \mathrm{diag}(X_{11}, X_{22})$, Y_1, and Y_2, and is uniquely related to the previous one as*

$$X_{11} := P_{B,11} > 0, \; X_{22} := P_{B,22} > 0,$$

$$Y_1 := P_{B,22} K, \; Y_2 := P_B G L,$$

satisfying the following LMI:

$$W_T \left(X, Y_1, Y_2 \mid \alpha, \varepsilon, \rho \right) := \begin{bmatrix} \bar{\Gamma}_1 & Y_2 C \left(G^\mathsf{T} \right)^{-1} \\ \left(G^\mathsf{T} \right)^{-1} C^\mathsf{T} Y_2^\mathsf{T} & \bar{\Gamma}_2 \end{bmatrix} < 0, \tag{12.27}$$

where

$$\bar{\Gamma}_1 = X G A_\alpha \left(G^\mathsf{T} \right)^{-1} + \left(G^\mathsf{T} \right)^{-1} A_\alpha^\mathsf{T} G^\mathsf{T} X + \begin{bmatrix} 0 \\ Y_1 \end{bmatrix} \left(G^\mathsf{T} \right)^{-1} +$$

$$\left(G^\mathsf{T} \right)^{-1} \begin{bmatrix} 0 \\ Y_1 \end{bmatrix}^\mathsf{T} + \varepsilon C_2 \left(1 + \rho \right) \left(G^\mathsf{T} \right)^{-2}$$

$$\bar{\Gamma}_2 = X G A_\alpha \left(G^\mathsf{T} \right)^{-1} + \left(G^\mathsf{T} \right)^{-1} A_\alpha^\mathsf{T} G^\mathsf{T} X - Y_2 C \left(G^\mathsf{T} \right)^{-1} - \left(G^\mathsf{T} \right)^{-1} C^\mathsf{T} Y_2^\mathsf{T} +$$

$$\varepsilon C_2 \left(1 + \rho^{-1} \right) \left(G^\mathsf{T} \right)^{-2}. \tag{12.28}$$

Proof. The matrix W^* defined by (12.25) can be represented in a block format as

$$W^* := \begin{bmatrix} W_{11} & W_{12} \\ W_{21} & W_{22} \end{bmatrix} < 0, \; W_{11} := \begin{bmatrix} \Omega_1 & \left(P_A \right)^{-1} L C \\ C^\mathsf{T} L^\mathsf{T} \left(P_A \right)^{-1} & \Omega_2 \end{bmatrix},$$

$$W_{22}^\mathsf{T} = \begin{bmatrix} -\varepsilon I_{n\times n} & 0_{n\times p} \\ 0_{p\times n} & -\varepsilon I_{p\times p} \end{bmatrix}, \; W_{12}^\mathsf{T} = \begin{bmatrix} 0_{n\times n} & \left(P_A \right)^{-1} L \\ \left(P_A \right)^{-1} & -\left(P_A \right)^{-1} L \end{bmatrix} = W_{21},$$

where Ω_1 and Ω_2 are defined by (12.26). By Schur's complement, we obtain

$$W_{11} < W_{12}^\mathsf{T} W_{22}^{-1} W_{12} = -\varepsilon^{-1} \begin{bmatrix} P_A^{-2} & -P_A^{-2} L \\ -L^\mathsf{T} P_A^{-2} & 2 L^\mathsf{T} P_A^{-2} L \end{bmatrix} = -\varepsilon^{-1} F Q F^\mathsf{T} < 0$$

with

$$F := \begin{bmatrix} I & 0 \\ -L^\mathsf{T} & L^\mathsf{T} \end{bmatrix} \text{ and } Q := \begin{bmatrix} P_A^{-2} & 0 \\ 0 & P_A^{-2} \end{bmatrix}.$$

To solve the optimization problem with BMI constraints (12.11), we propose a transformation based on a characterization of the matrix $B \in R^{n \times m}$. Since the rank of the matrix B is m, this matrix can be partitioned in regular form: using the definition of matrices G and B, the inequality (12.11) can be converted to

$$W_{11} = \begin{bmatrix} \Omega_1 & G^\mathsf{T} P_B G F C \\ C^\mathsf{T} F^\mathsf{T} G^\mathsf{T} P_B G & \Omega_2 \end{bmatrix} < -\varepsilon^{-1} L Q L^\mathsf{T} < 0.$$

Since

$$GB = \begin{bmatrix} 0_{(n-m) \times m} \\ I_{m \times m} \end{bmatrix},$$

we have

$$(P_A)^{-1} BK = G^\mathsf{T} \begin{bmatrix} 0 \\ P_{B,22} K \end{bmatrix}$$

and

$$\Omega_1 = G^\mathsf{T} P_B G A_\alpha + A_\alpha^\mathsf{T} G^\mathsf{T} P_B G +$$

$$G^\mathsf{T} \begin{bmatrix} 0 \\ P_{B,22} K \end{bmatrix} + \begin{bmatrix} 0 \\ P_{B,22} K \end{bmatrix}^\mathsf{T} G + \varepsilon C_2 (1 + \rho) I.$$

Observe that $W_{11} < 0$ if and only if

$$W_T := T W_{11} T^\mathsf{T} < 0$$

for a nonsingular matrix T. Define T as the block-diagonal matrix

$$T^\mathsf{T} := \begin{bmatrix} G^{-1} & 0_{n \times n} \\ 0_{n \times n} & G^{-1} \end{bmatrix}.$$

Taking into account that $P_A > 0$, we may conclude that $W_{11} < 0$ if and only if (12.27) and (12.28) are satisfied. This completes the proof of the lemma. ∎

In the new variables X_{11}, X_{22}, Y_1, Y_2, the optimization problem (12.10) can be formulated as follows:

$$\frac{\alpha}{\varepsilon C_1} \operatorname{tr} \{X^{-1}\} \rightarrow \inf_{W_T(X,Y_1,Y_2,\Lambda | \alpha, \varepsilon, \rho) < 0, \ X > 0}.$$

If $X^*, Y_1^*, Y_2^*, \Lambda^*$ is a solution of the considered constrained optimization problem, then the optimal gain matrices K^* and L^* can be uniquely found as

$$K^* = \left(X_{22}^*\right)^{-1} Y_1^*, \quad L^* = G^{-1}\left(X^*\right)^{-1} Y_2^*.$$

12.2.11 Numerical Aspects

The numerical values X^*, Y_1^*, Y_2^* can be obtained (approximately estimated) recursively by a two-step recurrent procedure:

- First, fixing some initial values of the scalar parameters $\alpha = \alpha_0$, $\varepsilon = \varepsilon_0$, and $\rho = \rho_0$, we apply the MATLAB toolbox SeDuMi in order to solve the corresponding optimization problem with LMI constraints (12.27). As a result, we obtain the matrices $P_{B,11}^0$, $P_{B,22}^0$, K_0, and L_0.
- From the obtained matrices $P_{B,11}^0$, $P_{B,22}^0$, K_0, and L_0, we augment the parameters α taking

$$\alpha_1 = \alpha_0 + \Delta\alpha$$

where $0 < \Delta\alpha \ll 1$, and we decrease ε, ρ, and ρ, which yields

$$\varepsilon_1 = \varepsilon_0 - \Delta\varepsilon, \quad \rho_1 = \rho_0 - \Delta\rho,$$

$$0 < \Delta\varepsilon \ll 1, \ 0 < \Delta\rho \ll 1.$$

- Then these steps are repeated until the SeduMi toolbox "informs" us that the current LMI has no solution. The last admissible parameters are declared optimal.

12.2.12 Illustrative Example

In this section, we consider as an illustrative example the underactuated double inverted pendulum (well known as a *Pendubot system*). The control to be designed is intended to maintain (stabilize) the pendulum in the vertical position using only shoulder torque in the first (lowest) joint. The experimental results were obtained using the PendCon system. A Digital Signal Processor C6713DSK board control system was integrated on a 16-bit expansion bus-slot of a personal computer. The real-time process was given by a Win-Con compiler (MATLAB SIMULINK) providing the programming environment. The control input was transmitted to a 24-Volt DC Motor with 1000 Cnt/Rev Optical Encoder from Maxon Inc., a 24VDC at 2.1AMP power supply was employed.

A mathematical model of the considered systems is given in Section 11.5.1.

The problem to be solved is to stabilize this system at the upper equilibrium point of the desired upper unstable position.[2] By construction, the denominator is not equal to zero. In view of this, we have

$$\dot{x}(t) = Ax(t) + Bu + \xi\,(x(t), t)\,,$$

$$\xi\,(x(t), t) := f(x(t)) - Ax(t) + \zeta\,(t)\,.$$

The physically measured states are the positions (angles) $q_1 + \zeta_1(t)$ and $q_2 + \zeta_2(t)$ disturbed by noise. So we have

$$y(t) = Cx(t) + \zeta(t), \ C = \begin{bmatrix} 1 & 0 & 0 & 0 \\ 0 & 1 & 0 & 0 \end{bmatrix}.$$

Select

$$A = \begin{bmatrix} 0 & 0 & 1 & 0 \\ 0 & 0 & 0 & 1 \\ 55.62 & -46.76 & 0 & 0 \\ -80.47 & 155.30 & 0 & 0 \end{bmatrix}, \ C_0 = 1.209, \ C_1 = 1.025.$$

Notice that the pairs (A, B) and (C, A) constitute respectively a controllable and observable pair. Applying the suggested technique for the Pendubot system, for $k_1 = 0.5$ and $k_2 = 0.03$, we obtain (after five recurrent steps)

$$\alpha^* = 1.5, \ \varepsilon^* = 0.0056, \ \rho^* = 0.7, \ \Lambda^* = 0.5 I_{n \times n}$$

and

$$P^* = 10^{-3} \begin{bmatrix} 1.6040 & -0.2689 & -1.5289 & 0.4251 \\ -0.2689 & 0.5205 & -0.0864 & -0.0341 \\ -1.5289 & -0.0864 & 2.3613 & -0.6556 \\ 0.4251 & -0.0341 & -0.6556 & 0.3135 \end{bmatrix},$$

$$K^* = \begin{bmatrix} 11.8093 & 10.0139 & 9.2711 & 5.6993 \end{bmatrix},$$

$$L^* = 10^4 \begin{bmatrix} 3.8735 & 1.3459 & 2.1939 & 1.6105 \\ 1.7817 & 2.9592 & 2.1499 & 1.6103 \end{bmatrix}^{\mathsf{T}}.$$

The plots are shown in Figs. 12.1, 12.2, and 12.3.

---

[2] We use the variable coordinate change $x_1 = q_1$, $x_2 = q_2$, $x_3 = \dot{q}_1$, and $x_4 = \dot{q}_2$. The Pendubot top position is $(x_1, x_2, x_3, x_4) = (\pi/2, 0, 0, 0)$.

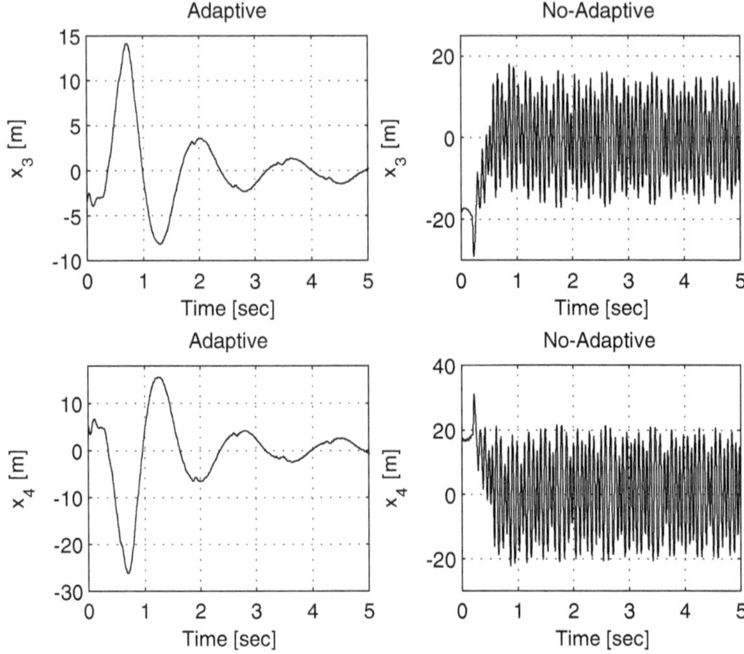

Fig. 12.1 Link velocity of the Pendubot system for the nonadaptive and adaptive cases

The advantage of the adaptive controller is obvious: *the ratio of the corresponding "ellipsoid size" is less than* 1 *in the nonadaptive case, about* 80%, *i.e., for the system we have*

$$r_i^{ad} / r_i = 0.1084.$$

12.3 A-Adaptation in the Attractive Ellipsoid Method

Adjustment (or adaptation) is suggested to be made only in some fixed sample times, so that the corresponding gain matrix of the controller remains constant within every interval between two neighboring sample times. Namely,

$$u(t) = K_{t_i} \hat{x}(t), \ t \in (t_{i-1}, t_i], \tag{12.29}$$

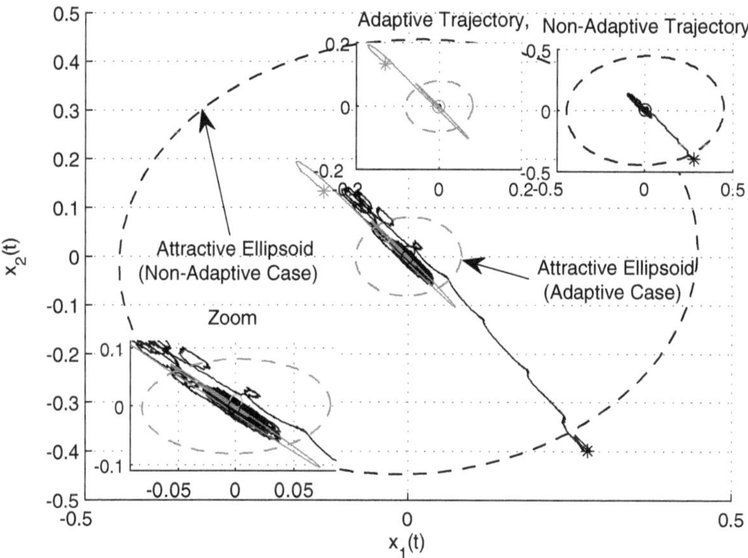

Fig. 12.2 The attractive ellipsoid corresponding to $\hat{x}_1$ and $\hat{x}_2$ for the Pendubot system

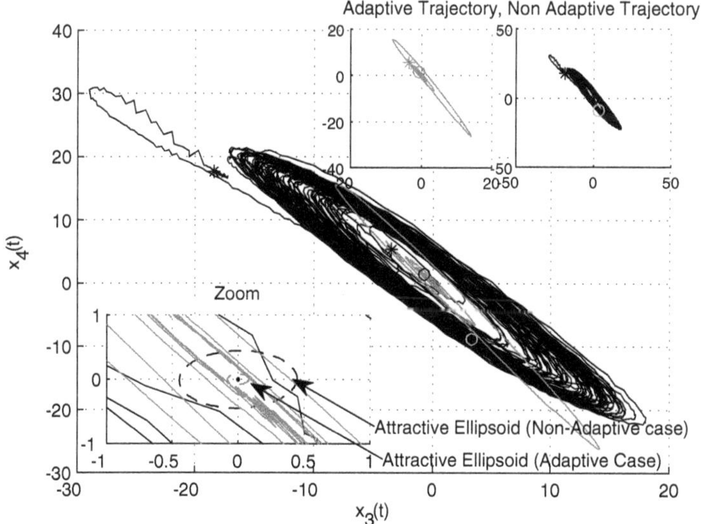

Fig. 12.3 The attractive ellipsoid corresponding to $\hat{x}_3$ and $\hat{x}_4$ for the Pendubot system

where

$$t_0 < t_1 < \cdots < t_i < \cdots$$

is a given sequence of fixed correction moments. The method proposed below
guarantees that under a specific "regularized persistent excitation condition," the

controlled system trajectories converge to an ellipsoid of *minimal size* (the minimal size is referred to as the minimal trace of the corresponding ellipsoidal matrix).

12.3.1 Quasilinear Model with Adjusted Feedback and Problem Formulation

Observe that for the system (12.1), closed by the *switched feedback* (12.29), we have

$$\dot{x}(t) = Ax(t) + BK_{t_i}\hat{x}(t) + \xi_x(x(t),t), \; t \in (t_{i-1}, t_i]. \tag{12.30}$$

Here we are interested in designing the feedback matrices K_{t_i} and L_{t_i} providing the stabilization (or the boundedness) of the trajectories of the closed-loop system (12.30). Since every bounded trajectory can be imposed on some convex set (here we will select an ellipsoid), here we suggest selecting the parameters K_{t_i} and L_{t_i} in an "optimal manner," making the "size" of this set (ellipsoid) as small as possible.

Now we are ready to formulate the problem that we will consider and solve here.

Problem formulation. *Based on the available information*

$$\{y(t), \hat{x}(t), \dot{\hat{x}}(t), u(t)\}_{t \geq 0},$$

designate the sequences $\{K_{t_i}\}_{i=1,2,...}$ *and* $\{L_{t_i}\}_{i=1,2,...}$ *of the gain matrices* $K_{t_i} \in \mathbb{R}^{m \times n}$, $L_{t_i} \in \mathbb{R}^{n \times p}$ *that for every plant with uncertainties from the quasi-Lipschitz class, provide the existence of an attractive ellipsoid of minimal possible "size":*

$$\mathrm{tr}\left\{\bar{P}_{t_i}^{-1}\right\} \to \inf_{K_{t_i}\, L_{t_i},\, (i=1,2,...)}. \tag{12.31}$$

The sequence $\{K_{t_i}\}_{i=1,2,...}$ of the gain matrices in (12.29), realizing (12.31), guarantees the so-called *stability zone* for the class of uncertain systems (12.9).

12.3.2 "A"-Adaptation

The control design technique based on the traditional AEM requires knowledge of the matrices B in (12.30) and the matrix A, as well as the scalars d_0, d_1 defining the class of quasi-Lipschitz systems to be controlled. As a result, the designed robust control guarantees the boundedness of the corresponding system trajectories within an ellipsoid of minimal size for every nonlinearity from the given quasi-Lipschitz class $\mathcal{C}$. But when we intend to control a real process with an unknown nonlinear model, we are not interested in designing a robust control for the entire class $\mathcal{C}$:

our aim is to control the given uncertain plant, and we do not worry about other plants from this class. Obviously, within a neighborhood of the point $x = 0$, a more precise selection of the matrix A is $A \simeq \dfrac{\partial f}{\partial x}(0)$ (if, of course, the vector function f is differentiable at the origin). Unfortunately, we cannot realize this choice, since the mapping $f : \mathbb{R}^n \to \mathbb{R}$ is unknown a priori! That is why we need to try to approximate this matrix A online, and using this information, make the attractive ellipsoid for the given plant of smallest possible size in comparison with one serving for the description of the class $\mathcal{C}$. Below, we present the approach that realizes online the so-called "A"-approximation (identification) process.

The Extended System with "A"-Adaptation

To obtain the "best" estimate of the matrix A_{t_i} at time t_i, we take into account that during the previous time interval $(t_{i-1}, t_i]$, the feedback gain matrix in $u(t)$ was equal to K_{t_i}. Note that for such a time interval ($i = 1, 2, \ldots$), the dynamic term in (12.30) satisfies the following ODE:

$$\dot{x}(t) = \dot{\hat{x}}(t) + \dot{e}(t) = Ax(t) + BK_{t_i}\hat{x}(t) + \xi_x[x(t), t],$$

$$\dot{\hat{x}}(t) + [A - L_{t_i}C]e(t) + \xi_x[x(t), t] - L_{t_i}\xi_y[x(t), t] =$$

$$Ax(t) + BK_{t_i}\hat{x}(t) + \xi_x[x(t), t],$$

$$\dot{\hat{x}}(t) - A\hat{x}(t) - L_{t_i}y(t) - L_{t_i}\xi_y[x(t), t] = BK_{t_i}\hat{x}(t),$$

implying that the uncertain term $\left(L_{t_i}e(t) - L_{t_i}\xi_y[x(t), t]\right)$ is given by

$$L_{t_i}e(t) - L_{t_i}\xi_y[x(t), t] =$$

$$\dot{\hat{x}}(t) - [A + BK_{t_i}]\hat{x}(t), \quad t \in (t_{i-1}, t_i], \quad i = 1, 2, \ldots . \tag{12.32}$$

Note that here, $\dot{\hat{x}}(t)$ is not available (but may be estimated) as well as the "best" (for the given plant) matrix A. Our objective here is to find the "best" estimate of the matrix A, minimizing (in some average sense, for example, mean-square) the effects of $\xi_x[x(t), t]$.

Taking into account that the only measurable vector at time t is the system output $y(t)$, let us approximate the state $x(t)$ by the estimate $\hat{x}(t)$, defined by (12.2). Observe that in view of (12.29), it follows that

$$\dot{\hat{x}}(t) = [A + BK_{t_i}]\hat{x}(t) + L_{t_i}(y - C\hat{x}). \tag{12.33}$$

Define the state estimation error

$$e(t) = x(t) - \hat{x}(t), \tag{12.34}$$

for which we have

$$\dot{e}(t) = [A - L_{t_i} C] e(t) + \xi_x[x(t), t] - L_{t_i}\xi_y[x(t), t], \ e(0) = e_0. \tag{12.35}$$

Combining (12.30), (12.33), and (12.35) for the extended vector

$$z^\mathsf{T}(t) = [x^\mathsf{T}(t), \ \hat{x}^\mathsf{T}(t), \ e^\mathsf{T}(t)],$$

we derive the following model:

$$\dot{z}(t) = \mathcal{A}_{t_i} z(t) + F_{t_i}\xi\left[x\left(t\right), t\right], \ z(0) = z_0, \tag{12.36}$$

where

$$\mathcal{A}_{t_i} = \begin{bmatrix} A + BK_{t_i} & 0_{n \times n} & -BK_{t_i} \\ 0_{n \times n} & A + BK_{t_i} & L_{t_i}C \\ 0_{n \times n} & 0_{n \times n} & A - L_{t_i}C \end{bmatrix},$$

$$F_{t_i} = \begin{bmatrix} I_{n \times n} & 0_{n \times p} & 0_{n \times n} \\ 0 & L_{t_i} & 0_{n \times p} \\ I & -L_{t_i} & 0_{n \times n} \end{bmatrix}, \ \xi\left[x\left(t\right), t\right] = \begin{bmatrix} \xi_x[x(t), t] \\ \xi_y[x(t), t] \\ 0_{(2n-p)n} \end{bmatrix}.$$

Below, we will suppose that

$$\|\xi\|^2 \le b_0 + b_1\|x(t)\|^2. \tag{12.37}$$

Using (12.32), we derive the relation

$$\delta\left[\hat{x}(t), \dot{\hat{x}}(t)\right] = L_{t_i}\left[y(t) - C\hat{x}(t)\right] - BK_{t_i}\hat{x}(t) = \dot{\hat{x}}(t) - A\hat{x}(t),$$

where $y(t)$, $\hat{x}(t)$, $\dot{\hat{x}}(t)$ are the measurable vectors available at every time $t \in (t_{i-1}, t_i]$.

Definition 12.3. Define the least squares estimate A_{t_i}, obtained from the measurements within the time interval $(t_{i-1}, t_i]$, as

$$A_t := \arg\min_A \int_{t_{i-1}}^{t} \left\|\delta\left[\hat{x}(\tau), \dot{\hat{x}}(\tau)\right]\right\|^2 d\tau. \tag{12.38}$$

This means that A_t minimizes the quadratic-integral error of the "joint uncertainty effect."

Proposition 12.1 (On the "A"-approximation procedure). *The "best" matrix approximation A_t, defined by (12.38), is given by the following set of differential equations:*

$$\dot{A}_t = -\left[A_t\hat{x}(t) - \dot{\hat{x}}(t)\right]\hat{x}^\mathsf{T}(t)\Gamma_t,$$

$$\dot{\Gamma}_t = -\Gamma_t\hat{x}(t)\hat{x}^\mathsf{T}(t)\Gamma_t, \ t \in (t_{i-1}, t_i],$$

(12.39)

where the initial condition $\Gamma_{t_{i-1}}$ satisfies the regularized persistent excitation condition

$$\Gamma_{t_{i-1}} = \left[\int_{t_{i-2}}^{t_{i-1}} \hat{x}(t)\hat{x}^\mathsf{T}(t)dt + \rho I_{n\times n}\right]^{-1}, \ 0 < \rho \ll 1.$$

(12.40)

The initial conditions for A_0 may be selected as a matrix A defining the class $\mathcal{C}$ of quasi-Lipschitz uncertainties.

Proof. To find A_t defined on (12.38) in an analytical form, let us follow Chap. 12 of Poznyak (2008). The functional to be minimized, (12.38), can be represented in the following form:

$$J_i^t(A) := \int_{t_{i-1}}^t \left\|\delta\left[\hat{x}(\tau), \dot{\hat{x}}(\tau)\right]\right\|^2 d\tau = \int_{t_{i-1}}^t \left\|\dot{\hat{x}}(\tau) - A\hat{x}(\tau)\right\|^2 d\tau$$

$$\int_{t_{i-1}}^t \left(\operatorname{tr}\left\{\dot{\hat{x}}(\tau)\dot{\hat{x}}^\mathsf{T}(\tau)\right\} - 2\operatorname{tr}\left\{\hat{x}(\tau)A\dot{\hat{x}}^\mathsf{T}(\tau)\right\} + \operatorname{tr}\left\{A\hat{x}(\tau)\hat{x}^\mathsf{T}(\tau)A^\mathsf{T}\right\}\right)d\tau =$$

$$\operatorname{tr}\left\{\int_{t_{i-1}}^t \dot{\hat{x}}(\tau)\hat{x}^\mathsf{T}(\tau)d\tau\right\} - 2\operatorname{tr}\left\{\left(\int_{t_{i-1}}^t \dot{\hat{x}}^\mathsf{T}(\tau)\hat{x}(\tau)d\tau\right)A\right\} +$$

$$\operatorname{tr}\left\{A\left(\int_{t_{i-1}}^t \hat{x}(\tau)\hat{x}^\mathsf{T}(\tau)d\tau\right)A^\mathsf{T}\right\}.$$

Using the formulas

$$\frac{\partial}{\partial A}\operatorname{tr}(BAC) = B^\mathsf{T}C^\mathsf{T}, \quad \frac{\partial}{\partial A}\operatorname{tr}(ABA^\mathsf{T}) = AB^\mathsf{T} + AB,$$

we find that the estimate A_t satisfies

$$\frac{\partial}{\partial A}J_i^t(A)\,|_{A=A_t} = -2\int_{t_{i-1}}^t \dot{\hat{x}}(\tau)\hat{x}^\mathsf{T}(\tau)d\tau + 2A_t\int_{t_{i-1}}^t \hat{x}(\tau)\hat{x}^\mathsf{T}(\tau)d\tau = 0_{n\times n},$$

and hence

$$A_t = Z_t \Gamma_t, \quad Z_t := \int_{t_{i-1}}^{t} \dot{\hat{x}}(\tau)\hat{x}^{\mathsf{T}}(\tau)d\tau,$$

$$\Gamma_t := \left[\int_{t_{i-1}}^{t} \hat{x}(\tau)\hat{x}^{\mathsf{T}}(\tau)d\tau\right]^{-1}.$$

(12.41)

Certainly, the representation (12.41) makes sense if the so-called "persistent excitation condition" holds, namely, when

$$\Gamma_t^{-1} = \int_{t_{i-1}}^{t} \hat{x}(\tau)\hat{x}^{\mathsf{T}}(\tau)d\tau > 0.$$

In view of the relations

$$\Gamma_t^{-1}\Gamma_t = I_{n\times n} \implies \dot{\Gamma}_t^{-1}\Gamma_t + \Gamma_t^{-1}\dot{\Gamma}_t = 0_{n\times n},$$

$$\dot{\Gamma}_t = -\Gamma_t \dot{\Gamma}_t^{-1}\Gamma_t = -\Gamma_t x(t)x^{\mathsf{T}}(t)\Gamma_t,$$

the time differentiation of (12.41) leads to the following differential form of the matrix A_t:

$$\dot{A}_t = Z_t \dot{\Gamma}_t + \dot{Z}_t \Gamma_t = -Z_t \Gamma_t \hat{x}(t)\hat{x}^{\mathsf{T}}(t)\Gamma_t + \dot{\hat{x}}(t)\hat{x}^{\mathsf{T}}(t)\Gamma_t =$$

$$-\left[A_t \hat{x}(t) - \dot{\hat{x}}(t)\right]\hat{x}^{\mathsf{T}}(t)\Gamma_t.$$

Observe that by soft modification of the initial condition with

$$0 < \rho \ll 1,$$

we obtain (12.40). The specific "persistence excitation condition" provides the existence of this solution, which contains the small regularization term $\rho I_{n\times n}$, permitting to avoid the problems with the inversion of $\Gamma_{t_{i-1}}$ even the matrix $\int_{t_{i-2}}^{t_{i-1}} \hat{x}(t)\hat{x}^{\mathsf{T}}(t)dt$ is singular. This regularization practically does not change the dynamics of the adaptation procedure (12.39). This completes the proof. ∎

12.3.3   Closed-Loop Representation and Storage Function

Observe that in the original process (12.36), the term $\xi[x(t), t]$ is unavailable, and the matrix "A" participating in the quasilinear format (12.30) is unknown as well. But the adaptation procedure (12.39) provides its "best" approximation A_t for every $t \in (t_{i-1}, t_i]$. So to realize the adaptive robust control, let us maintain the matrix A_t

as the constant matrix at each interval, namely, put $A_t = A_{t_{i-1}}$ for all $t \in (t_{i-1}, t_i]$ and select the feedback controller as (12.29). Then the system (12.36), where the identification procedure (12.39) is applied, can be represented as

$$\dot{z}(t) = \hat{\mathcal{A}}_{t_i} z(t) + \hat{F}_{t_i} \xi \left[x(t), t \right], \quad z(0) = z_0,$$

$$\hat{\mathcal{A}}_{t_i} = \begin{bmatrix} A_{t_i} + BK_{t_i} & 0_{n \times n} & -BK_{t_i} \\ 0_{n \times n} & A_{t_i} + BK_{t_i} & L_{t_i} C \\ 0_{n \times n} & 0_{n \times n} & A_{t_i} - L_{t_i} C \end{bmatrix}, \tag{12.42}$$

$$\hat{F}_{t_i} = \begin{bmatrix} I_{n \times n} & 0_{n \times p} & 0_{n \times n} \\ 0 & L_{t_i} & 0_{n \times p} \\ I & -L_{t_i} & 0_{n \times n} \end{bmatrix}.$$

Observe that to find K_{t_i} and L_{t_i}, one has to use the measurements from the previous interval, that is, at time t_i, we use data up to t_{i-1}. Then the time approximation matrix A_{t_i} is not the best matrix estimation for the current time interval. Thus the problem formulation consists in finding the "best" matrix estimation defining the $\mathcal{C}$ class in the time interval referred below as to $\bar{A}_{t_i}$.

Proposition 12.2 (On the storage function). *If the collection* $(P_{t_i}, \alpha_i, \varepsilon_i, K_{t_i}, L_{t_i})$ *satisfy the matrix constraints*

$$W_i(P_{t_i}, \alpha_i, \varepsilon_i, K_{t_i}, L_{t_i}) := \begin{bmatrix} P_{t_i} \bar{\mathcal{A}}_{t_i} + \bar{\mathcal{A}}_{t_i}^{\mathsf{T}} P_{t_i} + \alpha_i P_{t_i} + R_{t_i} & P_{t_i} \hat{F} \\ \hat{F}^{\mathsf{T}} P_{t_i} & -\varepsilon_i I_{n \times n} \end{bmatrix} < 0,$$

$$P_{t_i} := diag\left[P_{t_i}^A, P_{t_i}^A, P_{t_i}^A \right], \quad P_{t_i}^A = (P_{t_i}^A)^{\mathsf{T}} > 0, \; 0 < \alpha_i \; and \; 0 < \varepsilon_i,$$

$$\bar{\mathcal{A}}_{t_i} := \begin{bmatrix} \bar{A}_{t_i} + BK_{t_i} & 0_{n \times n} & -BK_{t_i} \\ 0_{n \times n} & \bar{A}_{t_i} + BK_{t_i} & L_{t_i} C \\ 0_{n \times n} & 0_{n \times n} & \bar{A}_{t_i} - L_{t_i} C \end{bmatrix},$$

$$R_{t_i} = \begin{bmatrix} \varepsilon b_1 I_{n \times n} & 0_{n \times n} & 2P_{t_i}^A \left(A_{t_i} - \bar{A}_{t_i} \right) \\ 0_{n \times n} & \varepsilon b_1 I_{n \times n} & 0_{n \times n} \\ 2\left(A_{t_i} - \bar{A}_{t_i} \right)^{\mathsf{T}} P_{t_i}^A & 0_{n \times n} & 0_{n \times n} \end{bmatrix}, \tag{12.43}$$

under the solution of the following ODE:

$$\frac{d}{dt} \bar{A}_t = -\left[A_{t_i} \hat{x}(t) - \dot{\hat{x}}(t) \right] \hat{x}^{\mathsf{T}}(t),$$

$$-\frac{\alpha}{2} \left(A_{t_i} - \bar{A}_{t_i} \right) + \frac{4}{k_1} P_{t_i}^A \hat{x}(t) \hat{x}^{\mathsf{T}}(t), \tag{12.44}$$

for each time interval $(t_{i-1}, t_i]$ Then the "storage" (energetic) function

$$V_{t_i}[z(t)] := z^{\mathsf{T}}(t) P_{t_i} z(t) + \theta_i,$$

$$\theta_i := k\operatorname{tr}\left\{\left[A_t - \bar{A}_t\right]^{\mathsf{T}}\left[A_t - \bar{A}_t\right]\right\}, \quad k > 0,$$

$$(12.45)$$

defined for every $t \in (t_{i-1}, t_i]$, satisfies the following differential inclusion:

$$\dot{V}_{t_i}[z(t)] \le -\alpha_i V_{t_i}[z(t)] + b_0 \varepsilon_i, \quad \forall \, i = 1, 2, \ldots \tag{12.46}$$

[b_0 is defined in (12.37)].

Proof. Consider the system given by (12.42) in the time interval $t \in (t_{i-1}, t_i]$, and calculate the derivative of the storage function (12.45) along the trajectories of the system (12.36):

$$\dot{V}(z(t)) = \left[\begin{array}{c} z(t) \\ \xi \end{array}\right]^{\mathsf{T}} \left[\begin{array}{cc} P_{t_i} \mathcal{A}_{t_i} + \mathcal{A}_{t_i}^{\mathsf{T}} P_{t_i} & P_{t_i} F_{t_i} \\ F_{t_i}^{\mathsf{T}} P_{t_i} & 0_{n \times n} \end{array}\right] \left[\begin{array}{c} x(t) \\ \xi(x(t), t) \end{array}\right]$$

$$+ 2k\operatorname{tr}\left\{\left[\frac{d}{dt}A_t - \frac{d}{dt}\bar{A}_t\right]^{\mathsf{T}}\left[A_t - \bar{A}_t\right]\right\}.$$

For each time interval, add and subtract on the right-hand side of the last equation the terms

$$\alpha_i V(z(t)), \quad \varepsilon_i \xi^{\mathsf{T}}\xi, \quad 2x^{\mathsf{T}}(t) P_{t_i} \bar{A}_t x(t), \quad 2\hat{x}^{\mathsf{T}}(t) P_{t_i} \bar{A}_t \hat{x}(t), \quad 2e^{\mathsf{T}}(t) P_{t_i} \bar{A}_t e(t).$$

In view of the upper estimate

$$\xi^{\mathsf{T}}\xi \le b_0 + b_1 \|x(t)\|^2$$

and using the definition of the "A"-approximation procedure (12.39), we obtain

$$\dot{V}(z(t)) \le y^{\mathsf{T}}(t) \left[\begin{array}{cc} P_{t_i} \bar{A}_{t_i} + \bar{A}_{t_i}^{\mathsf{T}} P_{t_i} + \alpha_i P_{t_i} + R_{t_i} & P_{t_i} \hat{F} \\ \hat{F}^{\mathsf{T}} P_{t_i} & -\varepsilon_i I_{n \times n} \end{array}\right] y(t) - \alpha_i V_t(x(t))$$

$$+ \varepsilon_i d_0 + 2\operatorname{tr}\left\{\left[\frac{d}{dt}A_t - \frac{d}{dt}\bar{A}_t + \frac{\alpha}{2}\left(A_{t_i} - \bar{A}_{t_i}\right) + \frac{4}{k_1} P_{t_i}^A \hat{x}(t)\hat{x}^{\mathsf{T}}(t)\right]^{\mathsf{T}}\left[A_t - \bar{A}_t\right]\right\}$$

with the matrices $\bar{A}_{t_i}$, $\hat{F}_{t_i}$, and R_{t_i} defined by (12.42). Here the extended vector $y(t)$ is given by

$$y^{\mathsf{T}} := [z^{\mathsf{T}}(t), \, \xi^{\mathsf{T}}].$$

Selecting the variations of $\bar{A}_t$ satisfying

$$\frac{d}{dt}\bar{A}_t = -\left[A_{t_i}\hat{x}(t) - \dot{\hat{x}}(t)\right]\hat{x}^{\mathsf{T}}(t) - \frac{\alpha}{2}\left(A_{t_i} - \bar{A}_{t_i}\right) + \frac{4}{k_1}P_{t_i}^A\hat{x}(t)\hat{x}^{\mathsf{T}}(t)$$

$$\bar{A}_{t_{i-1}} = A_{t_{i-1}}, \ t \in (t_{i-1}, t_i],$$

we get

$$\dot{V}[z(t)] \leq y^{\mathsf{T}}W_i(P_{t_i}, \alpha_i, \varepsilon_i)y - \alpha_i V[z(t)] + \varepsilon_i b_0.$$

Since by the assumption of the theorem we have $W_i < 0$, we directly obtain (12.43). Note that the functions $\dot{V}_i$ satisfy (12.46) on the time intervals $t \in (t_{i-1}, t_i]$ for all $i := 1, 2, \ldots$, and then

$$V_{t_i}(x(t_i)) \leq \beta_i/\alpha_i + [V_{t_i}(x(t_{i-1})) - \beta_i/\alpha_i]e^{-\alpha_i \tau_i} =$$

$$\frac{\beta_i}{\alpha_i} - \frac{\beta_i}{\alpha_i}e^{-\alpha_i \tau_i} + e^{-\alpha_i \tau_i}V_{t_{i-1}} \tag{12.47}$$

with

$$\beta_i := d_1\varepsilon_i$$

and

$$\tau_i := t_i - t_{i-1}.$$

Note that in Theorem 12.1, additionally the energetic function (12.45) exhibits the property

$$\dot{V}_t[z(t)] \leq -\alpha_i V_t[z(t)] + b_0\varepsilon_i, \ t \in (t_{i-1}, t_i], \ \forall i = 1, 2, \ldots,$$

which completes the proof.   ∎

12.3.4   Stability Analysis

Here we just note that if there exists a set of solutions $(P_{t_i}, \alpha_i, \varepsilon_i, K_{t_i}, L_{t_i})$ over the current time interval such that (12.43) holds, then the storage function (12.45) is not necessarily a monotonically decreasing function. This means that V_{t_i} is not a Lyapunov function for the considered system, and one may obtain only zone convergence. Below, we suggest a construction of a Lyapunov function for zone-convergence stability analysis.

Lyapunov Function and the Convergence Zone

Let us consider the function

$$
G(t) := \sum_{i=1}^{\infty} \chi_i(t)\, \mathcal{G}_i(t),
$$

$$
\chi_i(t) := \begin{cases} 1 \ if \ t \in (t_{i-1}, t_i], \\ 0 \ if \ t \notin (t_{i-1}, t_i], \end{cases} \quad \sum_{i=1}^{\infty} \chi_i(t) = 1,
$$

(12.48)

where

$$
\mathcal{G}_i(t) = \left(\left[\sqrt{V_i[z(t)]} - \sqrt{\frac{\beta_i}{\alpha_i}} \right]_+ \right)^2, \ t \in (t_{i-1}, t_i],
$$

$$
\beta_i := d_1 \varepsilon_i, \ [\gamma]_+ := \begin{cases} \gamma \ if \ \gamma \geq 0, \\ 0 \ if \ \gamma < 0. \end{cases}
$$

(12.49)

Observe that the function $[\gamma]_+$ is not differentiable at the point $\gamma = 0$, but the function $([\cdot]_+)^2$ is differential everywhere. In (12.49), the process $z(t)$ is governed by (12.36), and therefore the function $G(t)$ is well defined on all possible trajectories of (12.36).

Proposition 12.3 (Convergence zone). *If*

1. the collection

$$
(P_{A,t_i}, Q_{t_i}, K_{t_i}, L_{t_i}, \varepsilon_{1,i}, \varepsilon_{2,i}, \alpha_i)
$$

satisfies the set of matrix inequalities in Proposition 12.2 within each time interval $(t_{i-1}, t_i]$;

2. the additional dynamic constraints are satisfied at each stage $i = 1, 2, \ldots$:

$$
P_{i-1} \leq P_i, \ \frac{\beta_{i-1}}{\alpha_{i-1}} \geq \frac{\beta_i}{\alpha_i};
$$

(12.50)

then the function $G(t)$ is a Lyapunov function of the dynamical system (12.42) satisfying

$$
\frac{d}{dt} G(t) \leq 0
$$

(12.51)

for all $t \geq 0$ and

$$
\frac{d}{dt} G(t) < 0
$$

if

$$\sqrt{V_{i(t)}[z(t)]} > \mu_{i(t)} := \sqrt{\beta_{i(t)}/\alpha_{i(t)}}, \tag{12.52}$$

$$i(t) := \{i : t \in (t_{i-1}, t_i]\}.$$

Moreover, the Lyapunov function G (t) (12.48) has the "attraction property"

$$\left[\sqrt{V_{i(t)}[z(t)]} - \mu_{i(t)}\right]_+ \to 0 \ as \ t \to \infty.$$

Proof. Recall that the "generalized" derivative of the Heaviside function

$$\theta(t) := \begin{cases} 1 \ \text{if} \ t \geq 0, \\ 0 \ \text{if} \ t < 0, \end{cases}$$

is the Dirac delta function $\delta(t) = \theta'(t)$ with the property

$$\int_{t=-\infty}^{\infty} \delta(t)\, f(t)\, dt = f(0)$$

for every function $f(t)$ right-continuous at the origin. Since

$$\chi_i(t) = \theta(t - t_{i-1}) - \theta(t - t_i),$$

differentiation of (12.48) leads to

$$\frac{d}{dt} G(t) = \sum_{i=1}^{\infty} \frac{d}{dt} [\chi_i(t)\, \mathcal{G}_i(t)] =$$

$$\sum_{i=1}^{\infty} \chi_i(t) \frac{d}{dt} \mathcal{G}_i(t) + \sum_{i=1}^{\infty} [\delta(t - t_{i-1}) - \delta(t - t_i)]\, \mathcal{G}_i(t).$$

This is a singularly perturbed differential equation that in the equivalent integral form can be represented as follows:

$$G(t) - G(0) = \int_{s=0}^{t} \sum_{i=1}^{\infty} [\delta(s - t_{i-1}) - \delta(s - t_i)] \mathcal{G}_i(s) ds + \int_{s=0}^{t} \left[\sum_{i=1}^{\infty} \chi_i(s) \tfrac{d}{ds} \mathcal{G}_i(s) \right] ds$$

$$= \sum_{i=1}^{\infty} [\mathcal{G}_i(t_{i-1}) - \mathcal{G}_i(t_i)] + \int_{s=0}^{t} \left[\sum_{i=1}^{\infty} \chi_i(s) \left[\sqrt{V_i(s)} - \sqrt{\frac{\beta_i}{\alpha_i}} \right]_+ \frac{\frac{d}{ds} V_i(s)}{\sqrt{V_i(s)}} \right] ds \leq$$

$$\mathcal{G}_1(t_0) + [-\mathcal{G}_1(t_2) + \mathcal{G}_2(t_2)] + \cdots + [-\mathcal{G}_{i-1}(t_{i(t)}) + \mathcal{G}_i(t_{i(t)})]$$

$$+ \cdots + \int_{s=0}^{t} \left[\sum_{i=1}^{\infty} \chi_i(s) \left[\sqrt{V_i(s)} - \sqrt{\frac{\beta_i}{\alpha_i}} \right]_+ \frac{(-\alpha_i V_i(s) + \beta_i)}{\sqrt{V_i(s)}} \right] ds.$$

By the "monotonicity condition" (12.50), it follows that

$$-\mathcal{G}_{i-1}\left(t_{i(t)}\right) + \mathcal{G}_i\left(t_{i(t)}\right) \leq 0,$$

and we obtain

$$G(t) - G(0) \leq \mathcal{G}_1(t_0) + I(t),$$

where

$$
I(t) := \int_{s=0}^{t} \left[\sum_{i=1}^{\infty} \chi_i(s) \left[\sqrt{V_i(s)} - \sqrt{\frac{\beta_i}{\alpha_i}} \right]_+ \frac{(-\alpha_i V_i(s) + \beta_i)}{\sqrt{V_i(s)}} \right] ds =
$$

$$
-\int_{s=0}^{t} \sum_{i=1}^{\infty} \alpha_i \chi_i(s) \left[\sqrt{V_i(s)} - \sqrt{\frac{\beta_i}{\alpha_i}} \right]_+ \frac{\left(V_i(s) - \frac{\beta_i}{\alpha_i} \right)}{\sqrt{V_i(s)}} ds =
$$

$$
-\int_{s=0}^{t} \sum_{i=1}^{\infty} \alpha_i \chi_i(s) \left[\sqrt{V_i(s)} - \sqrt{\frac{\beta_i}{\alpha_i}} \right]_+ \frac{\left(\sqrt{V_i(s)} - \sqrt{\frac{\beta_i}{\alpha_i}} \right)\left(\sqrt{\frac{\beta_i}{\alpha_i}} + \sqrt{V_i(s)} \right)}{\sqrt{V_i(s)}} ds \qquad (12.53)
$$

$$
= -\int_{s=0}^{t} \sum_{i=1}^{\infty} \alpha_i \chi_i(s) \left[\sqrt{V_i(s)} - \sqrt{\frac{\beta_i}{\alpha_i}} \right]_+^2 \frac{\sqrt{\frac{\beta_i}{\alpha_i}} + \sqrt{V_i(s)}}{\sqrt{V_i(s)}} ds =
$$

$$
-\int_{s=0}^{t} \sum_{i=1}^{\infty} \alpha_i \chi_i(t) G_i(t) \frac{\left(\sqrt{V_i(t)} + \sqrt{\frac{\beta_i}{\alpha_i}} \right)}{\sqrt{V_i(t)}} ds \leq 0.
$$

We now introduce the so-called "dominating process" $\tilde{G}(t)$ satisfying

$$\tilde{G}(t) - \tilde{G}(0) = \mathcal{G}_1(t_0) + I(t). \qquad (12.54)$$

It is clear that if $G(0) = \tilde{G}(0)$, then $G(t) \leq \tilde{G}(t)$. Differentiation of the last identity implies

$$\frac{d}{dt}\tilde{G}(t) = -\sum_{i=1}^{\infty} \alpha_i \chi_i(t) G_i(t) \frac{\left(\sqrt{V_i(t)} + \sqrt{\frac{\beta_i}{\alpha_i}} \right)}{\sqrt{V_i(t)}} \leq 0.$$

The right-hand side of the last expression is strictly negative if $\sqrt{V_i(t)} > \mu_{i(t)}$. Moreover, since $\tilde{G}(t)$ is a nonnegative monotonically nonincreasing function, it follows by Weierstrass's theorem that $\tilde{G}(t)$ has a limit:

$$\lim_{t \to \infty} \tilde{G}(t) = \tilde{G}^*.$$

From (12.54) it follows that

$$0 \leq \tilde{G}(t) + |I(t)| = \mathcal{G}_1(t_0) + \tilde{G}(0) = \text{const.}$$

Taking $t \to \infty$, we obtain

$$0 \leq \tilde{G}^* + \limsup_{t \to \infty} |I(t)| < \infty,$$

implying $\limsup\limits_{t \to \infty} |I(t)| < \infty$. This means that there exists a time sequence $\{s_k\}_{k=1,2,\dots}$ such that

$$\sum_{i=1}^{\infty} \alpha_i \chi_i(s_k) \left[\sqrt{V_i(s_k)} - \sqrt{\frac{\beta_i}{\alpha_i}} \right]_+^2 \frac{\left(\sqrt{V_i(s_k)} + \sqrt{\frac{\beta_i}{\alpha_i}} \right)}{\sqrt{V_i(s_k)}} \xrightarrow[k \to \infty]{} 0,$$

and as a result

$$\sum_{i=1}^{\infty} \alpha_i \chi_i(s_k) \left[\sqrt{V_i(s_k)} - \sqrt{\frac{\beta_i}{\alpha_i}} \right]_+^2 = G(s_k) \xrightarrow[k \to \infty]{} 0.$$

Hence from the continuity of $\tilde{G}(t)$, it follows that $\tilde{G}(s_k) \xrightarrow[k \to \infty]{} 0$. But the sequence $\tilde{G}(t)$ converges, and hence all its subsequences have the same limit point, yielding $G^* = 0$, and as a result,

$$0 = G^* = \lim_{t \to \infty} \tilde{G}(t) \geq \lim_{t \to \infty} G(t) \geq 0,$$

implying $G(t) \xrightarrow[t \to \infty]{} 0$. This completes the proof of the proposition.   ■

Corollary 12.2. *If in Proposition 12.3, the numerical sequence $\{\mu_{i(t)}\}$ and the matrix sequence $\{P_i^{-1}\}$ are monotonically nonincreasing, that is,*

$$\mu_{i(t')} \geq \mu_{i(t'')} \text{ for } t' < t'', \ P_{i-1}^{-1} \geq P_i^{-1}, \tag{12.55}$$

then by Weierstrass's theorem, both have their limits

$$\lim_{t \to \infty} \mu_{i(t)} = \tilde{\mu}, \ \lim_{i \to \infty} P_i^{-1} = \tilde{P}^{-1},$$

which means that the ellipsoid $\mathcal{E}\left(0, \dfrac{1}{\tilde{\mu}^2} \tilde{P} \right)$ is attractive.

12.3.5 On the "Minimal Size" of the Attractive Ellipsoid

Attractive Ellipsoid in the Extended z-Space

To guarantee that at each time interval $(t_{i-1}, t_i]$, the proposed adaptive robust controller provides an ellipsoid of "minimal size," one needs to obtain the solution of the following optimization problem:

$$
\mathrm{tr}\left\{ \frac{\alpha_{i(t)}}{\beta_{i(t)}} P_i^{-1} \right\} \;\rightarrow\; \inf_{P_{A,t_i},\, Q_{t_i},\, K_{t_i},\, L_{t_i},\, \varepsilon_{1,i},\, \varepsilon_{2,i},\, \alpha_i}, \tag{12.56}
$$

subject to the constraints (12.43) and (12.55). Denote the solution of this optimization problem (12.56) at each step $i = 1, 2, \dots$ by

$$
(P_{t_i}^A)^*,\, Q_{t_i}^*,\, K_{t_i}^*,\, L_{t_i}^*,\, \varepsilon_{1,i}^*,\, \varepsilon_{2,i}^*,\, \alpha_i^*.
$$

By the monotonicity conditions (12.55), we may conclude that the limits

$$
\lim_{i \to \infty} (P_{t_i}^A)^* := \tilde{P}_A^*, \quad \lim_{t \to \infty} \mu_{i(t)}^* := \tilde{\mu}^*
$$

exist, so that we can consider the ellipsoid $\mathcal{E}\left(0, (\mu^2)^* \tilde{P}^*\right)$ with

$$
\tilde{P}^* = \mathrm{diag}\left[P_{t_i}^A, P_{t_i}^A, P_{t_i}^A \right]
$$

as an asymptotically attractive ellipsoid of "minimal size" in the extended space of the variable $z(t)$.

Attractive Ellipsoid in the State Space

To estimate the size of the "optimal attractive ellipsoid" in the state space of the system (12.42), note that

$$
x_t = H z_t, \quad H := \left[I_{n \times n}\; 0_{n \times n}\; 0_{n \times n} \right] = \left[0_{n \times n}\; I_{n \times n}\; I_{n \times n} \right],
$$

and therefore

$$
x_t^\mathsf{T} P_x x_t = z_t^\mathsf{T} H^\mathsf{T} P_x H z_t.
$$

On the other hand, the attractive ellipsoid in the extended z-space is $\dfrac{1}{(\mu^2)^*} \left(\tilde{P}^*\right)^{-1}$, so the equality

$$
\frac{1}{(\mu^2)^*} \left(\tilde{P}^*\right)^{-1} = H^\mathsf{T} P_x^{-1} H
$$

must be satisfied. Unfortunately, in general, this identity cannot be satisfied, since the size of the matrix P_x in which we are interested is less than that of the matrix $\tilde{P}^*$. Therefore, we propose estimating the "minimal ellipsoid" in the x-space as the solution P_x^* of the following optimization problem:

$$\left\| \frac{1}{(\mu^2)^*} \left(\tilde{P}^*\right)^{-1} - H^{\mathsf{T}} P_x^{-1} H \right\|^2 \to \min_{P_x \geq 0}, \tag{12.57}$$

where the norm in the Hilbert space of finite-dimensional matrices is defined as

$$\|A\|^2 := \langle A, A \rangle = \mathrm{tr}\left\{ A A^{\mathsf{T}} \right\}.$$

The solution P_x^* of the optimization problem (12.57) is as follows:[3]

$$\left(P_x^*\right)^{-1} = \frac{1}{(\mu^2)^*} \left(H^{\mathsf{T}}\right)^+ \left(\tilde{P}^*\right)^{-1} H^+.$$

Since in our case,

$$H^+ = \left[0_{n \times n} \; I_{n \times n} \; I_{n \times n} \right]^+ = \frac{1}{2} H^{\mathsf{T}},$$

we obtain

$$\left(P_x^*\right)^{-1} = \frac{1}{2(\mu^2)^*} \left(\tilde{P}_A^*\right)^{-1}. \tag{12.58}$$

Hence the attractive ellipsoid of "minimal size" in the state space that guarantees the property $\limsup_{t \to \infty} x_t^{\mathsf{T}} \left(P_x^*\right)^{-1} x_t \leq 1$ is defined by the ellipsoidal matrix P_x^* (12.58). So the attractive ellipsoid in x-space is *twice the size* of the corresponding minimal ellipsoid in z-space. To make this ellipsoid of minimal size, it suffices to find the solution of the optimization problem (12.57) under the restrictions (12.43).

12.3.6   Numerical Aspects

Observe that this optimization problem is a nonlinear optimization problem, subject (with fixed α_i, ε_i) to the BMI (12.43). This bilinear (under fixed scalars) optimization problem can be solved using the MATLAB toolbox Tomlab and SeDuMi (Peaucelle et al. 2002).

---

[3] H^+ is the pseudoinverse matrix to H defined in the Moore–Penrose sense (Poznyak 2008).

Algorithm 12.1 (Numerical procedure). *To obtain the numerical solution of (12.56) and (12.43) for the time interval* $t \in (t_{i-1}, t_i]$, *we select the initial condition for the adaptation procedure* $\bar{A}_{t_0} = A$ *(where* A *is the given matrix defining the class of uncertainties* $C(A, c_1, c_2)$*). The next steps of calculation,*

$$K_{t_i}, P_{t_i} \, (i = 1, 2, \ldots),$$

are obtained recursively by the following recurrent procedure:

1. *First, fixing some initial values of the scalar parameters* $\alpha_i = \alpha_{0,i}$ *and* $\varepsilon_i = \varepsilon_{0,i}$, *we apply the MATLAB toolbox SeDuMi to solve the corresponding constraint optimization problem. As a result, we obtain the matrices* P_{t_0} *and* K_{t_0}.
2. *Fixing the obtained matrices* P_{t_0} *and* K_{t_0}, *we suggest the augmentation of the parameters* α_i *and* ε_i, *taking*

$$\alpha_{1,i} = \alpha_{0,i} + \Delta\alpha_i, \ 0 < \Delta\alpha_i \ll 1,$$
$$\varepsilon_{1,i} = \varepsilon_{0,i} + \Delta\varepsilon_i, \ 0 < \Delta\varepsilon_i \ll 1.$$

3. *Then these steps are iterated in time, maintaining the property (12.55). When the SeduMi toolbox "informs" us that the current LMI has no solution, we stop the procedure. The last admissible parameters are declared the optimal ones* $(\alpha_i^*, \varepsilon_i^*, K_{t_i}^*, P_{t_i}^*)$ *for the time interval* $(t_{i-1}, t_i]$.
4. *Applying the switched controller* $u(t) = K_{t_i}^* x(t)$ *in (12.42), we realize online the algorithm (12.44) and obtain the matrix* A_{t_i} *for the current time interval* $(t_{i-1}, t_i]$.
5. *Increase* $i = i + 1$ *and return to step 1, verifying the following conditions:*

   (a) *If the condition (12.55) holds, we may conclude that the matrix* P_{t_i}, α_i, *and* ε_i *are the final solution, which may be declared the optimal collection* $(\alpha_i^*, \varepsilon_i^*, K_{t_i}^*, P_{t_i}^*)$ *for each time interval* $(t_{i-1}, t_i]$.
   (b) *If (12.55) does not hold, return to step 1 with*

$$\alpha_{0,i} = \alpha_i^*, \varepsilon_{0,i} = \varepsilon_i^*.$$

Numerical Example

In this section, we consider a benchmark example. The mathematical model of the considered systems can be represented in standard Cauchy affine (with respect to the control) form:

$$\dot{x}(t) = f(x) + Bu + \xi_x(x, t), \ y(t) = Cx + \xi_y(x, t),$$

with the model governed by following nonlinear dynamics:

$$f(x) := \begin{bmatrix} -5x1+x_2\mathrm{sign}(x2) \\ -x_2-4x_1\mathrm{sign}(x_1) \end{bmatrix}, \ B := \begin{bmatrix} 0 \\ 1 \end{bmatrix}, \ C = \begin{bmatrix} 1 & 0 \end{bmatrix},$$

$$\xi_x(t) := \begin{bmatrix} 0.02\sin(60t) \\ 0.1\sin(60t) \end{bmatrix}, \ \xi_y(t) := C\xi_x(t).$$

(12.59)

Numerical Data

Applying the suggested technique for this benchmark system (12.59) and using (12.42), we get

$$A_{t_0} = -\begin{bmatrix} 0.7 & 0 \\ 0 & 1.3 \end{bmatrix}, \ x_0 = \begin{bmatrix} -0.12 \\ 0.3 \end{bmatrix},$$

$$\hat{x}_0 = \begin{bmatrix} -0.1 \\ 0 \end{bmatrix}, \ d_0 = 0.52, \ d_1 = 1.65.$$

After eleven recurrent iterations of Algorithm 12.1 for $k = 1$, we obtained

$$K_{t_0} = \begin{bmatrix} -8.5318 & -0.6598 \end{bmatrix}, \ L_{t_0} = \begin{bmatrix} 1.6752 \\ -0.0191 \end{bmatrix}, \ P^A_{t_0} = \begin{bmatrix} 2.3051 & 0 \\ 0 & 5.0915 \end{bmatrix},$$

with $\alpha_i = 0.1$ and $\varepsilon_i = 0.01$. Applying Algorithm 12.1, stopping at iteration $i = 20$, we obtained the following results:

$$K_{t_{20}} = \begin{bmatrix} -0.0021 & -5.0317 \end{bmatrix}, \ L_{t_{20}} = \begin{bmatrix} 24.829 \\ 0.0059 \end{bmatrix}, \ P^A_{t_{20}} = \begin{bmatrix} 0.3192 & 0 \\ 0 & 1.0141 \end{bmatrix}.$$

The identification procedure led to

$$A_{t_{20}} = \begin{bmatrix} 0.0012 & 0.0098 \\ -0.0005 & -0.9890 \end{bmatrix}, \ \bar{A}_{t_{20}} = \begin{bmatrix} 0.0497 & 0.0042 \\ -0.0001 & 0.0517 \end{bmatrix}.$$

The time intervals $(t_{i-1}, t_i]$ were equal to 0.5 s uniformly.

Simulation Results

To illustrate the corresponding numerical results, we have divided the plots into two subfigures: Figures 12.4 and 12.5 present the real trajectories and velocities of the system. Figure 12.6 represent the results of the identification procedure (12.39). Finally, Figs. 12.7 and 12.8 present the corresponding system trajectories and attractive ellipsoids.

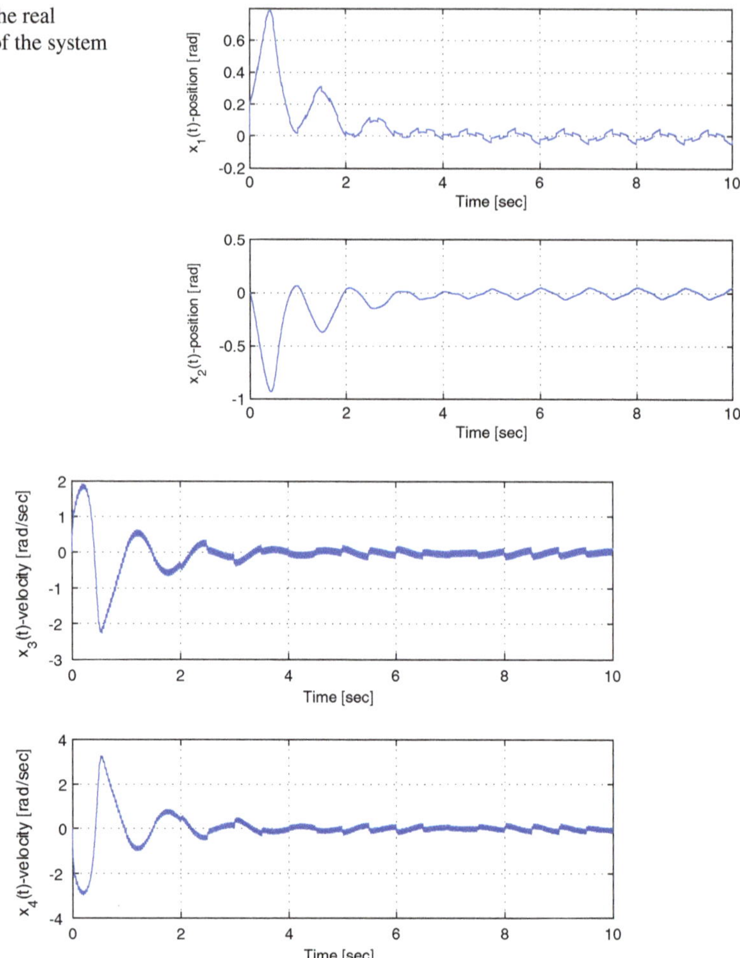

Fig. 12.4 The real trajectories of the system

Fig. 12.5 The real velocities of the system

12.4   Conclusion

- The suggested technique combines the observer and the controller, whose gain matrices are adjusted in time intervals by a conventional AEM procedure supplemented by two learning (adaptive) procedures. The gain matrices in the control actuator and in the Luenberger-type observer are corrected online at the beginning of each time interval (KL-adaptation).
- The "best" A-approximation (identification), denoted here by A_{t_i}, minimizes the quadratic-integral error of the "joint uncertainty" effect. The use of such recurrent procedures together with the resolution of the constraint optimization problem (realizing the minimization of the attractive ellipsoid at each iteration) constitutes

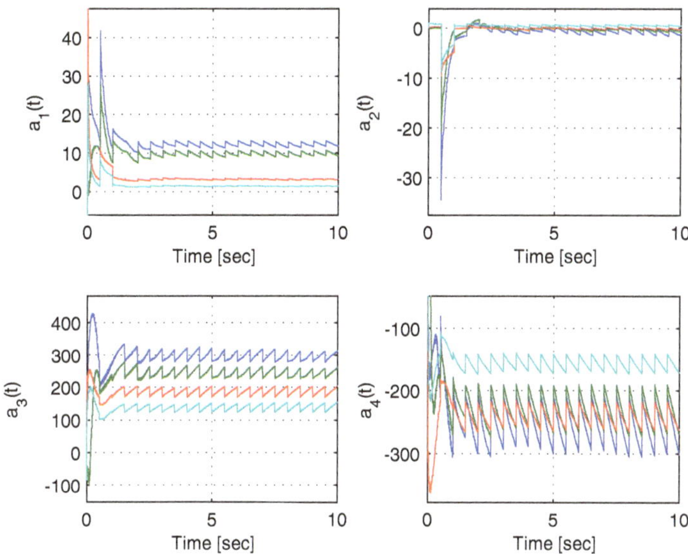

Fig. 12.6 A_t estimates

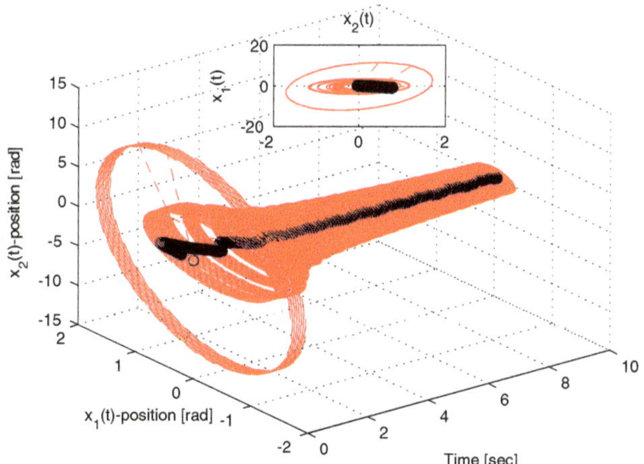

Fig. 12.7 The corresponding system trajectories x_1, x_2 and attractive ellipsoids

the main feature of the AEM with "A"-adaptation. A numerical example dealing with a benchmark system stabilization in unstable steady state was given to demonstrate the high effectiveness (more than 91%) of the suggested approach in comparison to the traditional AEM without adaptation.

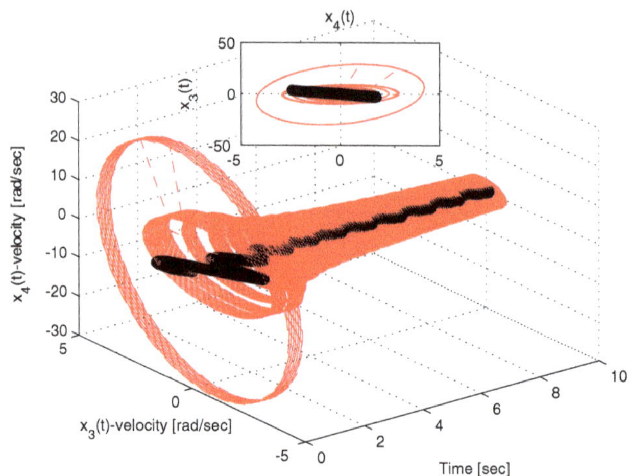

Fig. 12.8 The corresponding system trajectories x_3, x_4 and attractive ellipsoids

Bibliography

Aliprantis, C., & Border, K. (1999). *Infinite dimensional analysis*. Berlin: Springer.

Anderson, B. D. O. (1985). Adaptive systems, lack of persistency of excitation and bursting phenomena. *Automatica, 21*, 247–258.

Antsaklis, P. (2000). Special issue on hybrid systems: Theory and applications—a brief introduction to the theory and applications of hybrid systems. *Proceedings of the IEEE, 88*(7), 879–887.

Artstein, Z. (1982). Linear systems with delayed controls: A reduction. *IEEE Transactions on Automatic Control, 27*(4), 869–879.

Atkinson, K., & Han, W. (2005). *Theoretical numerical analysis*. New York: Springer.

Aubin, J., & Celina, A. (1984). *Differential inclusions*. Berlin: Springer.

Azhmyakov, A. (2006). Stability of differential inclusions: A computational approach. *Mathematical Problems in Engineering*, 1–15.

Azhmyakov, V. (2011). On the geometric aspects of the invariant ellipsoid method: Application to the robust control design. In *Proceedings of the 50th Conference on Decision and Control* (pp. 1353–1358).

Azhmyakov, V., Poznyak, A., & Gonzalez, O. (2013). On the robust control design for a class of nonlinearly affine control sustems: The attractive ellipsoid approach. *Journal of Industrial and Managment Optimization, 9*(3), 579–593.

Azhmyakov, V., Poznyak, A., & Juárez, R. (2013). On the practical stability of control processes governed by implicit differential equations: The invariant ellipsoid based approach feedback stabilization of networked control systems with packet dropouts. *Journal of the Franklin Institute, 350*, 2229–2243.

Balandin, D., & Kogan, M. (2007). *Control laws synthesis* (in Russian). Moscow: Fismatlit.

Barabanov, A., & Granichin, O. (1984). Optimal controller for linear plants with bounded noise. *Automation and Remote Control, 45*, 39–46.

Barkhordari-Yazdi, M., & Jahed-Motlagh, M. (2009). Stabilization of a CSTR with two arbitrarily switching modes using modal state feedback linearization. *Chemical Engineering Journal, 155*(3), 838–843.

Bartolini, G., Fridman, L., Pisano, A., & Usai, E. (2008). *Modern sliding mode control theory*. Lecture Notes in Control and Information Sciences (Vol. 375). Berlin: Springer.

Basar, T., & Olsted, G. J. (1999). *Dynamic noncooperative game theory*. Philadelphia: SIAM.

Basin, M., & Calderon-Alvarez, D. (2010). Optimal filtering over linear observations with unknown parameters. *Journal of the Franklin Institute, 347*, 988–1000.

Basin, M., Rodriguez-Gonzalez, J., & Fridman, L. (2007). Optimal and robust control for linear state-delay systems. *Journal of the Franklin Institute, 344*, 830–845.

© Springer International Publishing Switzerland 2014

A. Poznyak et al., *Attractive Ellipsoids in Robust Control*, Systems & Control: Foundations & Applications, DOI 10.1007/978-3-319-09210-2

Bekiaris-Liberis, N., & Krstic, M. (2012). Compensation of time-varying input and state delays for nonlinear systems. *Journal of Dynamic Systems, Measurement, and Control, 134*(1), paper 011009.

Bellman, R. (1956). *Dynamic programming and modern control theory.* New York: Academic.

Ben-Abdallah, A. (2009). On the practical output feedback stabilization for nonlinear uncertain systems. *Nonlinear Analysis: Modelling and Control, 14*(2), 145–153.

Ben-Tal, A., & Nemirovski, A. (2001). *Lectures on modern convex optimization.* Philadelphia: SIAM.

Berkovitz, L. (1974). *Optimal control theory.* New York: Springer.

Blanchini, F. (1999). Set invariance in control: A survey. *Automatica, 35*, 1747–1767.

Blanchini, F., & Miami, S. (2008). *Set theoretic methods in control. Systems and control: Foundations and applications.* Boston, MA: Birkhäuser.

Boiko, I. (2009). *Discontinuous control systems frequency-domain analysis and design.* New York: Birkhäuser.

Boltyanski, V., Gamkrelidze, R., & Pontryagin, L. (1956). On the theory of optimal processes. *Doklady AN USSR, 110*(1), 7–10 (in Russian).

Boltyanski, V., & Poznyak, A. (2012). *Robust maximum principle: theory and applications.* Boston: Birkhäuser.

Boskovic, J. D., & Mehra, R. K. (2000). Multi-mode switching in flight control. In *Proceedings of the 19-th Digital Avionics Systems Conference* (Vol. 2, pp. 6F2/1–6F2/8).

Boyd, S., Ghaoui, E., Feron, E., & Balakrishnan, V. (1994). *Linear matrix inequalities in system and control theory.* Philadelphia: SIAM.

Boyd, S., & Vandenberghe, L. (1997). Semidefinite programming relaxations of non-convex problems in control and combinatorial optimization. In A. Paulraj, V. Roychowdhury, & C. D. Schaper, *Communications, computation, control and signal processing: A tribute to Thomas Kailath* (pp. 279–288). Boston: Kluwer.

Branicky, M. S. (1998). Multiple Lyapunov functions and other analysis tools for switched and hybrid systems. *IEEE Transactions on Automatic Control, 43*, 475–482.

Brockett, R. W., & Liberzon, D. (2000). Quantized feedback stabilization of linear systems. *IEEE Transactions on Automatic Control, 47*, 1279–1289. info/brockett2000.pdf.

Burton, T. (2006). *Stability by fixed point theory for functional differential equations.* New York: Dover.

Chen, P., Qin, H., & Huang, J. (2001). Local stabilization of a class of nonlinear systems by dynamic output feedback. *Automatica, 37*, 969–981.

Chistyakov, V. (2007). Nonlocal theorems on existence of solutions of differential-algebraic equations of index 1. *Russian Mathematics, 51*, 71–76.

Choi, H.-L., & Lim, J.-T. (2010). Output feedback regulation of a chain of integrators with an unknown time-varying delay in the input. *IEEE Transactions on Automatic Control, 55*(1), 263–268.

Choi, S.-B., & Hedrick, J. (1996). Robust throttle control of automotive engines. *ASME Journal of Dynamic Systems, Measurement and Control, 118*, 92–98.

Coddington, E., & Levinson, N. (1955). *Theory of ordinary differential equations.* New York, NY: McGraw-Hill.

Corless, M. (1990). Guaranteed rates of exponential convergence for uncertain systems. *Journal of Optimization Theory and Applications, 64*(3), 481–494.

Coutinho, D., Trofino, A., & Barbosa, K. (2003). Robust linear dynamic output feedback controllers for a class of nonlinear systems. In *Proceedings of the Conference on Decision and Control* (pp. 374–379).

Daafouz, J., Riedinger, P., & Iung, C. (2002). Stability analysis and control synthesis for switched systems: A switched Lyapunov function approach. *IEEE Transactions on Automatic Control, 47*(11), 1883–1887.

Dahleh, M., Pearson, J., & Boyd, J. (1988). Optimal rejection of persistent disturbances, robust stability, and mixed sensitivity minimization. *IEEE Transactions on Automatic Control, 33*, 722–731.

Dai, L. (1989). *Singular control systems*. Berlin: Springer.

Davila, J., & Poznyak, A. (2011). Dynamic sliding mode control design using attracting ellipsoid method. *Automatica, 47*, 1467–1472.

DeCarlo, R., Branicky, M., Petterson, S., & Lennartson, S. (2000). Perspectives and results on the stability and stabilizability of hybrid systems. *Proceedings of the IEEE: Special Issue Hybrid Systems, 88*, 1069–1082.

Deimling, K. (1992). *Multivalued differential equations*. Berlin: W. de Gruyter.

Doyle, J. (1983). Synthesis of robust controllers and filters. In *Proceedings of IEEE Conference on Decision and Control* (pp. 109–114).

Duncan, G., & Schweppe, F. (1971). Control of linear dynamic systems with set constrained disturbances. *IEEE Transactions on Automatic Control, 16*, 411–423.

Edwards, C., & Spurgeon, S. (1998). *Sliding mode control: Theory and applications*. London, UK: Taylor & Francis.

Edwards, R. (1995). *Functional analysis*. New York: Dover.

Filippov, A. (1988). *Differential equations with discontinuous right-hand sides*. Dordrecht: Kluwer.

Flower, A. C., & Mackey, M. C. (2002). Relaxation oscillations in a class of delay differential equations. *SIAM Journal of Applied Mathematics, 63*, 299–323.

Francis, B., Helton, J., & Zames, G. (1984). H^∞-optimal feedback controllers for linear multivariable systems. *IEEE Transactions on Automatic Control, 29*, 888–900.

Fridman, E. (2001). New Lyapunov–Krasovskii functionals for stability of linear retarded and neutral type systems. *Systems and Control Letters, 43*, 309–319. delay/fridman2001.pdf.

Fridman, E. (2006). Descriptor discretized Lyapunov functional method: analysis and design. *IEEE Transactions on Automatic Control, 51*, 890–897.

Fridman, E. (2010). A refined input delay approach to sampled-data control. *Automatica, 46*, 421–427.

Fridman, E., & Dambrine, M. (2009). Control under quantization, saturation and delay: An LMI approach. *Automatica, 45*, 2258–2264. delay/fridman2009.pdf.

Fridman, E., & Niculescu, S.-I. (2008). On complete Lyapunov–Krasovskii functional techniques for uncertain systems with fast-varying delays. *International Journal of Robust and Nonlinear Control, 18*, 364–374. delay/fridman2008.pdf.

Fridman, E., & Orlov, Y. (2007). On stability of linear parabolic distributed parameter systems with time-varying delays. In *Proceedings of the 46th Conference on Decision and Control* (pp. 1597–1602), New Orlean, USA.

Fridman, E., Seuret, A., & Richard, J.-P. (2004). Robust sampled-data stabilization of linear systems: An input delay approach. *Automatica, 40*(8), 1441–1446.

Fridman, E., & Shaked, U. (2002). An improved stabilization method for linear time-delay systems. *IEEE Transaction on Automatic Control, 47*(11), 1931–1937.

Fridman, L., Acosta, P., & Polyakov, A. (2001). Robust eigenvalue assignment for uncertain delay control systems. In *IFAC Workshop Time Delay Systems* (pp. 239–244), Santa Fe, USA.

Fu, M., & Xie, L. (2005). The sector bound approach to quantized feedback control. *IEEE Transactions on Automatic Control, 50*, 1698–1711. info/fu2005.pdf.

Gao, H., & Chen, T. (2008). A new approach to quantized feedback control systems. *Automatica, 44*, 534–542. info/gao2008.pdf.

Gel'fand, I., & Shilov, G. (1968). *Generalized functions*. New York: Academic.

Ghaoui, L., & Niculescu, S. (2000). *Advances in linear matrix inequalities in control*. Philadelphia: SIAM.

Glover, J. D., & Schweppe, F. C. (1971). Control of linear dynamic systems with set constrained disturbance. *IEEE Transaction on Automatic Control, 16*, 411–423. ellipsoids/glover1971.pdf.

Gonzalez-Garcia, S., Polyakov, A., & Poznyak, A. (2009). Output linear controller for a class of nonlinear systems using the invariant ellipsoid technique. In *American Control Conference* (pp. 1160–1165).

Gonzalez-Garcia, S., Polyakov, A., & Poznyak, A. (2011). Using the method of invariant ellipsoids for linear robust output stabilization of spacecraft. *Automation and Remote Control, 72*, 540–555.

Grimble, M. (1994). *Robust industrial control.* Hemel Hempstead, UK: Prentice Hall International.

Gu, K., Kharitonov, V., & Chen, J. (2003). *Stability of time-delay systems.* New York: Birkhäuser.

Haddad, W., & Chellaboina, V. (2008). *Nonlinear Dynamical Systems and Control.* Princeton: Princeton University Press.

Hale, J. (1969). *Ordinary differential equations.* New York: Wiley.

Han, X., Fridman, E., & Spurgeon, S. (2012). Sliding mode control in the presence of input delay: A singular perturbation approach. *Automatica, 48*(8), 1904–1912.

Haykin, S. (2009). *Neural networks and learning machines.* Upper Saddle River, NJ: Prentice Hall.

Henrion, D., Loefberg, J., Kocvara, M., & Stingl, M. (2006). Solving polynomial static output feedback problems with PENBMI. In *Proceedings of the IEEE Conference on Decision and Control* (pp. 7581–7586).

Hespanha, J. P., & Morse, A. S. (1999). Stability of switched systems with average dwell-time. In *Proceedings of the 38th IEEE Conference on Decision and Control* (pp. 2655–2660), Phoenix, AZ.

Himmelberg, C. (1975). Measurable relations. *Fundamenta Mathematicae, 87*, 53–72.

Holmstrom, K., Goran, A., & Edvall, M. (2009). *User's Guide for TOMLAB/CPLEX v12.1.*

Hong, Y. (2008). *A globally uniformly ultimately bounded adaptive robust control approach to a second-order nonlinear motion system with input saturation.* West Lafayette: Purdue University.

Ioannou, P., & Sun, J. (1996). *Robust adaptive control.* Upper Saddle River, NJ: Prentice Hall.

Isidori, A. (1995). *Nonlinear control systems.* New York, Springer.

Isidori, A., Teel, A., & Praly, L. (2000). A note on the problem of semiglobal practical stabilization of uncertain nonlinear systems via dynamic output feedback. *System and Control Letters, 39*, 165–171.

Jong, M. L., & Jay, H. L. (2005). Approximate dynamic programming-based approaches for input–output data-driven control of nonlinear processes. *Automatica, 41*, 1281–1288.

Kabamba, P. T., & Hara, S. (1993). Worst-case analysis and design of sampled-data control systems. *IEEE Transactions on Automatic Control, 38*(9), 1337–1358.

Karafyllis, I., & Jiang, Z.-P. (2011). *Stability and stabilization of nonlinear systems.* London: Springer.

Khalil, H. (2002). *Nonlinear systems.* Upper Saddle River: Prentice Hall.

Krstic, M. (2009). *Delay compensation for nonlinear, adaptive, and PDE systems.* Boston: Birkhäuser.

Kruszewski, A., Jiang, W. J., Fridman, E., Richard, J.-P., & Toguyeni, A. (2012). A switched system approach to exponential stabilization through communication network. *IEEE Transactions on Control Systems Technology, 20*(4), 887–900.

Kunkel, P., & Mehrmann, V. L. (2006). *Differential-algebraic equations: analysis and numerical solutions.* Switzerland: EMS Publishing House.

Kurzhanski, A., & Valyi, I. (1997). *Ellipsoidal calculus for estimation and control.* Boston, MA: Birkhäuser.

Kurzhanski, A. B., & Varaiya, P. (2006). Ellipsoidal techniques for reachability under state constraints. *SIAM Journal on Control and Optimization, 45*, 1369–1394. ellipsoids/kurzhanski2006.pdf.

Kurzhanski, A., & Veliov, V. (1994). *Modeling techniques and uncertain systems.* New York: Birkhäuser.

Lakshmikantham, V., Leela, S., & Martynyuk, A. (1990). *Practical stability of nonlinear systems.* New Jersey, London, Singapore, Hong Kong: World Scientific.

Leonhard, W. (1966). *Control of electrical drives.* Berlin: Springer.

Li, X., & Yurkovitch, S. (1999). Sliding mode control of systems with delayed states and controls. In *Variable structure systems, sliding mode and nonlinear control* (pp. 93–108). Berlin: Springer.

Liberzon, D. (2003). *Switching in systems and control*. Boston: Birkhäuser.

Liberzon, D., & Morse, A. S. (1999). Basic problems in stability and design of switched systems. *IEEE Control Systems Magazine, 19*, 59–70.

Lin, H., & Antsaklis, P. (2009). Stability and stabilizability of switched linear systems: A survey of recent results. *IEEE Transactions on Automatic Control, 54*(2), 308–322.

Lin, W., & Qian, C. (2001). Semi-global robust stabilization of mimo nonlinear systems by partial state and dynamic output feedback. *Automatica, 37*, 1093–1101.

Ljung, L. (1999). *Sistem identification: Theory for the user*. New Jersey: Prentice Hall PTR.

Loukianov, A. G., Espinosa-Guerra, O., Castillo-Toledo, B., & Utkin, V. A. (2006). Integral sliding mode control for systems with time delay. In *International Workshop on Variable Structure Systems* (pp. 256–261), Alghero, Italy.

Lunze, J., & Lamnabhi-Lagarrigue, F. (2009). *Handbook of hybrid systems control: Theory, tools and applications*. New York: Cambridge University Press.

Maciejowski, J. (1989). *Multivariable feedback design*. New York: Addison Wesley.

Mahmoud, M. S. (2000). *Robust control and filtering for time-delay systems*. New York: Marcel Dekker.

Malek-Zaverei, M., & Jamshidi, M. (1987). *Time delay systems analysis, optimization and applications, systems and control*. Amsterdam, Netherlands: North Holland.

Manitius, A., & Olbrot, A. (1979). Finite spectrum assignment problem for systems with delay. *IEEE Transactions on Automatic Control, 24*, 541–553.

Masubuchi, I., Kamitane, Y., Ohara, A., & Suda, N. (1997). h_∞ control for descriptor systems: A matrix inequalities approach. *Automatica, 33*(4), 669–673.

Matveev, A. S., & Savkin, A. V. (2007). An analogue of Shannon information theory for detection and stabilization via noisy discrete communication channels. *SIAM Journal on Control and Optimization, 46*, 1323–1367. info/matveev2007.pdf.

Mazenc, F., Niculescu, S., & Krstic, M. (2012). Lyapunov-Krasovskii functionals and application to input delay compensation for linear time-invariant systems. *Automatica, 48*(7), 1317–1323.

Mera, M., Poznyak, A., & Azhmyakov, V. (2011). On the robust control design for a class of continuous-time dynamical systems with a sample-data output. In *Proceedings of the IFAC World Congress*.

Mera, M., Poznyak, A., Azhmyakov, V., & Fridman, E. (2009). Robust control for a class of continuous-time dynamical systems with sample-data outputs. In *Proceedings of International Conference on Electrical Engineering, Computing Science and Automatic* (pp. 1–7), Toluca, Mexico.

Michel, A., Hou, L., & Liu, D. (2007). *Stability of dynamical systems*. New York: Birkhäuser.

Naghshtabrizi, P., Hespanha, J., & Teel, A. (2008). Exponential stability of impulsive systems with application to uncertain sampled-data systems. *Systems and Control Letters, 57*, 378–385.

Nair, G. N., & Evans, R. J. (2003). Exponential stabilisability of finite-dimensional linear systems with limited data rates. *Automatica, 39*, 585–593. info/nair2003.pdf.

Narendra, K. S., & Annaswamy, A. M. (2005). *Stable adaptive systems*. New York: Dover Publications Inc.

Nazin, S., Polyak, B., & Topunov, M. (2007). Rejection of bounded exogenous disturbances by the method of invariant ellipsoids. *Automation and Remote Control, 68*, 467–486.

Ordaz, P., Alazki, H., & Poznyak, A. (2013). A sample-time adjusted feedback for robust bounded output stabstazation. *Kybernetika, 49*(6), 911–934.

Ordaz, P., & Poznyak, A. (2012). The Furuta's pendulum stabilization without the use of a mathematical model: Attractive ellipsoid method with KL-adaptation. In *51-th IEEE Conference on Decision and Control*, Maui, Hawaii.

Ordaz, P., & Poznyak, A. (2014). "KL"-gain adaptation for attractive ellipsoid method. *IMA Journal of Mathematical Control and Information*. doi:10.1093/imamci/dnt046.

Orlov, Y. (2008). *Discontinuous systems: Lyapunov analysis and robust synthesis under uncertainty conditions*. New York: Springer.

Peaucelle, D., Henrion, D., Labit, Y., & Taitz, K. (2002). *User's guide for SeDuMi Interface 1.04*.

Peng, C., & Tian, Y.-C. (2007). Networked h_∞ control of linear systems with state quantization. *Information Sciences, 177*, 5763–5774. info/peng2007.pdf.

Peng, C., Tian, Y.-C., & Yue, D. (2011). Output feedback control of discrete-time systems in networked environments. *IEEE Transactions on Systems, Man and Cybernetics, Part A : Systems and Humans, 41*, 185–190. info/peng2011.pdf.

Phat, V. N., Jiang, J., Savkin, A. V., & Petersen, I. R. (2004). Robust stabilization of linear uncertain discrete-time systems via a limited capacity communication channel. *Systems and Control Letters, 53*, 347–360. info/phat2004.pdf.

Polik, I., & Terlaky, T. (2007). A survey of the s-lemma. *SIAM Review, 49*(1), 371–418.

Polyak, B. (2001). Convexity of nonlinear image of a small ball with applications to optimization. *Set-Valued Analysis, 9*, 159–168.

Polyak, B. T., Nazin, S., Durieu, C., & Walter, E. (2004). Ellipsoidal parameter or state estimation under model uncertainty. *Automatica, 40*, 1171–1179. ellipsoids/polyak2004.pdf.

Polyak, B. T., & Topunov, M. V. (2008) Suppression of bounded exogenous disturbances: Output feedback. *Automation and Remote Control, 69*, 801–818. ellipsoids/polyak2008.pdf.

Polyakov, A. (2010). On practical stabilization of the systems with relay delay control. *Automation and Remote Control, 71*(11), 2331–2344.

Polyakov, A. (2012). Minimization of disturbances effects in time delay predictor-based sliding mode control systems. *Journal of the Franklin Institute, 349*(4), 1380–1396.

Polyakov, A., Efimov, D., Perruquetti, W., & Richard, J.-P. (2013). Output stabilization of time-varying input delay systems using interval observation technique. *Automatica, 49*(11), 3402–3410.

Polyakov, A., & Poznyak, A. (2009). Lyapunov function design for finite-time convergence analysis: "Twisting" controller for second-order sliding mode realization. *Automatica, 45*, 444–448.

Polyakov, A., & Poznyak, A. (2011). Invariant ellipsoid method for minimization of unmatched disturbances effects in sliding mode control. *Automatica, 47*, 1450–1454.

Pontryagin, L. S., Boltyansky, V. G., Gamkrelidze, R. V., & Mishenko, E. F. (1969). *Mathematical theory of optimal processes* (Translated from Russian). New York: Interscience.

Poznyak, A. (2004). *Variable structure systems: From Principles to Implementation, chapter 3*. IEE Control Series (Vol. 66, pp. 45–78). London, UK: The IET.

Poznyak, A. (2008). *Advanced mathematical tools for automatic control engineers: Deterministic techniques*. Amsterdam: Elsevier.

Poznyak, A., Azhmyakov, V., & Mera, M. (2011). Practical output feedback stabilisation for a class of continuous-time dynamic systems under sample-data outputs. *International Journal of Control, 84*, 1408–1416.

Poznyak, A., Sanchez, E., & Yu, W. (2001). *Differential neural networks: Identification, state estimation and trajectory tracking*. Singapore: World Scientific.

Pytlak, R. (1999). *Numerical methods for optimal control problems with state constraints*. Berlin: Springer.

Reis, T., & Stykel, T. (2011). Lyapunov balancing for passivity-preserving model reduction of RC circuits. *SIAM Journal of Applied Dynamic Systems, 10*, 1–34.

Rheinbolt, W. (1981). On the existence and uniqueness of solutions of nonlinear semi-implicit differential-algebraic equations. Nonlinear analysis. *Theory, Methods and Applications, 16*, 642–661.

Rheinbolt, W. (1988). Differential-algebraic systems as differential equations on manifolds. *Mathematics of Computation, 43*, 473–482.

Richard, J.-P. (2003). Time-delay systems: An overview of some recent advances and open problems. *Automatica, 39*(10), 1667–1694.

Rockafellar, R. (1970). *Convex analysis*. Princeton: Princeton University Press.

Roh, Y. H., & Oh, J. H. (1999). Robust stabilization of uncertain input delay systems by sliding mode control with delay compensation. *Automatica, 35*, 1861–1865.

Rudin, W. (1991). *Functional analysis* (2nd ed.). New York: MacGraw-Hill Inc.

Sastry, S. (1999). *Nonlinear systems*. New York: Springer.

Sastry, S., & Bodson, M. (1994). *Adaptive control: stability, convergence, and robustness.* New York: Prentice-Hall.

Scherer, C., & Weiland, S. (2000). *Course on linear matrix inequalities in control.* Lecture Notes DISC. Berlin: Springer.

Schweppe, F. (1973). *Uncertain dynamic systems.* Englewood Cliffs, NJ: Prentice Hall.

Seuret, A., Floquet, T., Richard, J.-P., & Spurgeon, S. K. (2007). A sliding mode observer for linear systems with unknown time varying delay. In *American Control Conference* (pp. 4558–4563).

Shannon, C. E. (1948). A mathematical theory of communication. *Bell System Technical Journal, 27,* 379–423. info/shannon1948.pdf.

Shorten, R., & Cairbre, F. O. (2001). A proof of global attractivity for a class of switching systems using non-quadratic Lyapunov approach. *IMA Journal of Mathematical Control and Information, 18*(3), 341–353.

Shorten, R. N. (1996). A Study of Hybrid Dynamical systems with application of automotive control. PhD thesis, Department of Electrical and Electronic Engineering, University College Dublin, Republic of Ireland.

Sontag, E. (1998). *Mathematical control theory.* New York: Springer.

Stephanopoulos, G. (1984). *Chemical process control: An introduction to theory and practice.* New York: Prentice-Hall.

Sussmann, H. J., Sontag, E. D., & Yang, Y. (1994). A general result on the stabilization of linear systems using bounded controls. *IEEE Transactions on Automatic Control, 39*(12), 2411–2425.

Takaba, K. (2002). Stability analysis of interconnected implicit systems based on passivity. In *Proceedings of the IFAC 15th World Congress,* Barcelona, Spain.

Tatikonda, S., & Mitter, S. (2004). Control under communication constraints. *IEEE Transactions on Automatic Control, 49,* 1056–1068. info/tatikonda2004.pdf.

Teel, A. R. (1996). A nonlinear small gain theorem for the analysis of control systems with saturation. *IEEE Transactions on Automatic Control, 41*(9), 1256–1270.

Teel, A. R., Nesic, D., & Kokotovic, P. V. (2001). A note on the robustness of input-to-state stability. In *Proceedings of the 40th IEEE Conference on Decision and Control* (pp. 875–880), Orlando, USA.

Tian, E., Yue, D., & Chen, P. (2008). Quantized output feedback control for networked control systems. *Zeitschrift für Angewandte Mathematik und Physik, 178,* 2734–2749. info/tian2008.pdf.

Ting Shu, H., & Zong Li, L. (2001). *Control systems with actuator saturation: Analyze and design.* Boston: Birkhäuser.

Toker, O., & Ozbay, H. (1995). On the NP-hardness of solving bilinear matrix inequalities and simultaneous stabilization with static output feedback. In *Proceedings of the American Control Conference.*

Usoro, P., Schweppe, F., Gould, L., & Wormley, D. (1982). A Lagrange approach to set-theoretic control synthesis. *IEEE Transactions on Automatic Control, 27*(2), 393–399.

Utkin, V. (1992). *Sliding modes in control optimization.* Berlin: Springer.

Utkin, V. (1993). Sliding mode control design principles and applications to electric drives. *IEEE Transactions on Industrial Electronics, 40*(1), 23–26.

Utkin, V., Guldner, J., & Shi, J. (1999). *Sliding modes in electromechanical systems.* London: Taylor & Francis.

Watanabe, K. (1986). Finite spectrum assignment and observer for multivariable systems with commensurate delays. *IEEE Transactions on Automatic Control, 31*(6), 543–550.

Wicks, M. A., Peleties, P., & DeCarlo, R. A. (1994). Construction of piecewise Lyapunov functions for stabilizing switched systems. In *Proceedings of the 33rd Conference on Decision and Control* (pp. 3492–3497), Lake Buena Vista, FL.

Yakubovich, E. (1976). Solution of the optimal control problem for the linear discrete systems. *Automation and Remote Control, 36,* 1447–1453.

Yakubovich, V. (1977). S-procedure in nonlinear control theory. *Vestnik Leningrad University, 4,* 73–93.

Yefremov, M., Polyakov, A., & Strygin, V. (2006). An algorithm for the active stabilization of a spacecraft with viscoelastic elements under conditions of uncertainty. *Journal of Applied Mathematics and Mechanics, 70*, 723–733.

Zabczyk, J. (1995). *Mathematical control theory: An introduction.* Boston: Birkhäuser.

Zhai, G. S., Hu, B., Yasuda, K., & Michel, A. N. (2001). Disturbance attenuation properties of time-controlled switched systems. *Journal of the Franklin Institute, 338*(7), 765–779.

Zhang, W.-A., & Yu, L. (2007). Output feedback stabilization of networked control systems with packet dropouts. *IEEE Transactions on Automatic Control, 52*, 1705–1710. info/zhang2007.pdf.

Zheng, G., Barbot, J.-P., Boutat, D., Floquet, T., & Richard, J.-P. (2011). On observability of nonlinear time-delay systems with unknown inputs. *IEEE Transactions on Automatic Control, 56*(8), 973–1978.

Zhou, K., Doyle, J., & Glover, K. (1996). *Robust and optimal control.* Upper Saddle River, NJ: Prentice Hall.

Zubov, V. (1962). *Mathematical methods for the study of automatic control systems.* New York: Pergamon Press.

Index

© Springer International Publishing Switzerland 2014
A. Poznyak et al., *Attractive Ellipsoids in Robust Control*, Systems & Control:
Foundations & Applications, DOI 10.1007/978-3-319-09210-2